NATURAL QUIET and NATURAL DARKNESS

NATURAL QUIET and NATURAL DARKNESS

THE "NEW" RESOURCES OF THE NATIONAL PARKS

EDITED BY

**Robert Manning, Peter Newman,
Jesse Barber, Christopher Monz,
Jeffrey Hallo, and Steven Lawson**

UNIVERSITY PRESS OF NEW ENGLAND
Hanover and London

University Press of New England
www.upne.com
© 2018 University Press of New England
All rights reserved
Manufactured in the United States of America
Designed by Vicki Kuskowski
Typeset in Scala by Copperline Book Services

Cover photos show an old growth forest at Muir Woods National Monument
and the night sky above Arches National Park. These photos and places are
emblematic of the growing importance of natural quiet and natural darkness—
both ecologically and experientially—as national park resources.
Left: NPS/Muir Woods Arch. *Right*: PS/Jacob W. Frank

For permission to reproduce any of the material in this book, contact Permissions,
University Press of New England, One Court Street, Suite 250, Lebanon NH 03766;
or visit www.upne.com

Library of Congress Cataloging-in-Publication Data available upon request

Hardcover ISBN: 978-1-5126-0188-6
Paperback ISBN: 978-1-5126-0189-3

5 4 3 2 1

CONTENTS

Part 4: The Experience of Natural Darkness   225

ACKNOWLEDGMENTS

This book was supported by grants from the Natural Sounds and Night Skies Division of the National Park Service. Additional support was provided by the National Science Foundation, Grant CNH 1414171. Valuable administrative, logistical, and technical support was provided by William Valliere at the University of Vermont. Graphic design and document editing support were provided by Kristen Neilson and Joseph Wildey of RSG. We also thank the Utah Agricultural Experiment Station for their support of this book. Special appreciation is expressed to Jane Rodgers and the Division of Science and Resource Management at Grand Canyon National Park for its Scholar in Residence program. Time spent at the park offered the opportunity to help shape and complete this book.

Most of the chapters in this book are edited versions of papers published in the scientific and professional literature. Appreciation is expressed to the following authors and publishers (listed in the order of their appearance in the book).

Clinton Francis and Jesse Barber. "A Framework for Understanding Noise Impacts on Wildlife: An Urgent Conservation Priority." *Frontiers in Ecology and the Environment* 11, no. 6 (2013): 305–13. Publisher: John Wiley & Sons, Inc. © 2017 Ecological Society of America. All rights reserved.

Heidi Ware, Christopher McClure, Jay Carlisle, and Jesse Barber. "A Phantom Road Experiment Reveals Traffic Noise Is an Invisible Source of Habitat Degradation." *Proceedings of the National Academy of Sciences* 112, no. 39 (2015): 12105–9.

Clinton Francis, Nathan Kleist, Catherine Ortega, and Alexander Cruz. "Noise Pollution Alters Ecological Services: Enhanced Pollination and Disrupted Seed Dispersal." *Proceedings of the Royal Society B: Biological Sciences* 279, no. 1739 (July 22, 2012): 2727–35. Reprinted by permission of the Royal Society.

Stephen Simpson, Andrew Radford, Sophie Nedelec, Maud Ferrari, Douglas Chivers, Mark McCormick, and Mark Meekan. "Anthropogenic Noise Increases Fish Mortality by Predation." *Nature Communications* 7 (2016): 1–7.

Taegan McMahon, Jason Rohr, and Ximena Bernal. "Light and Noise Pollution Interact to Disrupt Interspecific Interactions." *Ecology* 98, no. 5 (May 1, 2017): 1290–99. doi:10.1002/ecy.1770. John Wiley & Sons, Inc. © 2017 The Ecological Society of America. Ecology (ISSN 0012-9658) is published monthly on behalf of the Ecological Society of America by Wiley Subscription Services, Inc., a Wiley Company, 111 River St., Hoboken, NJ 07030-5774. Periodicals postage paid at Hoboken, NJ and additional offices. Postmaster: Send all address changes to ECOLOGY, John Wiley & Sons Inc., c/o The Sheridan Press, PO Box 465, Hanover, PA 17331. ESA Headquarters: 1990 M Street NW, Suite 700, Washington DC 20036 USA. Tel 1-202-833-8773, Fax 1-202-833-8775, Email esahq@esa.org www.esa .org/journals.

David Weinzimmer, Peter Newman, Derrick Taff, Jacob Benfield, Emma Lynch, and Paul Bell. "Human Responses to Simulated Motorized Noise in National Parks." *Leisure Sciences* 36, no. 3 (2014): 251–67. Reprinted by permission of the publisher (Taylor and Francis Ltd, http://www.tandfonline.com).

Jacob Benfield, Derrick Taff, Peter Newman, and Joshua Smyth. "Natural Sound Facilitates Mood Recovery from Stress." *Ecopsychology* 6, no. 3 (2014): 183–88.

Robert Manning, Peter Newman, Kurt Fristrup, Dave Stack, and Ericka Pilcher. "A Program of Research to Support Management of Visitor-Caused Noise at Muir Woods National Monument." *Park Science* 26, no. 3 (2010): 54–58.

Derrick Taff, Peter Newman, Steven R. Lawson, Alan Bright, Lelaina Marin, Adam Gibson, and Tim Archer. "The Role of Messaging on Acceptability of Military Aircraft Sounds in Sequoia National Park." *Applied Acoustics* 84 (2014): 122–28. Reprinted with permission from Elsevier.

Logan Park, Steven Lawson, Ken Kaliski, Peter Newman, and Adam Gibson. "Modeling and Mapping Hikers' Exposure to Transportation Noise in Rocky Mountain National Park." *Park Science* 26, no. 3 (2010): 59–64.

Kevin Gaston, Jonathan Bennie, Thomas Davies, and John Hopkins. "The Ecological Impacts of Nighttime Light Pollution: A Mechanistic Appraisal." *Biological Reviews* 88, no. 4 (November 2013): 912–27. © 2013 The Authors. Biological Reviews © 2013 Cambridge Philosophical Society. Publisher: John Wiley & Sons, Inc.

Florian Altermatt, and Dieter Ebert. "Reduced Flight-to-Light Behaviour of Moth Populations Exposed to Long-Term Urban Light Pollution." *Biology Letters* 12, no. 4 (2016): 20160111. Reprinted by permission of the Royal Society.

Bart Kempenaers, Pernilla Borgstroem, Peter Loes, Emmi Schlicht, and Mihai Valcu. "Artificial Night Lighting Affects Dawn Song, Extra-Pair Siring Success, and Lay Date in Songbirds." *Current Biology* 20, no. 19 (October 12, 2010): 1735–39. Reprinted with permission from Elsevier.

Daniel Lewanzik and Christian C. Voigt. "Artificial Light Puts Ecosystem Services of Frugivorous Bats at Risk." *Journal of Applied Ecology* 51, no. 2 (2014): 388–94. Publisher: John Wiley & Sons, Inc. © 2014 The Authors. Journal of Applied Ecology © 2014 British Ecological Society.

Richard ffrench-Constant, Robin Somers-Yeates, Jonathan Bennie, Theodoros Economou, David Hodgson, Adrian Spalding, and Peter K. McGregor. "Light Pollution Is Associated with Earlier Tree Budburst across the United Kingdom." In *Proc. R. Soc. B*, 283:20160813. The Royal Society, 2016.

Dan Duriscoe. "Preserving Pristine Night Skies in National Parks and the Wilderness Ethic." In *George Wright Forum*, 18, no. 4 (2001): 30–36.

Brandi Smith and Jeffrey Hallo. "A System-Wide Assessment of Night Resources and Night Recreation in the U.S. National Parks: A Case for Expanded Definitions." *Park Science* 29, no. 2 (2013): 54–59.

Robert Manning, Ellen Rovelstad, Chadwick Moore, Jeffrey Hallo, and Brandi Smith. "Indicators and Standards of Quality for Viewing the Night Sky in the National Parks." *Park Science* 32, no. 2 (2015): 1–9.

Britton Mace, and Jocelyn McDaniel. "Visitor Evaluation of Night Sky Interpretation in Bryce Canyon National Park and Cedar Breaks National Monument." *Journal of Interpretation Research* 18, no. 1 (2013): 39–57.

Fredrick Collison, and Kevin Poe. "'Astronomical Tourism': The Astronomy and Dark Sky Program at Bryce Canyon National Park." *Tourism Management Perspectives* 7 (2013): 1–15.

INTRODUCTION

THE "NEW" RESOURCES
OF THE NATIONAL PARKS

Robert Manning, Peter Newman, Jesse Barber,
Christopher Monz, Jeffrey Hallo, and Steven Lawson

We begin this book by introducing the subject of natural quiet and natural darkness in the national parks. We trace the evolution of national park resources and conclude that natural quiet and natural darkness are the latest resources to be "discovered" in the national parks. Natural quiet is generally defined as the sounds of nature with no interference by human-caused noise (or noise pollution), and natural darkness is defined as darkness undiminished by human-caused light (or light pollution). Recent research has found that natural quiet and natural darkness are vital to the integrity of many of the planet's ecological relationships, and that they can add substantially to the quality of visiting and appreciating the national parks and related protected areas. A growing body of research has given rise to a program of management by the National Park Service (NPS) and other agencies designed to help protect natural quiet and natural darkness from rapidly increasing noise and light pollution. Finally, we introduce several conceptual frameworks that can help guide the management and protection of natural quiet and natural darkness in national parks and protected areas more broadly.

THE EVOLUTION OF NATIONAL PARK RESOURCES

National parks have been established to protect a variety of natural and cultural resources. But how do we determine what resources should be protected in national parks? Indeed, what is a "resource"? The classic definition of a resource is something that is useful or valuable, but increasingly scarce. Thus, in determining what constitutes a national park resource, the values and beliefs of society may be just as important as the physical characteristics of a resource. It is not surprising, then, that the thinking of society about the

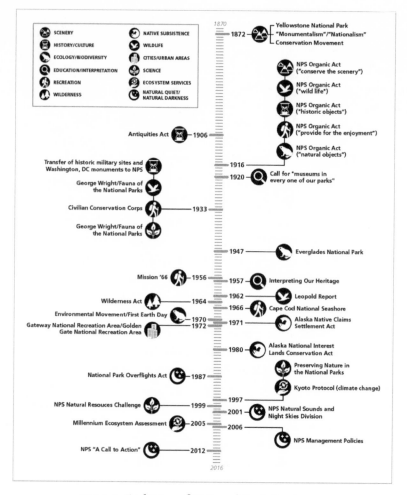

FIG I.1: Evolution of National Park Resources

natural and cultural resources of national parks has evolved over time, and this process continues today. Figure i.1 outlines the evolution of national park resources over the nearly 150 years in which the parks have been in existence, and the first two parts of this introduction describe this evolution.

Of course, most lands designated as national parks were originally occupied and used by Native Americans. These people were unfairly driven from their lands, places that they considered home and that often had spiritual meaning, as well as being vital sources of food and shelter. The subsistence value of these lands existed long before they were designated as national parks, but the subsistence value of national parks was not officially recognized until passage of the Alaska Native Claims Settlement Act of 1971 and the Alaska National Interest Lands Conservation Act of 1980. These laws recognized subsistence uses and rights of Alaska Natives in several large Alaskan national parks that now carry the name "preserve."

The focus of the first national parks (established in the last half of the nineteenth century) was on the vast, sublime landscapes of the American West. Yosemite and Yellowstone are iconic examples. These parks were manifestations of "romanticism" and "nationalism"—the parks were seen, respectively, as refuges from the pollution and crowds of growing urban areas and the social ills of the Industrial Revolution and as important markers of American pride in the monumental wilderness of the US frontier. This was the beginning of the nation's conservation movement, a period in which the nation began to save much of its remaining natural landscape. The Organic Act of 1916 that created the NPS (39 Stat. 535, 16 U.S.C. §1) specifically directed the agency to "conserve the scenery" of the national parks. Thus, the first widely appreciated resource of the national parks was scenery: the wild and awe-inspiring beauty of the landscape.

In the early twentieth century, the significance and meaning of national parks was extended to include history and culture. By then the nation had accumulated a proud and substantive history, including a foundational legacy of independence and freedom. Moreover, the country's astounding prehistory was being discovered at Native American–related sites such as Mesa Verde. History and culture became vital park resources, and more national parks were established to honor these increasingly important elements of US society. Many of these parks were created under the auspices of the 1906 Antiquities Act, which allowed the president to establish national monuments. The Organic Act of 1916 specifically noted the historical values of national parks, directing the NPS to preserve the parks' "historic objects." In 1933, President Franklin Roosevelt signed an executive order that substantially expanded the role of the national parks in historic preservation by transferring most of the War Department's historic and military sites—Gettysburg, Antietam, and many others—and the national memorials of Washington, D.C., to the national park system. It was during this period that history and culture became firmly established as vital resources of the national parks.

Wildlife was considered a national park resource from the very beginning of the parks. The great herds of elk and bison and the grizzly and black bears of Yellowstone helped create excitement about establishing Yellowstone National Park, the first national park, in 1872. However, only with passage of the Organic Act of 1916 was wildlife formally recognized as an important park resource. The act specified that the NPS was to conserve the "wild life" that resided in the national parks. Managing wildlife has proved challenging, and park managers have gone through periods of intense study and introspection, first in the 1930s, under the leadership of George Wright—an NPS biologist and coauthor of the influential report *Fauna of the National Parks of the United States* (G. Wright, Dixon, and Thompson 1933)—and then in the 1960s, with preparation of what was popularly called the Leopold Report (Leopold et al. 1963), which attempted to establish a more science-based approach to wildlife management.

The national parks began to be valued as an educational resource very shortly after passage of the Organic Act of 1916. In 1920, the director of the NPS called for development of "museums in every one of our parks" (quoted in National Park Service n.d.c) and many parks established museums that included artifacts of both natural and human history that proved popular with visitors. Moreover, the NPS established its own brand of education called "interpretation," in an attempt to provide scientific and technical information about the parks to visitors in an engaging and understandable manner. The NPS philosophy of interpretation was first presented in a comprehensive way by the NPS interpreter Freeman Tilden in *Interpreting Our Heritage* (1957).

The twentieth century also witnessed the birth of the modern science of ecology, which played a primary role in the rise of the US environmental movement in the 1960s. One of the most visible manifestations of this growing environmental consciousness was the observation in 1970 of the first Earth Day. The focus on "natural objects" specified in the Organic Act of 1916 was a nascent expression of the ecological values of national parks. However, from a more modern vantage point, the natural components of national parks—such as tall mountains, wild rivers, and iconic wildlife—were more than just natural objects; they were also increasingly important ecological resources. Officials at existing national parks began to recognize and manage these natural resources more carefully as important areas of biological diversity and ecological reserves. Everglades National Park, established in 1947, was the first national park created explicitly to protect biological diversity and ecological integrity. More modern interpretations of ecology suggested that the landscapes of national parks consist of geological and biological resources that are intertwined to form complex and dynamic ecosystems. This ecological reality implied that national parks should be established and managed in a more holistic way and that the national park system be extended to encompass the full array of North American ecosystems and their associated biodiversity. Thus, the natural environment, in the form of landscapes that consist of geologic landforms and the species that inhabit them, has evolved as a foundational national park resource.

Opportunities for outdoor recreation were recognized very early as an important national park resource; the Organic Act of 1916 directed the NPS to "provide for the enjoyment" of the national parks by the people. During the Depression, the Civilian Conservation Corps conducted many public works projects in the national parks, creating trails, roads, and facilities designed to promote recreation and public appreciation of the parks. In the latter half of the twentieth century, the recreational values of national parks were given growing emphasis. The period after World War II in America was marked by sustained growth in the population, economic well-being, increased leisure time, and transportation, and these trends collectively sent millions of Americans to the national parks in search of outdoor recreation opportunities, often overwhelming the capacity of the national parks. The NPS responded in 1956 with its Mission '66 program, a ten-year investment in national park infrastructure—roads, trails, campgrounds, and visitor centers—designed to accommodate rapidly increasing numbers of visitors to the national parks and to celebrate the agency's fiftieth anniversary in 1966. It was during this period that recreation came to be recognized as an increasingly important national park resource. In fact, many new parks were established, such as Cape Cod National Seashore in 1966, to offer more recreational opportunities to the public. Many other recreation-related units were added to the national park system in the following decades, including national seashores, lakeshores, and recreation areas.

The Wilderness Act was passed by Congress in 1964 after a protracted national debate. The act mandated the establishment of wilderness on many federal lands in an effort to preserve these places. Many wilderness areas have been established in the national parks, and now over half of the acreage of the national park system has been declared wilderness by Congress. Thus, wilderness—places that have been saved from economic development or been restored to their natural condition—is one of the most important resources in the national parks.

As unintuitive as it might first seem, urban areas have become a vital national park resource. Many historical and cultural sites are in urban areas—the monuments and memorials of Washington, D.C., for example. But in the 1970s there was a deliberate effort to expand the presence of national parks into some of the nation's great, but often troubled, cities. Social policy suggested that residents of growing cities deserved better access to national parks. Gateway National Recreation Area in New York City and Golden Gate National Recreation Area in San Francisco, both established in 1972, were the first of these national parks. Today, nearly a third of the units of the national park system are in or close to urban areas.

The NPS's relationship to science has had a checkered history, with the agency often lacking a strong commitment to science-based management of the national parks. This is ironic, given that the early government-sponsored expeditions of the Yellowstone area had an explicitly scientific orientation. However, this initial scientific interest quickly faded with the creation of Yellowstone and subsequent early national parks. Interest in science in the national parks was occasionally rekindled—for example, by George Wright and the Leopold Report (Leopold et al. 1963), as noted above. But a more sustained interest in science in the national parks did not appear until the latter part of the twentieth century, largely influenced by the environmental movement and rapid developments in the science of ecology. In 1997, the NPS historian Richard Sellars published *Preserving Nature in the National Parks*, a book that was strongly critical of the NPS's lack of commitment to science. Moreover, the national parks were seen by many scientists as the ideal places to conduct ecological research on natural environments, since they were often relatively undisturbed. In other words, the national parks were seen as an important scientific resource. These ideas coalesced in 1999 with a collaborative effort by Congress and the NPS to establish the agency's Natural Resources Challenge, a program that brought greater emphasis on, visibility of, and funding for science in the national parks. This program is conducted by NPS staff members and a growing core of university-based scientists.

More recently, national park scientists and managers have been paying increasing attention to an expanding array of ecosystem services and their associated benefits. Ecosystem services include all the ways in which human health and well-being depend on a healthy environment. High-profile examples include the physical fitness derived from using the parks, the mental health benefits associated with being in a natural environment, and the sequestration of carbon by the forests preserved in many of the national parks—which helps slow climate change. The Kyoto Protocol of 1997 attracted increasing global attention to the issue of climate change and the role of natural environments, including national parks and related reserves, in moderating this phenomenon. The 2005 Millennium Ecosystem Assessment was an important effort by thousands of scientists and policy makers to compile the multiple ways in which the natural environment contributes to human welfare. Growing scientific and public concern about climate change has placed special emphasis on carbon sequestration as an ecosystem service. The National Park Service's *Healthy Parks Healthy People US* program was established in 2011 to emphasize the benefits of the natural environment to both physical and mental health. Ecosystem services constitute new and important park resources that Americans need to recognize more fully, use more actively, and protect more consciously.

NATURAL QUIET AND NATURAL DARKNESS AS "NEW" PARK RESOURCES

Clearly, the notion of what constitutes important national park resources has evolved over the nearly 150-year history of us national parks. And this evolution is continuing. Two of the newly recognized resources of the national parks are natural quiet and natural darkness. Of course, both of these resources have been present in the national parks from their inception, but they have only recently been added to the agenda of scientists and park managers, and the public is just now becoming more aware of them. As the world becomes a noisier place and more widely bathed in the glare of human-caused light, natural quiet and natural darkness are rapidly disappearing. The national parks are some of the last refuges for these increasingly scarce and valuable resources.

The 1987 National Parks Overflight Act was an early sign of growing awareness of the need for natural quiet and natural darkness in the national parks. This legislation was intended to protect the parks' quiet from the growing practice of scenic air tours over the national parks, often called "flightseeing." In 2000, the NPS established its Natural Sounds and Night Skies Division, to take a broader and more coordinated approach to protecting and managing natural quiet and natural darkness in the national parks. The values of natural quiet and natural darkness received more formal recognition in 2006, when they were codified in the NPS's management policies. Moreover, as the NPS approached its centennial in 2016, the preservation of night skies was singled out in an important report, *A Call to Action* (National Park Service 2011).

Natural Quiet

Many of the sounds of nature are iconic manifestations of the national parks—wolves howling in Yellowstone National Park, elk bugling in Rocky Mountain National Park, the thundering crescendo of the waterfalls of Yosemite National Park, great claps of thunder at Great Smoky Mountains National Park, and the cathedral-like quiet of the groves of ancient trees at Redwood National and State Park and Sequoia National Park. "Soundscapes" are a vital component of the integrity and authenticity of national parks, and these sounds of nature are beginning to be more widely recognized as an important addition to the conventional focus of park managers on preserving landscapes. An increasingly meaningful objective of park managers is to maintain or restore natural quiet. However, it is important to recognize that in some parks, certain human-caused sounds are also integral and appropriate components of the environment. Examples include cannons booming at Gettysburg National Military Park and the tolling of the church bells at San Antonio Missions National Historical Park. Scientists and park managers recognize a subtle, but important, difference between sound and noise: noise is defined as unwanted or inappropriate sound.

Excessive human-caused noise has become a pervasive problem throughout US society, and this noise is increasingly drifting into many national parks. Transportation (cars and other vehicles on park roads and aircraft overhead) is a prime source of noise in many national parks. In fact, over the past forty years, the US population has increased by approximately one-third, while traffic on US roads has tripled. Several measures of aircraft traffic grew by a factor of three or more in approximately the same period. Depending on terrain, weather, and vegetation, transportation-related noise can travel deep into the wilderness portions of the parks. The NPS is instituting public transit systems in many national parks, often using vehicles that rely on propane and other alternative fuels, and this can substantially reduce noise. Good examples of this development include the shuttle bus systems at Zion National Park and Acadia National Park. Since the NPS does not control the airspace above parks, park managers are working with the Federal Aviation Administration to regulate the number of flightseeing aircraft over the national parks, establish minimum flight altitudes, and define acceptable flight routes and flight-free zones. Recent estimates suggest that there are as many as 200,000 tour flights over national parks each year, nearly 100,000 at Grand Canyon National Park alone. Human-caused noise can substantially detract from the quality of visitors' experiences in the national parks: it is an unwanted distraction and can mask the sounds of nature.

Noise can also have important biological and ecological impacts. An increasing number of studies have found that acoustical cues play important roles in animal behavior, including predator-prey interactions, reproductive cycles, and territoriality. Masking of these cues by human-caused noise can threaten wildlife. Recent research suggests that noise-related impacts on wildlife may be especially pronounced in marine environments. Human-caused noise has also been found to have a detrimental effect on people, including reduced cognitive function and sleep disturbance.

Natural Darkness

Natural darkness is also receiving more attention in the national parks by scientists, managers, and visitors. The growing awareness that natural darkness is a national park resource is a result of the expanding recognition of its value and its rapid disappearance. For millennia, our ancestors gazed upon the cosmos in their enduring efforts to understand the physical and metaphysical worlds, which suggests that night skies are an important cultural resource. Human culture is conventionally organized around the rhythms of the sun, moon, and stars; observations of the night sky are embodied in the religions and mythology of cultures around the globe; and the celestial world has been the inspiration for art, literature, and other forms of cultural expression. But natural darkness is being lost as populations grow and related development occurs. Human-caused light is pervasive in a world where commerce and other human activities occur twenty-four hours a day. An increasing number of park visitors cannot see the Milky Way from their home communities and relish the opportunity to experience natural darkness and to see and experience a pristine night sky in the national parks. And park managers are responding with more interpretive programs focused on the night sky and other activities in natural darkness, such as observing nocturnal wildlife.

Traditional night-dependent recreation in national parks—like camping, making campfires, and stargazing—may become even more important to visitors as increasing urbanization makes these activities less common. But even within parks, lights from vehicles, roadways, buildings, the activities of visitors and managers, and other human-caused sources may lead to diminished enjoyment of darkness and night-dependent recreation activities. However, it should be noted that like sound, light itself is not inherently deleterious in the national parks. In some cases, well-designed and appropriate lighting allows some park resources to be viewed at night and more fully appreciated. For example, lighting allows visitors to appreciate the caves of Carlsbad Caverns National Park and Mammoth Cave National Park and adds drama and aesthetic appeal to Statue of Liberty National Monument and the monuments and memorials of the National Mall and Memorial Parks in Washington, D.C.

Contemporary science has begun to recognize the importance of natural darkness by demonstrating the vital role of darkness in the biological world. Many species—estimated to represent at least half of all animals—rely on the absence of light for breeding and feeding, seasonal migrations, and other vital behaviors. For example, the flower of the queen of the night cactus opens for only a few hours on one night each year and is pollinated by bats and moths. The absence of natural darkness could affect the behavior of these pollinators, thus endangering the ability of the cactus to reproduce. Artificial light can also affect humans through sleep disturbance and related health effects.

THE GROWING SCIENCE OF
NATURAL QUIET AND NATURAL DARKNESS

Recent research has dramatically demonstrated that both natural quiet and natural darkness are rapidly disappearing, even in the national parks. For example, a study of many US national parks and protected areas found noise pollution to be pervasive, often surpassing levels known to disrupt wildlife behavior, threaten animals' fitness, and degrade the quality of visitors' experiences (Buxton et al. 2017). Some park roads have sound levels that in certain circumstances have been found to be more than four times higher than natural conditions (Barber et al. 2009). Even remote backcountry areas are not immune, especially due to aircraft overflights. During peak traffic hours, aircraft can be heard at the Snow Flats backcountry site of Yosemite National Park nearly 70 percent of the time (Barber et al. 2009). At Grand Canyon National Park, where air tours have become especially popular, aircraft noise can be heard nearly 80 percent of the time in certain locations (Bell et al. 2009).

Night skies are endangered primarily by light pollution that reduces the brightness of the stars and other celestial objects and prevents the human eye from fully adapting to natural darkness (Bogard 2013). "Light trespass"— when light extends beyond one person's property onto another's—is another term for this issue. It is conventional in our society to object to any form of pollution imposed unfairly on another person, and artificial light is increasingly considered a pollutant. Some observers have even likened light trespass to secondhand smoke. Outdoor lighting that is excessive, inefficient, or ineffective can produce light pollution that degrades the quality of natural darkness and the night sky by creating glare and sky glow (light that is reflected back to earth by the atmosphere). It has been estimated that by 2000 two-thirds of Americans could no longer see the Milky Way from their homes (Cinzano et al. 2001). Light pollution is caused partly by increasing development, but it may be more related to lighting that is oriented upward or sideways, thus spilling into the atmosphere, rather than down at the intended target. Light from urban areas can reduce the brightness of the night sky over two hundred miles away (National Park Service 2017b).

Biology and Ecology

A growing body of scientific and professional literature is beginning to document the impacts of human-caused noise and artificial light on the ecology of national parks and the quality of visitors' experiences there and in related areas. As noted above, an increasing number of studies have found that human-caused noise can interfere with a variety of animal behaviors such as predator-prey relationships, breeding and reproductive success, territoriality, communication with young, and assessment of habitat quality (Barber et al. 2009; Rabin et al. 2006). For example, a study of the effect of anthropogenic noise on boreal songbirds showed that breeding success was significantly reduced in noisy environments compared to control sites, noisy areas were disproportionally populated with younger males that were less successful in attracting mates, and bird populations were 1.5 times denser in quiet control sites than in nearby noisy areas (Bayne et al. 2008). Studies have also shown significant effects of noise on ungulates (for example, deer and elk), particularly when these studies have been conducted on a large landscape scale (Barber et al. 2010). For example, within three miles of human infrastructure or activities, caribou reduced their use of the habitat by 50–95 percent. Noise-related impacts on wildlife may be especially pronounced in marine environments, and this warrants more research and management attention.

By demonstrating the importance of darkness in the biological world, research is also raising substantial concerns about the ecological impacts of light pollution. Many species are nocturnal and rely on the absence of light for breeding, feeding, migration, and other behaviors. A classic example is the marine turtle. Turtle hatchlings emerge at night from their nests on ocean beaches and instinctively crawl toward the lighted horizon. Traditionally, this horizon has been the sea, as it has reflected the light of celestial bodies. But now the land is often lighted from nearby development, and this attracts many turtle hatchlings that ultimately die from predation, exhaustion, or desiccation. Hundreds of thousands of hatchlings die each year in Florida alone. Gulf Islands National Seashore uses a core of volunteers who "nest sit" active turtle reproductive sites, but there is a limit to the volunteers' effectiveness. The emerging scientific literature suggests that natural darkness is as important as the light of day to the functioning of ecosystems. And there is a growing literature on the effects of light pollution on humans through sleep disorders and related health effects.

Experience

As noted above, noise and light pollution can affect the quality of visitors' experiences in national parks and related areas. Studies spanning many years have documented that escaping noise and enjoying the sounds of nature are important motivations for visiting national parks and related reserves (see, for example, Driver et al. 1991). In fact, a national survey found that 72 percent of Americans reported that opportunities to experience natural quiet and the sounds of nature are a very important reason for preserving national parks (Haas and Wakefield 1998). In another survey, 91 percent of park visitors considered enjoying both natural quiet and the sounds of nature to be very important reasons for visiting these areas (C. McDonald et al. 1995). Moreover, a number of studies have documented conflicts between motorized and nonmotorized recreation, and the noise associated with motorized recreation is one of the causes of this conflict (Manning 2011). A series of focus group sessions with visitors and other stakeholders at Yosemite National Park found that participants reported a number of noise-related issues as detracting from the quality of their park experiences, including noise from tour buses, automobiles, recreational vehicle generators, aircraft, machinery, construction, and radios (Manning 1998). A recent program of noise-related research conducted at Muir Woods National Monument identified a number of things that visitors felt increased the quality of their experiences at the park, including the sounds of water flowing in Redwood Creek, birds calling, and wind blowing in the trees. Factors that decreased the quality of their experiences included visitor-caused noise, such as loud talking and boisterous behavior (Pilcher et al. 2009; Stack et al. 2011). The same program of research also identified the potential maximum level of visitor-caused noise acceptable to visitors and found that these noise thresholds were sometimes violated in the park (Manning et al. 2010).

A similar program of research on night skies was recently conducted at Acadia National Park (Manning et al. 2015). A large majority of survey respondents reported that the quality of the night sky at Acadia was important to them and that the NPS should manage the park to protect the quality of the night sky. Most respondents reported that they had not seen many of the celestial objects included in the survey (for example, moon, stars, and constellations), but for those who had, seeing the object substantially added to the quality of their experience. Similarly, most respondents said that they had not seen many of the sources of human-caused light included in the survey (such as car headlights and lighted buildings) and that this had substantially added to the quality of their experience. Minimum thresholds for night sky conditions were also identified in this program of study.

MANAGEMENT OF NATURAL QUIET AND NATURAL DARKNESS IN THE NATIONAL PARKS

National parks, especially those far from urban areas, are some of the last refuges of both natural quiet and natural darkness, and attention to these resources is increasingly reflected in NPS policy and management. For example, it has been argued that the night sky should be recognized as an important and increasingly scarce resource that must be managed and preserved, and that this is a natural extension of the Organic Act of 1916 as well at the Wilderness Act of 1964 (Public Law 88577, 16 U.S.C. 1131–1136, September 3, 1964), both of which call for preservation of important natural and cultural resources (Duriscoe 2001). This argument can be extended to natural quiet as well. Current NPS management policies explicitly address natural quiet and natural darkness (National Park Service 2006), and the NPS created a Natural Sounds and Night Skies Division in 2000 with the following mission: "The Natural Sounds and Night Skies Division works to restore, maintain, and protect acoustical environments and naturally dark skies throughout the National Park System. We work in partnership with parks and others to increase scientific understanding and inspire public appreciation of the value and character of undiminished soundscapes and star-filled skies" (National Park Service 2017b). This division is monitoring natural quiet and natural darkness in the national park system and working with park staff members to manage these resources.

Night sky interpretive programs are now conducted at an increasing number of national parks, as shown by the night sky festivals and "star parties" at Acadia National Park, Death Valley National Park, Yosemite National Park, and a number of other national parks; the creation of a night sky ranger position at Bryce Canyon National Park; and the development of an observatory at Chaco Culture National Historical Park. The NPS established its Night Sky Team—a small group of scientists—in 1999, and this has led to rigorous measures of night sky quality and associated monitoring in the national park system. Night sky quality is included as a vital sign by many of the thirty-two NPS Inventory and Monitoring Networks that cover the national park system. The recent influential NPS report, "A Call to Action," includes a recommendation that the agency "Lead the way in protecting natural darkness as a precious resource and create a model for dark sky protection" (National Park Service 2014). A recent survey of managers across the national park system found that night skies (and night resources more broadly, including the opportunity to observe nocturnal species) are frequently appreciated by visitors and that managers are interested in doing more to identify and manage night resources (B. Smith and Hallo 2013).

An example of these management programs is an effort at Muir Woods National Monument designed to reduce visitor-caused noise in the Cathedral Grove portion of the park, a place where natural quiet is especially important. Through signs and interpretive programs, visitors are asked to reduce the noise they make by turning off cell phones and car alarms and talking in low voices, and this has quieted the area dramatically (Manning et al. 2010; Pilcher et al. 2009; Stack et al. 2011). As noted above, an increasing number of parks are holding annual star parties or night sky festivals designed to sensitize visitors to the importance of night skies. In addition, many national parks are retrofitting park lighting, including reorienting the lighting downward, reducing unneeded light, and installing lights activated by motion sensors to reduce light in the parks.

Like an increasing number of other national park management issues, management of natural quiet and natural darkness is inherently challenging because the NPS cannot directly control noise and light generated outside the parks. However, the NPS has been able to work with some outside communities and other partners to help address these issues. For example, park managers at Chaco Culture National Historical Park worked with a number of stakeholder groups to support passage of the New Mexico Night Sky Protection Act, which regulates outdoor lighting throughout the state, and managers at Acadia National Park worked with officials in the gateway town of Bar Harbor to pass a progressive lighting ordinance.

National parks are vital refuges for people seeking a respite from the din and glare of everyday life. It can be remarkably rewarding to hear the sounds of nature, and an increasing amount of research suggests that these sounds can even be therapeutic (Hartig et al. 2003). Moreover, having a clear view of the Milky Way—the edge of our galaxy—can offer a comforting and thought-provoking perspective of our place in the universe. Plants and animals need these refuges as well. While challenging, managing these issues can have multiple benefits—improving the habitat for wildlife, increasing visitors' enjoyment and appreciation, and lowering costs for the public by reducing the amount of energy needed and reliance on the automobile. Moreover, human-caused noise and light are two types of pollution that can be greatly reduced with little financial cost to society. For example, there can be substantial long-term cost savings when excessive and inefficient lighting is reduced and modernized.

FRAMEWORKS FOR MANAGING NATURAL QUIET AND NATURAL DARKNESS IN THE NATIONAL PARKS

With the increasing recognition and importance of natural quiet and natural darkness as resources in national parks and related areas, careful thought should be given to how these resources should be managed and protected there. Too much human-caused noise and light can threaten the ecological integrity of parks and degrade the quality of visitors' experiences. But how can natural quiet and natural darkness be protected in the national parks? A number of conceptual and organizational frameworks have been developed in the scientific and professional literature on parks and outdoor recreation that can help sharpen our thinking about managing natural quiet and natural darkness. These frameworks provide a foundation for helping managers meet the difficult but fundamentally important objectives of national parks and related areas.

The Dual Mission of National Parks

National parks are established for two sometimes competing purposes: to protect important natural and cultural resources and to offer the public opportunities to use, enjoy, and appreciate these areas. When parks are used for outdoor recreation, vital natural and cultural resources can be affected and degraded, as can the quality of visitors' experiences. Yellowstone National Park was established in 1872, but the NPS, the agency charged with managing the national parks, was created by Congress only in 1916. In a classic phrase, the legislation creating the NPS states that the national parks are to be managed "to conserve the scenery and the natural and historic objects and the wild life [sic] therein and to provide for the enjoyment of the same in such manner and by such means as will leave them unimpaired for the enjoyment of future generations." How can national parks be managed to achieve both objectives?

In the case of natural quiet and natural darkness, the recreational use of the national parks and the infrastructure needed to accommodate it contributes to human-caused sounds and light in and around the parks, which now collectively have more than 300 million visits annually. How much and what kinds of visitor use can be accommodated in the parks before their biological or ecological integrity is threatened and the quality of visitors' experiences is diminished to an unacceptable degree? Recognition of the dual and sometimes conflicting missions of national parks can help guide appropriate park management.

Common Property Resources

A classic paper in the environmental literature, titled "The Tragedy of the Commons," was published in the prestigious journal *Science* in 1968 (Hardin 1968). This paper identified a set of environmental problems—issues related to the commons, or lands and associated resources owned by society at large—that must be resolved through public policy and management action. Without explicit management, there is an inherent tendency to overuse common property resources. Garrett Hardin's ultimate prescription for managing the commons was through "mutual coercion, mutually agreed upon" (1968, 1247): without such collective action, environmental (and related social) tragedy is inevitable.

Hardin began his paper with an illustration that used perhaps the oldest and simplest example of an environmental commons, a shared pasture. Each herdsman is tempted to graze additional cattle on the commons because he reaps all the benefits but pays only a portion of the costs of resulting environmental degradation (these costs are shared by all herdsmen or even society at large). Hardin went on to identify and explore other examples of environmental commons, including national parks: "The National Parks present another instance of the working out of the tragedy of the commons. At present, they are open to all without limit. The parks themselves are limited in extent—there is only one Yosemite Valley—whereas population seems to grow without limit. The values that visitors seek in parks are steadily eroded. Plainly, we must soon cease to treat the parks as commons or they will be of no value to anyone" (1968, 1245).

The management of national parks represents an example of "mutual coercion, mutually agreed upon" that Hardin suggests is needed to protect parks, including their environmental and experiential resources. While this coercion (management and policy, in more contemporary terms) may be distasteful because it restricts freedom of choice—for example, by limiting the amount of use or types of use such as automobiles—it is ultimately needed to protect parks, the recreational experience in them, and the greater welfare of society (Manning 2007).

Natural quiet and natural darkness are quintessential examples of common property resources, as they are "owned" by society at large but "used" by individual people, businesses, communities, and so on. For example, a recreational vehicle owner can use a generator in a national park campground for his own benefit, but this may cause noise that degrades the experience of other park visitors. Similarly, the owner of a business near a national park might install nighttime lighting that serves the needs of her customers, but that diminishes the beauty of the night sky, thereby affecting the quality of the park experience. The concept of common property resources offers a foundation that can help guide informed management of national parks and protection of natural quiet and natural darkness.

Carrying Capacity

Carrying capacity (or visitor capacity, as it is sometimes called in the context of parks and other protected areas) has been an important part of natural resources and environmental management for decades. Its emergence can be traced to a historic publication titled *An Essay on the Principle of Population* (Malthus 1798). In this essay, Malthus reasoned that the human population tends to grow at an exponential rate, but that production of food and other vital resources tends to grow only linearly. In this way, the supply of food and other resources presents an ultimate limit to population growth, and if this limit is not respected, the carrying capacity of the earth (or selected geographic regions) may be exceeded, with highly detrimental effects. Malthus's ideas about carrying capacity and the limits of the earth to support human population growth have become foundational components of the contemporary environmental movement. In this sense, there is a strong similarity between carrying capacity as applied to parks and sustainability as applied to environmental management more broadly.

Scientific applications of carrying capacity were first advanced in the fields of fisheries, wildlife, and range management (Hadwen and Palmer 1922; Leopold 1933; Odum 1953), to calculate, for example, how many animals can ultimately be supported by a given area of range. Carrying capacity was first applied to parks and outdoor recreation in the 1960s (Lucas 1964; Wagar 1964). In this context, visitor capacity is generally defined as the amount and type of recreation that can be accommodated in a park without having unacceptable impacts on the park's resources and/or the quality of visitors' experiences (Manning 2007). Early research on visitor capacity sought to apply this concept exclusively to the environmental impacts of outdoor recreation, but this was soon supplemented by a consideration of experiential issues. In the preface to his influential monograph on carrying capacity, Alan Wagar (1964) wrote: "The study reported here was initiated with the view that the carrying capacity of recreation lands could be determined primarily in terms of ecology and the deterioration of areas. However, it soon became obvious that the resource-oriented point of view must be augmented by consideration of human values" (1964 n.p.).

Wagar's point was that as more people visit a park or related outdoor recreation area, not only are the environmental resources of the area affected, but so is the quality of visitors' recreation experiences, and park and outdoor recreation managers must address both resource and experiential concerns. Moreover, there are potentially important interactions between these components. For example, impacts on park resources can degrade the aesthetic quality of visitors' recreation experiences. Informed management of the visitor capacity of parks and related areas must take into account both of these components of parks and outdoor recreation and their potential interactions.

Finally, it is important to note that visitor capacity can be affected by the type and intensity of management (Manning 2007; Wagar 1968). For example, the durability of natural resources might be enhanced by fertilizing and irrigating vegetation, and the quality of recreation experiences might be enhanced by distributing recreation use more evenly across both space and time, thereby decreasing crowding.

Visitor capacity has remained an important, but challenging and often contentious, issue in the field of parks and outdoor recreation (Graefe et al. 1984; Manning 2007 and 2011; Shelby and Heberlein 1986; Stankey and Manning 1986; Whittaker et al. 2011). What is the ultimate capacity of parks for outdoor recreation? How can outdoor recreation be managed to ensure that it does not exceed a park's visitor capacity? These questions can and should be applied to natural quiet and natural darkness, just as visitor capacity is conventionally applied to other park resources and experiences (such as impacts on wildlife and the resulting diminished opportunities to observe wildlife in the national parks). Natural quiet and natural darkness are increasingly important park resources that have both environmental and experiential components, and management of these resources should be guided by consideration of the concept of visitor capacity. The way in which visitor capacity is conventionally understood and managed is outlined in the following two sections.

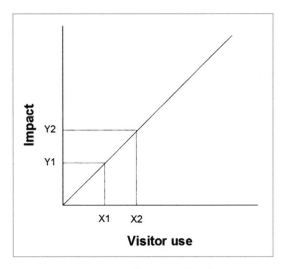

FIG I.2: Limits of Acceptable Change

Limits of Acceptable Change

Research on the application of visitor capacity to parks and outdoor recreation has documented a number of impacts that recreation can have on park resources and the quality of visitors' experiences (Hammitt et al. 2015; Manning 2011). For example, park visitors may trample fragile soils and vegetation and disturb wildlife. And as the number of visitors increase, parks may become crowded. With increasing use of parks comes increasing environmental and experiential impacts, and at some point these impacts may become unacceptable. But what determines the limits of acceptable change?

This issue is illustrated graphically in Figure i.2. This figure shows that increasing amounts of recreation result in increasing environmental and experiential impact. As the amount of recreation use increases from x1 to x2, the amount of impact increases from y1 to y2. (Figure i.2 is hypothetical; research suggests that such relationships may not be linear as suggested in the figure.) However, the limits of acceptable change are not clear from this relationship. Which of the points on the vertical axis—y1 or y2, or some other point—represents the maximum acceptable level of impact? Determining the limits of acceptable change is central to addressing the visitor capacities of parks for outdoor recreation (Frissell and Stankey 1972).

To emphasize and further clarify the limits of acceptable change and its relationship to visitor capacity, some writers have suggested distinguishing between descriptive and prescriptive components of visitor capacity (Shelby and Heberlein 1984 and 1986). The descriptive component focuses on factual, objective data such as the relationships in Figure i.2. For example, what is the relationship between the amount of visitor use and perceived crowding? The prescriptive component addresses the seemingly more subjective issue of how much impact or change is acceptable. For example, what level of perceived crowding should be allowed? Determining acceptable limits of change must be the foundation for managing the visitor capacity of parks, including the biological, ecological, and experiential impacts of humans on the resources of natural quiet and natural darkness.

Indicators and Thresholds

Contemporary approaches to determining the limits of acceptable change in parks and outdoor recreation are based largely on formulating management objectives and associated indicators and thresholds (also called standards) (Interagency Visitor Use Management Council 2016; Manning 2007 and 2011; Manning and Anderson 2012; Whittaker et al. 2011). Management objectives are statements about the desired conditions of parks and outdoor recreation, including the level of protection of park resources and the types and quality of visitors' experiences. Indicators are more specific, measurable, and manageable variables that reflect the meaning or essence of management objectives; they are quantifiable proxies of management objectives. Thresholds are numerical expressions of desired conditions for indicator variables; they are the minimum acceptable condition of indicator variables.

As an example, many national parks have developed networks of iconic roads that were designed and constructed for visitors to see and appreciate the parks from their cars. Thus, the management objective for this type of recreational opportunity might be called "scenic driving" or "driving for pleasure." Associated indicators and thresholds might focus on setting and enforcing a maximum level of traffic noise on these roads, as is implied by the management objective of "driving for pleasure." An indicator might be the loudness of traffic that can be heard on a given length of road, and a threshold might specify the maximum decibels of traffic noise. Indicators and thresholds must be developed to help inform the management of national parks, including the management of natural quiet and natural darkness.

Management objectives and indicators and thresholds have been incorporated as vital components of a management-by-objectives framework for addressing park carrying capacity and park management more broadly. This

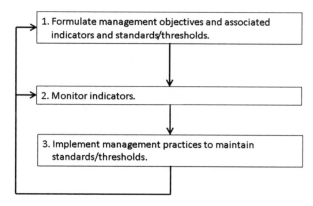

FIG I.3: Management-by-Objectives Framework

framework is illustrated in Figure i.3 and involves three steps. First, management objectives must be formulated. These are broad narrative statements that suggest the level of resource protection and the type of visitor experience that should be provided and maintained. Management objectives must then be expressed in a more quantitative manner, in the form of indicators and thresholds, so they can be monitored. Second, indicators must be monitored to determine their condition. Third, management actions must be implemented to maintain thresholds. This management-by-objectives framework is an adaptive process, in that monitoring provides periodic information about the condition of indicator variables as well as the effectiveness of management actions, and managers must adjust their management regime as needed. This framework can be used to address the more contemporary and generic environmental issue of sustainability.

A Landscape-Scale Approach to National Park Management

Many national parks and related protected areas are relatively large. For example, Yellowstone National Park is 2.2 million acres, and the largest national park—Wrangell–St. Elias National Park in Alaska—is more than 13.0 million acres. Yet nearly all national parks are incomplete geographic regions when viewed from an ecological or even a societal perspective. For example, Everglades National Park includes only a portion of the watershed necessary to protect its iconic wetlands and associated wildlife. Civil War parks such as Gettysburg National Military Park and Antietam National Battlefield protect the sites of iconic battles but are not nearly large enough to interpret the greater context of the battles fought at these sites and the related stories intimately associated with the Civil War. Important corollaries of this issue are that parks are not self-sufficient "islands," and that park managers must work with stakeholders outside park boundaries to protect the integrity of national parks.

There is a large and growing interest in taking a larger landscape-scale approach to national park management and conservation more broadly. For example, the conservation community is working at this larger geographic scale as it considers areas such as the Greater Yellowstone Ecosystem, Greater Grand Canyon, Yellowstone to Yukon, Crown of the Continent, and Path of the Pronghorn. In these areas and many others, national park staff members are developing working relationships with other public land managers, local communities, nonprofit organizations, and even private landowners to ameliorate the impacts on park resources that emanate from outside park boundaries.

The management and protection of natural quiet and natural darkness often require this type of landscape-scale approach. Many national parks are located close to population centers, and the noise and artificial light associated with these areas spill over into the parks. For example, light and noise from San Francisco can often be seen and heard on the beaches and bluffs that are a part of Golden Gate National Recreation Area. In fact, this human-caused noise and light extends for many miles to other national park areas, such as Muir Woods National Monument and Point Reyes National Seashore. Light from large cities can pollute the sky up to 200 miles away. As noted above, there are encouraging examples of park representatives having successfully worked with local communities and others to reduce the noise and light that filters into the parks. Working relationships among national parks and with other land management agencies, businesses, nonprofit organizations, and communities will be needed to effectively address noise- and light-related issues on a broad, landscape scale.

THE OBJECTIVES AND ORGANIZATION OF THE BOOK

A substantial body of research on natural quiet and natural darkness in parks and protected areas more broadly, including outdoor recreation in these areas, has been developed in the scientific and professional literature in recent years. However, this literature is widely scattered over a variety of academic and professional journals that cover both the natural and social sciences. The two primary objectives of this book are to identify, collect, organize, integrate, and synthesize representative and important components of this work, and to develop a series of principles or best management practices to help guide the study and management of parks to protect natural quiet and natural darkness.

The book is organized into four parts plus an introduction and conclusion. This introduction has traced the evolution of thought about what constitutes national park resources, described the emergence of natural quiet and natural darkness as increasingly important park environmental and experiential resources, and outlined a series of conceptual frameworks that might be used to help inform the management of natural quiet and natural darkness in the national parks. Parts 1 and 2 of the book contain a set of edited articles from the scientific and professional literature that addresses the biological or ecological (part 1) and experiential (part 2) components of natural quiet. Parts 3 and 4 use the same organizational approach for natural darkness: part 3 addresses the biological or ecological components of natural darkness, and part 4 addresses its experiential components. The conclusion draws on these and other articles to develop a series of principles or best practices for studying, managing, and protecting natural quiet and natural darkness in the national parks and related reserves.

Although the book focuses primarily on us national parks, its derivation and applications are considerably broader. Our emphasis on the national parks is mostly derived from our own programs of research, which have largely been conducted in a number of national parks across the country. Moreover, much of the scientific and professional literature we accessed on natural quiet and natural darkness is based on fieldwork conducted in the United States. However, the book includes studies conducted outside the United States, and several of the us-based studies described in the book were conducted in protected areas other than the national parks. All of the studies that we found in the scientific and professional literature—those in the us national parks, in other types of protected areas, and outside the United States—have contributed to the findings of this book. In particular, we have no reason to think that the principles we derive in the conclusion do not apply to parks and other protected areas everywhere.

In developing this book, our overarching objective has been to attract increased attention to both existing and needed science-based inquiries into natural quiet and natural darkness. Many of these efforts focus on national parks or other protected areas because these places are of such great importance to society, and they may be sites that can demonstrate how issues related to noise and light pollution can be best dealt with. We also hope to inspire students as well as academics and practitioners (both veteran and fledging) to expand their training, scientific inquiries, and efforts to include these topics. Relatively few people worldwide currently study or do other work directly related to the park resources of natural quiet and natural darkness whose importance is increasingly recognized. Yet many people have stories about the importance of a quiet place to enjoy the outdoors or experiences outdoors that were made better because of the darkness of night. We hope that you will use this book to begin (or refresh) your role as a person who cares about and helps improve natural quiet and natural darkness in parks and other protected areas.

THE ECOLOGY OF NATURAL QUIET

Part 1 of this book addresses the ecology of natural quiet. Natural quiet is a vital element of the biological and ecological integrity of national parks and protected areas more broadly. However, human-caused noise can affect natural quiet and other park resources as well. The five chapters in this part of the book document many of the wide-ranging impacts of human-caused noise. In addition to describing impacts on individual animals and plants, the authors note that these impacts can extend from individuals to species and even to communities of species and a suite of ecological services. For example, altered spatial and temporal patterns of noise-sensitive species that pollinate plants and disperse seeds can affect the foundational biological or ecological structure of forests.

1

A FRAMEWORK FOR UNDERSTANDING NOISE IMPACTS ON WILDLIFE

An Urgent Conservation Priority

Clinton Francis and Jesse Barber

A substantial literature has begun to uncover the costs of noise exposure for wildlife. This review article places over a decade of research into an integrated framework with the aim of understanding the mechanistic drivers of noise pollution's ecological effects on animals and ecosystems more broadly. The authors highlight several documented and possible routes through which a louder world might alter animal behavior, physiology, and, ultimately, fitness via changes in spatial distributions, temporal patterns, foraging efficiency, antipredator behavior, mate attraction, and territory defense. It is critical that we understand how noise influences wildlife so that protected area management strategies can be optimized to "protect the wildlife therein" (Organic Act 1916; 39 Stat. 535, 16 U.S.C. §1).

INTRODUCTION

An emerging aim in applied ecology and conservation biology is to understand how human-generated noise affects taxonomically diverse organisms in both marine (e.g., Slabbekoorn et al. 2010; Ellison et al. 2012) and terrestrial environments (e.g., Patricelli and Blickley 2006; Barber et al. 2010; Kight and Swaddle 2011). Noise is a spatially extensive pollutant and there is growing evidence to suggest that it may have highly detrimental impacts on natural communities, yet efforts to address this issue of emerging conservation concern lack a common framework for understanding the ecological consequences of noise. A conceptual scaffold is critical to scientific progress and to its ability to inform conservation policy. As more attention and resources are invested in understanding the full ecological effects of noise, it is important that investigators design research questions and protocols in light of the many possible costs associated with noise exposure and also that they properly link responses to several relevant features of noise, such as intensity, frequency and timing, that could explain wildlife responses.

Here we provide a framework using a mechanistic approach for how noise exposure can impact fitness at the level of the individual organism as a result of changes in behavior, and identify several acoustic characteristics that are relevant to noise exposure and ecological integrity. We provide representative examples of noise impacts, primarily from terrestrial systems; however, these issues are equally applicable to organisms in aquatic environments. We stress that many responses to noise exposure are less obvious than those that have typically been studied to date; these include signal modifications (e.g., changes in vocal frequency, amplitude, or vocalization timing) and decreases in site occupancy (Bayne et al. 2008; Francis et al. 2011c). Importantly, many likely behavioral responses to noise that merit further scientific study might be detrimental to individual fitness and have severe population-level consequences. As we show below, the presence of a species in a noisy area cannot be interpreted as an indication that it is not impacted by noise because there are many potential costs associated with noise exposure that have not been rigorously studied.

VARIATION IN RESPONSES TO THE SAME NOISE STIMULUS

Species differ in their sensitivities to noise exposure (Bayne et al. 2008; Francis et al. 2009a; Francis et al. 2011b); however, the degree to which individuals vary in sensitivity to noise during each life history stage or due to behavioral context has been underappreciated. For example, ovenbird (*Seiurus aurocapilla*) habitat occupancy appears uninfluenced by noise exposure (Habib et al. 2007; Bayne et al. 2008; Goodwin and Shriver 2011), yet males defending noisy territories are less successful in attracting mates (Habib et al. 2007). Reed buntings (*Emberiza schoeniclus*) also show reduced pairing success in noisy areas (Gross et al. 2010). Such examples should serve as a warning to biologists, land managers, and policy makers: the same noise stimulus can affect various response metrics in different ways. An organism might show little to no response to noise in terms of habitat occupancy or foraging rate, for example, but may experience strong negative impacts in terms of pairing success, number of offspring, physiological stress, or other measures of fitness (Figure 1.1). Because the various responses may range from

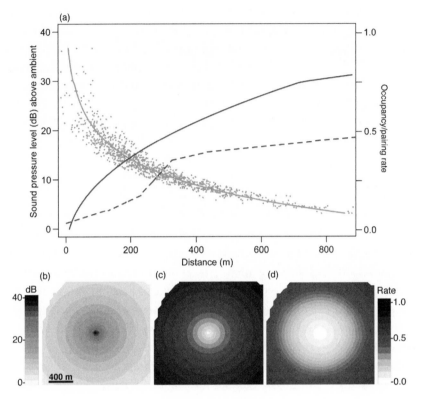

FIG 1.1: Responses to the same noise stimulus can take a variety of shapes. (a) The sound pressure level (SPL) of noise (gray) decreases with increasing distance from the source but may not reach "baseline" ambient levels until ˜1 kilometer away (this distance will vary depending on noise source and the environment). Response curves for species occupancy (black solid line) and pairing rates (black dashed line) in response to noise may have unique shapes, as might many other measures of species responses to noise stimuli. The relationship between SPL and distance is from Francis et al. (2011a) and Francis (unpublished data) with noise generated from gas well compressors. Behavioral responses are hypothetical but based on responses in Francis et al. (2011a). (b) Spatial propagation of elevated noise levels from a point source (such as a single car or an oil/gas compressor station), which decays at a spreading loss of 6 dB or more per doubling of distance, due to the geometry of the spherical wave front. It is important to note that line sources (such as a busy highway; not shown) lose only 3 dB per doubling of distance due to their cylindrical wave front. Clearly, knowledge of the geometry of anthropogenic noise stimuli is essential to understanding scale of exposure. (c) Spatial representation of species occupancy and (d) pairing success surrounding a point source of noise.

linear to threshold functions of noise exposure, investigators should take an integrative approach that incorporates several different metrics (e.g., density, pairing success, number of offspring), rather than using a single metric to describe how noise influences their study organism. But which alterations to behavior are most likely to occur and which are most detrimental? These are important questions because funding and logistical constraints ensure that measuring all of the potential impacts of noise is impossible. Fortunately, the nature of sound stimuli can guide investigators toward likely behavioral changes that may influence fitness.

CHARACTERIZING NOISE AND THE DISTURBANCE-INTERFERENCE CONTINUUM

Determining whether a particular noise stimulus is within an organism's sensory capabilities is foremost in importance; if a sound consists of frequencies that are outside of an organism's hearing range, it will not have a direct effect (Figure 1.2). Provided that an organism can hear the noise stimulus, its acoustic energy could cause permanent or temporary hearing loss, but this might only occur when the animal is extremely close to the source of the noise (Dooling and Popper 2007).

Instead, sounds may have their greatest influence on behavior, which then translates into fitness costs, but how and why noise elicits a response can vary greatly (Figures 1.2 and 1.3). At one extreme, noise stimuli that startle animals are perceived as threats and generate self-preservation responses (e.g., fleeing, hiding), which are similar to responses to real predation risk or non-lethal human disturbance (i.e., the risk-disturbance hypothesis, which posits that animal responses to human activities are analogous to their responses to real predation risk; Frid and Dill 2002). Noise stimuli at this end of the continuum are often infrequent, but are abrupt and unpredictable. At the other end of the continuum, noise can impair sensory capabilities by masking biologically relevant sounds used for communication, detection of threats or prey, and spatial navigation. These noise stimuli tend to be frequent or chronic and their spectral (i.e., frequency) content overlaps with biologically relevant sounds. Increases in noise intensity (loudness or amplitude) will increase the severity of the impacts, regardless of whether it is perceived as a threat or masks biologically relevant sounds. An important supplement to this dichotomy is that limited stimulus processing capacity could be responsible for some detrimental effects. For example, noise stimuli of various kinds may act as a distraction, drawing the animal's attention

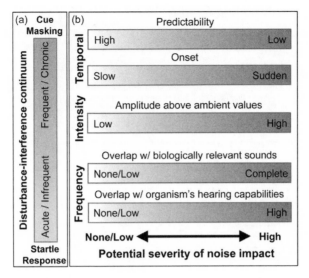

FIG 1.2: (a) The disturbance-interference continuum can range from acute or infrequent noise stimuli that will likely trigger startle or hide responses to frequent or chronic noises that interfere with cue detection. (b) The severity of an impact from a noise stimulus will depend on the temporal, intensity, and frequency features of the stimulus.

to a sound source and thereby impairing its ability to process information perceived through other sensory modalities (A. Chan et al. 2010b). Alternatively, noise may reduce auditory awareness, trigger increased visual surveillance, and compromise visually mediated tasks. The mechanistic details and ecological importance of such distractions still need to be fully explored. Regardless, the conservation implications of understanding the importance of noise as a distractor are not trivial; if distraction is a fundamental route for noise impacts, our concern might spread beyond those frequencies that overlap with biologically relevant signals.

BEHAVIORAL CHANGES

Although a limited number of laboratory studies have suggested that noise may affect gene expression, physiological stress, and immune function directly (Figure 1.3a; Kight and Swaddle 2011), the majority of noise-related impacts appear to involve behavioral responses across four categories: (1) changes in temporal patterns; (2) alterations in spatial distributions or

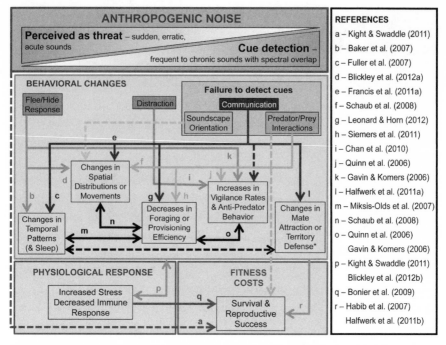

FIG 1.3: Conceptual framework for understanding how noise stimuli that are either perceived as a threat or interfere with cue detection (the disturbance-interference continuum) can elicit behavioral responses that have direct consequences for fitness, or via a physiological stress response, which can also feed back to behavioral changes. Startle/hide responses are more likely to occur in response to noise stimuli that are perceived as a threat (acute, erratic, or sudden sounds). Problems arising from a failure to detect cues are more likely to occur when noise stimuli are chronic and overlap with biologically relevant cues used for communication, orientation, and predator/prey detection. Problems arising from distraction may occur as a result of sounds with features ranging from those that interfere with cue detection to those that are perceived as threats. Letters = studies providing evidence for the link made for each arrow and are listed to the right. Dashed lines = a link we predict as important for which no current evidence exists. * = could result from a change in behavior or a failure to change behavior in response to noise.

movements; (3) decreases in foraging or provisioning efficiency coupled with increased vigilance and anti-predator behavior; and (4) changes in mate attraction and territorial defense (Figure 1.3). As we show below, these disturbance-, distraction-, and masking-mediated behavioral changes could directly impact individual survival and fitness or lead to physiological stress that may then compromise fitness.

Changes in Temporal Patterns

Sound stimuli that are perceived as threats can alter temporal patterns; for example, red foxes (*Vulpes vulpes*) cross busy roads when traffic rates are lower, suggesting noise cues might be affecting the timing of their movements (Figure 1.3b; P. Baker et al. 2007). Similarly, noise from boat traffic disrupts the timing of foraging by West Indian manatees (*Trichechus manatus*), potentially influencing foraging efficiency and energy budgets (Figure 1.3m; Miksis-Olds et al. 2007). Noise can also change behavior due to interference with cue detection. European robins (*Erithacus rubecula*) avoid acoustic interference from urban noise by singing at night, when noise levels are lower than during daylight hours (Figure 1.3c; Fuller et al. 2007). Although this example may appear to be an important behavioral adaptation that permits this species to overcome unfavorable acoustic conditions, the consequences of shifting the timing of song delivery are unknown. The effects of signal timing on mate attraction or territorial defense may be just as important to fitness as other signal features (e.g., frequency, syntax). For example, changes in the timing of song delivery of less than one hour can break down signaler-receiver coordination so that conspecific males may not recognize species-specific signals (Luther 2008). If signaler-receiver coordination is disrupted between singing males and responsive females, it is possible that the behavioral flexibility that permits shifts in signal timing in response to noise may be maladaptive.

Sleep is an important factor and follows a strong temporal profile. Although a substantial body of research has investigated the impact of noise on sleep in humans, scant information is available regarding other animals (reviewed in Kight and Swaddle 2011). Understanding the importance of sleep disruption on overall fitness is critical as we might expect detrimental influences even on species not typically described as dependent upon hearing (e.g., visually-oriented predators such as raptors).

Alterations in Spatial Distributions or Movements

Among the most obvious responses to noise are site abandonment and decreases in spatial abundance. These metrics may also be easiest and least costly to quantify, which perhaps explains why there are many such examples in the literature (e.g., Bayne et al. 2008; Eigenbrod et al. 2008; Francis et al. 2009). However, noise itself can affect an investigator's ability to measure responses to noise. For example, increases in continuous noise of 5–10 decibels (dB; A weighted) above baseline can reduce bird numbers during standard bird surveys by one half, greatly biasing measures of site occupancy and abundance (Ortega and Francis 2012). If not carefully considered, this detection problem could bias subsequent interpretations and management efforts.

Despite the known effects of noise on population sizes, there is still considerable evidence to suggest that animals may abandon areas when frequent or chronic noise stimuli interfere with cue detection or when more variable sounds are perceived as threats (Bayne et al. 2008; Goodwin and Shriver 2011; Blickley et al. 2012a). For example, birds with low-frequency vocalizations experience more acoustic interference from chronic low-frequency anthropogenic noise and therefore exhibit stronger negative responses to noise in their habitat use than birds with high-frequency vocalizations that experience less acoustic interference (Figure 1.3e; Francis et al. 2011b). These masking effects can be spatially extensive, potentially impairing communication at distances ranging from 0.5 to 1.0 kilometers or further from the noise source (Blickley and Patricelli 2012). Furthermore, changes in spatial distributions due to the effects of noise on cue detection are not restricted to intraspecific communication; for instance, greater mouse-eared bats (*Myotis myotis*), which use echolocation to search for prey on the ground, avoid hunting in noisy areas (Figure 1.3f; Schaub et al. 2008). In addition to disrupting cue detection at the intra- and interspecific level, ambient noise may also interfere with cue detection used for movement at larger spatial scales. For example, some frog species use conspecific calls to locate appropriate breeding habitat, while some species of newts use heterospecific calls for the same purpose (reviewed in Slabbekoorn and Bouton 2008). Whether noise exposure impedes animals from using such acoustic beacons to locate critical resources (e.g., water, food, habitat) is unknown and should be a focus of future research.

Site abandonment or decreases in population numbers can also occur in response to unpredictable, erratic, or sudden sounds, which are perceived as threats (Figure 1.3d). For example, greater sage grouse (*Centrocercus urophasianus*) lek attendance declines at a higher rate in response to experimentally introduced intermittent road noise than to continuous noise (Blickley et al. 2012a), suggesting that sage grouse site occupancy may depend more on perceived risk than on masking of acoustic cues. Nevertheless, masking of communication may have other consequences (Figure 1.1).

Species undoubtedly differ in their sensitivities to disruptive sounds, but individuals within a population also show such differences (Bejder et al. 2006). Individuals can vary greatly in their behavioral responses to stimuli, which may explain variations in their ability to cope with environmental change (Sih et al. 2004). The redistribution of sensitive and tolerant individuals across the landscape may not appear to be a problem. However, in the case of social animals, where group living provides protection from predation, the loss of sensitive individuals from the group through site abandonment could increase predation risk for the group as a whole, through the removal of the most vigilant group members. These sensitive individuals, who are now isolated from the group, lose the benefit of safety in numbers. Depending on population structure and the scale at which these individuals are displaced by noise, genetic diversity may be reduced because traits that govern risk-averse (shy/sensitive) and risk-prone (bold) behaviors can be heritable (Dingemanse et al. 2002).

Site abandonment and changes in abundance provide only a limited understanding of how noise can impact wildlife populations and communities. Importantly, abundance can also be misleading because areas where individuals are in high abundance do not always translate into high fitness for those individuals (e.g., R. Johnson and Temple 1986). Using such evidence to conclude that noise has no impact is problematic; individuals may not have alternative areas to occupy or other responses (survival, mating success, reproductive output) may be negatively affected by noise even when abundance is high (Figure 1.1a). These possibilities are especially likely when a noise stimulus is new and demographic processes have not had time to impact population size or when the population in an area that is exposed to noise is supplemented by individuals from elsewhere (i.e., source-sink dynamics).

Decreases in Foraging or Provisioning Efficiency and Increased Vigilance and Anti-Predator Behavior

Noise can impair foraging and provisioning rates directly (Figure 1.3, g and h) or indirectly as a consequence of increased vigilance and anti-predator behavior (Figure 1.3, i–k, o). When noise is perceived as a threat an organism may miss foraging opportunities ("missed opportunity cost"; Brown 1999) while hiding or as a result of maintaining increased vigilance (Figure 1.3k; Gavin and Komers 2006). Missed opportunities can also occur when noise interferes with cue detection. For instance, nestling tree swallows (*Tachycineta bicolor*) exposed to noise beg less in response to recorded playbacks of parents arriving at nests (e.g., calls, movement, sounds) than nestlings in quiet conditions, presumably because the ambient noise masks parent-arrival sounds (Figure 1.3g; Leonard and Horn 2012). Unfortunately, this study did not determine whether missed provisioning opportunities translated into costs, such as reduced nestling mass or fledging success.

Noise that interferes with cue detection can also hamper predators' hunting abilities. For example, among greater mouse-eared bats, search time for prey was shown to increase and hunting success decrease with exposure to experimental traffic noise (Figure 1.3h; Siemers and Schaub 2011). This decrease in foraging success may explain why some predators avoid noisy areas (Figure 1.3n; e.g., Schaub et al. 2008; Francis et al. 2009). Noise also impairs foraging in three-spined sticklebacks (*Gasterosteus aculeatus*), resulting in more unsuccessful hunting attempts (Purser and Radford 2011). It is also possible that noise interferes with the ability of prey species to hear approaching predators, which could impact fitness directly. Although likely, elevated predation risk due to noise has yet to be demonstrated, but some evidence does suggest that animals exposed to noise behave as though they are at greater risk of predation. For example, in the chaffinch (*Fringilla coelebs*), continuous noise impairs auditory surveillance, triggering increased visual surveillance, as a result of which the birds spend less time foraging (Figure 1.3j; Quinn et al. 2006). Noise that serves as a distraction may also lead to an increased latency in predator-escape response (Figure 1.3i; A. Chan et al. 2010b), potentially compromising survival. Both distraction and elevated vigilance could also cause a decrease in foraging rates and success (i.e., a tradeoff; Figure 1.3o; Gavin and Komers 2006; Quinn et al. 2006). Collectively, these studies suggest that both interference noise and noise perceived as a threat decrease the rate and frequency at which organisms obtain food. Studies aimed at understanding the extent to which these behavioral shifts represent a metabolic expense (relevant to survival and reproductive success) will help to reveal the hidden costs of noise exposure.

Changes in Mate Attraction and Territorial Defense

The most direct way in which noise may alter an individual's ability to attract mates or defend its territory is through energetic masking, in which potential receivers are simply unable to hear another individual's acoustic signals through noise that is frequent or continuous during important temporal signaling windows. Changes made to acoustic signals appear to be an adaptive behavioral adjustment that permits individuals to communicate under noisy conditions (e.g., Fuller et al. 2007; Gross et al. 2010; Francis et al. 2011c), yet these shifts could also incur a cost. In noisy areas, female great tits (*Parus major*) more readily detect male songs sung at higher frequencies than females typically prefer (Halfwerk et al. 2011a). However, males who sing predominantly at higher frequencies experience higher rates of cuckoldry (Figure 1.3). Great tits breeding in noisy areas also have smaller clutches and fewer fledglings (Halfwerk et al. 2011b); similarly, eastern bluebirds (*Sialia sialis*) experience decreased productivity when nesting in areas with elevated noise levels (Kight et al. 2012). Paired with patterns of decreased pairing success in noisy areas (Habib et al. 2007; Gross et al. 2010), these studies suggest that short-term signal adjustments in response to anthropogenic noise might function as evolutionary traps (e.g., Schlaepfer et al. 2002) in which behavioral responses to novel acoustic stimuli could be maladaptive. That is, behavioral shifts to be heard in noisy areas may come with the cost of compromising the attractiveness of the signal to potential mates. This possibility remains to be tested against other potential explanations for declines in pairing or reproductive success, but emphasizes why investigators should measure aspects of fitness in noise-impact studies rather than simply documenting changes in site occupancy or abundance.

Finally, although the list of species known to shift their signals in response to noise is growing, there is at least one frog and some bird species that do not alter their vocalizations in response to noise (e.g., Hu and Cardoso 2010; Love and Bee 2010; Francis et al. 2011c). More work is needed to provide a thorough understanding of the phylogenetic distribution of noise-dependent vocal change and researchers should strive to publish negative results, as knowledge of the apparent absence of these behavioral modifications are just as important as knowledge of their presence.

LINKING BEHAVIORAL CHANGES, PHYSIOLOGICAL RESPONSES, AND FITNESS COSTS

The behavioral changes mentioned above can have direct consequences for fitness (Figure 1.3r), such as reduced pairing success (Habib et al. 2007) or reduced reproductive success (Halfwerk et al. 2011b). However, behavior can influence, and be influenced by, physiological responses (Figure 1.3p; Kight and Swaddle 2011), which in turn can affect fitness (Figure 1.3q; Bonier et al. 2009). Kight and Swaddle (2011) review many links between noise, physiological stress, and behavioral change, so we only briefly mention them here.

It is well known that increased physiological stress affects fitness (Figure 1.3q); yet, to our knowledge, a direct link between increased physiological stress due to noise and decreased survival or reproductive success has not been shown in wild animals. The best evidence for this potential link comes from two studies. In one, Blickley et al. (2012b) found that greater sage grouse on leks exposed to experimental playback of continuous natural gas drilling noise or intermittent road noise had higher fecal glucocorticoid metabolites (fGM) than individuals on control leks. The authors suggest that masking of cues likely resulted in elevated stress levels, inhibiting social interactions, or leading to a heightened perception of predation risk. In the other, Hayward et al. (2011) showed that experimental exposure to motor-cycle traffic and motorcycle noise increased fGMs in northern spotted owls (*Strix occidentalis caurina*). In an observational component of the same study, spotted owls nesting in areas with higher levels of traffic noise fledged fewer offspring, even though they did not have elevated fGMs, suggesting that the effects of road noise may have been offset by greater prey availability in noisy areas. These two studies demonstrate that noise may lead to decreased fitness in sage grouse and spotted owls, and also clearly indicate that more research is needed to determine how noise exposure, physiological stress, and fitness are linked in wild populations.

SCALING UP BEHAVIORAL RESPONSES

Here, we have focused on effects of noise exposure at the level of the individual; however, studies that integrate individual behavior, population responses among multiple species, and species interactions are critical to understanding the cumulative, community-level consequences of noise. Measures of species richness are a good starting point, but may be misleading because species may respond negatively, positively, or not at all to sound stimuli (Bayne et al. 2008; Francis et al. 2009), individuals within a single species may respond quite differently to the same stimulus (Sih et al. 2004), and those that remain in noisy areas may suffer from one or more of the fitness costs discussed above. This variation within and among species in responses to noise guarantees that communities in noisy areas will not always be subsets of the species that make up communities in comparable quiet areas. Researchers should couple standard measures of richness and alpha (local) diversity with beta-diversity metrics that reflect variations in the composition of species within communities and among sites. Nevertheless, additional investigations will be needed to understand why species respond to sound stimuli as they do. Settlement patterns may not hinge on the intensity of noise, but are perhaps due to the presence or absence of cues indicating the presence of predators and heterospecific competitors (Francis et al. 2009). These other species (i.e., predators or competitors) may have unique settlement patterns in response to noise and will complicate efforts to measure how noise directly affects the species of interest. Disentangling these interactions will also be essential to understanding the consequences of noise exposure for organisms that are not directly impacted by noise, such as plants that depend on noise-sensitive faunal taxa (Francis et al. 2012a) or animals whose hearing range is not tuned to a particular frequency that makes up a sound stimulus.

CONCLUSIONS

Both policy and scientific literature have often oversimplified the effects of noise on wild animals, typically suggesting that species are either sensitive and abandon noisy areas or are not and remain. In our experience with stakeholders, habituation is an oft-cited reason for persistence and an absence of noise impacts, yet research on other stressors indicates that acclimation to a stressor might not release an organism from costs to fitness (Romero et al. 2009). Additionally, we have shown how behavioral modifications among individuals confronted with noise—even those individuals that outwardly appear to habituate—can lead to decreased fitness. Challenging the assumption that habituation to noise equals "no impact" will be difficult, but it will also be a critical component in revealing how a range of behavioral mechanisms link noise exposure to fitness costs. Ideally, we need to predict which combination of noise characteristics and behavioral contexts are most detrimental and under what circumstances behavioral changes affect fitness directly or indirectly. This will require an array of experimental and observational approaches and frameworks that complement the conceptual structure presented here (Figure 1.3). Other promising frameworks include the risk-disturbance hypothesis (Frid and Dill 2002), which provides an avenue for understanding energetic costs associated with wildlife responses to noise disturbances that are perceived as threats. Studies evaluating aspects of habitat selection and acoustic communication in response to noise may find it useful to frame questions in terms of ecological and evolutionary traps (Schlaepfer et al. 2002). Furthermore, investigators should strive to measure responses along a range of noise exposure levels to reveal the shape of response curves (e.g., threshold, linear) because these details will be indispensable to resource managers and policy makers when establishing and modifying regulatory limits that reflect the ecological effects of noise exposure.

An increase in anthropogenic noise levels is only one of many threats to biodiversity on which ecologists and policy makers should focus their attention. However, relative to other conservation problems, noise may also offer readily available solutions, which, if implemented, could lead to major, measurable improvements for both wildlife and people. For example, use of noise-attenuating walls could reduce the area of a landscape exposed to elevated noise levels from natural gas extraction activities by as much as 70% (Francis et al. 2011a) and similar solutions exist for mitigating noise from roadways and cities (Code of Federal Regulations 2010). These mitigation efforts could come with drawbacks; for instance, noise-attenuating walls near roads could restrict the movement of wildlife and impede gene flow. Nevertheless, as we develop a better understanding of the ecological effects of noise, implementation of mitigation efforts can begin in many well-studied and high-priority systems (e.g., oil and gas developments in natural areas, transportation networks in national parks), where benefits outweigh the potential costs. In addition to protecting contiguous natural habitat, reducing noise exposure in and around developed areas will not only benefit wildlife populations and diversity, but will also provide adjacent human populations with the suite of physiological benefits afforded by living in a quieter community.

2

A PHANTOM ROAD
EXPERIMENT REVEALS TRAFFIC
NOISE IS AN INVISIBLE SOURCE
OF HABITAT DEGRADATION

Heidi Ware, Christopher McClure, Jay Carlisle,
and Jesse Barber

Roads are a dominant source of noise in protected areas. These linear features are responsible for substantial declines in wildlife populations via myriad effects, including direct collisions, chemical pollution, invasive species, and noise pollution. This chapter presents clear experimental evidence that noise alone can be responsible for habitat degradation near roads. The authors recorded traffic noise in Glacier National Park and used an array of speakers to broadcast in roadless habitat the soundscape of that busy transportation corridor in a protected area, while monitoring migratory songbird distributions and physiological condition. The authors' results suggest that incorporating traffic mitigation into park and protected area management is important for maintaining habitat quality for wildlife.

INTRODUCTION

Human infrastructure shapes animal behaviors, distributions, and communities (Benítez-López et al. 2010; Fahrig and Rytwinski 2009). A meta-analysis of 49 datasets from across the globe found that bird populations decline within 1km of human infrastructure, including roads (Benítez-López et al. 2010). Observational studies of birds near roads implicate traffic noise as a primary driver of these declines (Francis and Barber 2013). Road ecology research has also shown negative correlations between traffic noise levels and songbird reproduction (Halfwerk et al. 2011b; Reijnen and Foppen 2006).

Birds that produce low frequency songs, likely masked by traffic noise, show the strongest avoidance of roads (Goodwin and Shriver 2011).

There is now substantial evidence that anthropogenic noise has detrimental impacts on a variety of species (Barber et al. 2010; Bunkley et al. 2015; Francis and Barber 2013; Kight and Swaddle 2011; Siemers and Schaub 2011). For example, work in natural gas extraction fields has demonstrated that compressor station noise alters songbird breeding distribution and species richness (Bayne et al. 2008; Habib et al. 2007; Francis et al. 2009a). However, explicit experiments would help to further rule out other characteristics of infrastructure, such as visual disturbance, collisions, chemical pollution, and edge effects, which might be driving these patterns (Francis and Barber 2013). In addition, although these studies implicate noise as a causal factor in population declines, many individuals remain despite noise exposure (Francis and Barber 2013), but at what cost? Proposed causes of decreased fitness for birds in noise include song masking, interference with mate evaluation, non-random distribution of territorial individuals, disruption of parent-chick communication, reduced foraging opportunities, and/or alterations in the foraging/vigilance trade-off (Francis and Barber 2013; Halfwerk et al. 2011b).

Here we parse the independent role of traffic noise from other aspects of roads experimentally by playing traffic sounds in a roadless area, creating a "phantom road." We focus on birds during migratory stopover, because energy budgets are streamlined; foraging, vigilance, and rest dominate activity (Hedenstrom 2008). To meet the amplified physiological needs of sustained nocturnal migratory flights, birds must increase foraging during periods of stopover while maintaining appropriate vigilance levels (Berthold 1996; Hedenstrom 2008). Any interference with foraging will decrease stopover efficiency and thus reduce migration speed—a likely surrogate for fitness (Hedenstrom 2008)—thereby increasing exposure to significant mortality risks during what can be the most perilous stage of a migratory bird's life cycle (Sillett and Holmes 2002). Anthropogenic noise might disrupt the foraging-vigilance tradeoff by acting as a form of perceived predation risk (Frid and Dill 2002; Shannon et al. 2014) or by reducing sensory awareness via distraction or acoustic masking (Francis and Barber 2013; Purser and Radford 2011). Using the "phantom road" experimental approach, we previously conducted count surveys of bird distributions at this site, finding a decrease in overall bird numbers of more than 25% (McClure et al. 2013). We hypothesized that the subset of birds choosing to stay at the site would experience other negative effects of traffic noise, and we predicted that the birds that remained would exhibit lower body condition and reduced ability to increase body condition (i.e., reduced stopover efficiency) in noise.

To test these predictions we used an array of speakers to re-create the soundscape of a ~0.5 km section of highway along a ridge in southwest Idaho, USA. This approach enabled us to turn the traffic noise *on* and *off* throughout fall migration at our phantom road site, and compare it with a nearby quiet control site, creating a modified before-after-control-impact design (Figure 2.1). Alternating noise on/off every four days, we sampled a different set of migrants during each block as birds arrived and departed from the stopover site. We measured sound levels (hourly level-equivalent, or LEQ) continuously during the season using acoustic recording units placed at mist net locations (Figure 2.1A). We compared mist-net capture rate (birds/net/hr) across site (control vs. phantom road) and noise treatment (on vs. off) to investigate whether birds were leaving or staying when exposed to traffic noise. Similar to our survey work (McClure et al. 2013), our best-fitting model indicated that capture rate decreased by 31% during phantom traffic noise playback, demonstrating that anthropogenic noise, independent of other road forces, fundamentally shapes bird distributions. However, 69% of birds remained despite the noise.

Focusing on birds exposed to a gradient of sound levels, we examined differences in body condition index (BCI) of newly captured birds. BCI is a size-adjusted metric of body mass calculated as mass (g)/natural wing chord (mm). Small changes in BCI represent large differences in condition (Winker et al. 1992). During migration, high body condition signifies birds with the energy stores needed for long migratory flights (Berthold 1996). The best-fitting model showed that as noise exposure increased, overall BCI of the bird community remaining at the road site decreased (Figure 2.1C). In fact, BCI in noise declined by a full standard deviation compared to the community mean in control conditions. In the absence of noise, BCI of the songbird community at the phantom road site did not differ from the values at the control site, indicating both were suitable stopover locations (Figure 2.1C). Models for individual species showed 5 of 21 species significantly decreased BCI in noise. Iterative exposure to noise during the multiple stopovers of saltatory migration may ultimately result in mortality (Sillett and Holmes 2002) or, in a better case scenario, reduced fitness manifested from slower migration speed (Hedenstrom 2008) which would likely impact fitness and survival in the subsequent life history stage (Marra et al. 1998).

Because we turned the phantom road off overnight to match typical diel traffic patterns, it is likely that nocturnal migrants (the majority of species in

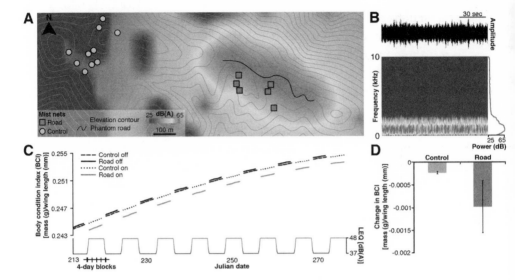

FIG 2.1: Phantom road playback causes songbird body condition decline. (A) Estimated sound levels (dB[A] 1 h LEQ: The level of a constant sound over a specified time period that has the same energy as the actual (unsteady) sound over the same interval) during periods when speakers were on: from August through October 2012–2013 in the Boise Foothills, Idaho. Sound level was modeled using NMSim (Wyle Laboratories, Inc., Arlington, VA) (McClure et al. 2013). Circles (control) and squares (road) represent capture sites. With the noise on, mean sound levels at the phantom road capture sites increased by 11 dB(A) to 48 dB(A) (s.e. = 0.3), while the control site averaged 2 dB(A) louder with noise on (mean ± s.e.; 41 dB[A] ± 0.2). With noise off, sound levels averaged 39 dB(A) (s.e. = 0.2) at the control capture sites and 37 dB(A) (s.e. = 0.3) at the phantom road. Elevation contours are 50m. (B) A two-minute sample of the phantom road file displayed as an oscillogram, a spectrogram and a power spectrum. (C) Predicted values for body condition index (BCI) as birds add fuel throughout fall migration. Estimates are based on the AIC-best model for BCI for all captures combined, with species as a random intercept. A consistent full standard deviation change in BCI is evident during each noise-on block (pattern of noise on blocks displayed along the x axis) throughout the migratory period. (D) Predicted mean change in BCI at the control and phantom road sites between noise off and noise on periods across the entire study. Error bars represent standard error. These differences in BCI (and associated error) are derived from the average of the predictions presented in panel C.

this study [see Poole 2005]) chose to land at our site when it was quiet, before the phantom road playbacks began in the morning. In effect, diurnally varying traffic noise might function as an ecological trap (Robertson and Hutto 2006) for migrants. Though staying in traffic noise has a cost, the energetic outlay for individuals to leave a given site might be even greater. Birds with low body condition are less likely to embark on migratory journeys than those in good condition, and depending on the suitability of surrounding habitat, it may not be worth the risk to disperse once landed (A. Smith and McWilliams 2014). We cannot differentiate whether the lower BCI we documented in traffic noise is the result of 1) higher body condition birds leaving the population or 2) birds losing body condition over the duration of noise exposure. We saw both reduced mean body condition and reduced bird numbers, suggesting that at least some birds with the energetic stores to migrate chose to leave the site and escape the costs of remaining in noise (A. Smith and McWilliams 2014).

To examine if the birds that remained in noise were suffering reduced ability to add migratory fuel (i.e., increase BCI), we regressed BCI of new captures against time of day to estimate stopover efficiency. Comparing stopover efficiency of individuals between sites provides an essential metric to compare the relative value of stopover habitat (Winker et al. 1992). The best-fitting model for the entire songbird community included noise intensity level [dB(A)] although the confidence intervals overlapped zero. For 9 individual species, the best-fitting model included a noise variable, however the confidence intervals overlapped zero for all but 3 of these species.

For MacGillivray's warblers, the best-fitting model showed that stopover efficiency substantially decreased with increasing decibel levels. MacGillivray's warblers did not show reduced capture rates in noise, and were the species that showed the strongest negative responses for both BCI and stopover efficiency, indicating that individuals stayed but did poorly in noise (Figure 2.2A). In contrast, Cassin's finches had significantly increased stopover efficiency in noise and a decreased capture rate (Figure 2.2B). This increase in stopover efficiency might reflect decreased competition for food resources in noise. Although stopover efficiency was increased in noise (Figure 2.2B), Cassin's finches showed lower initial BCI in traffic noise (Figure 2.2B), perhaps indicating individuals with higher BCI left the site during noise exposure. The best models for spotted towhees showed a reduced capture rate and also indicated different stopover efficiencies between on-off periods at the control and road sites with efficiency being negatively affected by noise along the phantom road (Figure 2.1).

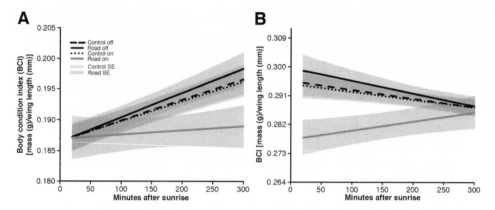

FIG 2.2: Stopover efficiency is altered in noise. Predicted values for stopover efficiency for MacGillivray's warblers (A) and Cassin's finches (B). Estimates were made using average day of season using the AIC-best model for BCI for all captures combined. Values were predicted by inputting average dB(A) levels for each site. Values are shown for the control site noise off [avg. 42dB(A)], control site noise on [43dB(A)], phantom site noise off [40dB(A)], and phantom site noise on [51dB(A)]. Darker gray shading represents standard error for the control site whereas lighter gray shading represents standard error for the phantom road.

It seems that for species impacted by noise, different strategies exist for managing the consequences, which might be based on differences in life history traits such as territoriality during stopover, migratory strategy, or flocking behavior. Our species-specific results show that birds may stay and incur a cost of remaining in noise (e.g., MacGillivray's warblers), or choose to leave (e.g., Cassin's finches). Leaving the noisy area may allow some species to avoid the costs of noise or a species may still experience the impacts of noise despite some individuals leaving (e.g., spotted towhees). Together, our observations of overall changes in the BCI of the entire bird community and of several individual species, as well as the changes in stopover efficiency of spotted towhee and MacGillivray's warbler, demonstrate that addition of traffic noise alone, without the other variables associated with actual roadways, can significantly decrease the value of a stopover site, thus degrading the quality of otherwise suitable habitat.

In support of our field results, we conducted a controlled laboratory study to test whether traffic noise alters the foraging-vigilance tradeoff in songbirds and could thus mechanistically underpin our field data. We focused on the second most common species from our field study, white-crowned sparrow (*Zonotrichia leucophrys*), a species that also decreased BCI in noise, to investigate the reduction in foraging and increase in vigilance implied by our community-wide body condition analysis. We quantified head-down duration (i.e., foraging) and head-up rate (i.e., vigilance), because these are known measures of avian visual vigilance that change when auditory surveillance is limited and that correlate with food intake and ability to detect predator attacks (Quinn et al. 2006). We also measured feeding duration ([# seconds per 8 minute trial spent feeding]) to quantify overall feeding bout duration. Using the same playback file as our field experiment, we played 61dB(A) and 55dB(A) traffic noise treatments, plus a silent control track (32dB(A)) to foraging sparrows (n = 20). White-crowned sparrows decreased foraging by ˜8%, increased vigilance levels by ˜21%, and decreased feeding duration by ˜30% when exposed to traffic noise [61dB(A); Figure 2.3]. Vigilance behavior of individuals did not change based on the number of trials experienced, indicating birds did not habituate to the noise. During energetically demanding periods in a bird's life, increasing vigilance can reduce survival because of increased starvation risk (Watson et al. 2007). In contrast to song masking, which can be partially overcome by frequency shifting (Mockford and Marshall 2009), release from masking is not possible for auditory cues necessary for aural vigilance (Barber et al. 2010). With limited auditory information, animals must resort to other methods such as visual scans to compensate for the increase in perceived predation risk, perhaps driven by masking of communication calls and predator-generated sounds (Gavin and Komers 2006 Quinn et al. 2006).

Our behavioral investigations in the lab offer compelling evidence that the body condition changes measured in the field were due at least in part to a change in foraging and vigilance behavior, but our field results could be due to a combination of factors that also deserve consideration. For example, noise might also increase physiological stress levels (Blickley et al. 2012b; Crino et al. 2011) that could cause additional declines in body condition. However, we view it as unlikely that noise can cause a stress response independent of a change in behavior. In addition, noise might indirectly change foraging rates through alterations in prey search time, sleep, or territoriality. For instance, our phantom road might have disrupted foraging behavior by reducing the acoustic detectability of insect prey (Montgomerie and Weatherhead 1997) or reducing insect numbers. We did not test for changes in insect abundance or distribution, but because we found noise impacts on a mixed community of both frugivorous and insectivorous birds, it seems unlikely that altered insect numbers explain a significant component of the observed

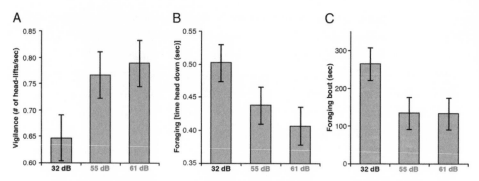

FIG 2.3: The foraging/vigilance trade-off is altered in noise. White-crowned sparrows foraging in traffic noise at 61 and 55 dB(A) had (A) reduced foraging rates, (B) increased vigilance, and (C) decreased foraging bout duration compared to trials in ambient conditions [32dB(A)]. Data are means ± standard error. [Mean head up rate (head lifts/s) for 61dB(A) = 0.79 ± 0.06, 55dB(A) = 0.77 ± 0.05, 32dB(A) = 0.65 ± 0.05. Mean head down duration (s): 61dB(A) = 0.41 ± 0.03, 55dB(A) = 0.44 ± 0.04, 32dB(A) = 0.50 ± 0.04. Mean foraging bout duration (s): 61dB(A) = 159.25 ± XX, 55dB(A) = 147 ± XX, 32dB(A) = 228 ± XX]. Birds showed more head lifts/s (β = 0.005 ± 0.002), decreased the amount of time spent with their heads down searching for seeds (β = -0.003 ± 0.001), and decreased total feeding duration (β = -4.589 ± 1.944) during noise playback compared to ambient conditions (Winker et al. 1992).

patterns. Effects were consistent between the 4-day noise-on blocks throughout migration, despite documented seasonal variation in fruit and arthropod availability at the site (Carlisle et al. 2012), so it is more likely that changes in bird behavior drove these responses. Our experimental design was not able to determine whether noise disrupts territoriality or dominance hierarchies during stopover. However, both territorial and non-territorial species showed negative effects of noise (Poole 2005). We expect that a subset of these indirect effects plus the behavioral changes quantified in the lab contributed to the body condition declines seen in our field experiment. Because provisioning is a constant requirement for birds throughout the year, other effects of noise that occur outside of migration (e.g., Halfwerk et al. 2011b; Reijnen and Foppen 2006) would be in addition to, rather than instead of, the impacts we document here.

Previous work that failed to find a change in animal distributions near roads or other infrastructure has assumed a lack of negative impacts from loud human activities (Benítez-López et al. 2010; Francis and Barber 2013).

Our results demonstrate that individuals may remain in an area with high levels of noise yet suffer significant costs. We found that different species chose different strategies: to either leave noisy areas, or stay and perhaps incur the costs of noise. We exposed the bird community at our phantom road to sound levels similar to some suburban neighborhoods (~55 dB[A] hourly LEQ) (Wayson 1998). Many protected areas and high-value habitats are currently exposed to these levels, and would benefit from noise relief measures (Barber et al. 2011; Lynch et al. 2011). The impact of noise reaches far beyond the physical footprint of human infrastructure. Unlike other aspects of roads, noise impacts can be minimized without removing the road itself. Substrate alteration and speed limit reduction on existing roads can significantly lower decibel levels (Wayson 1998).

Our results reveal the need for attention to noise impacts beyond distributional shifts (Francis and Barber 2013). For individuals that remain in areas disturbed by loud human activities, noise pollution represents an invisible source of habitat degradation that has been largely ignored—traffic noise degrades habitat value but leaves no physical signs of change. Stopover habitat loss and degradation have been identified as major contributing factors to migratory songbird declines worldwide (Robbins et al. 1989; Sanderson et al. 2006). Migrants are exposed to an unknown risk landscape at stopover sites and must therefore rely heavily on increased vigilance to compensate (Faaborg et al. 2010; Schmidt et al. 2010; Thomson et al. 2006). And unlike resident species, successful conservation of migratory species requires protection of habitats in breeding, wintering, and stopover locations (Faaborg et al. 2010). In addition, reduction in condition or delay in migration could have carry-over effects into the overwintering or breeding seasons (Reudink et al. 2009). Further understanding of anthropogenic noise's impact on body condition is key, as it is an important predictor of fitness across taxa and life stage (Marra et al. 1998). When managing natural systems, we should ensure that the habitat we protect remains of high quality, including the quality of the acoustic environment.

3

NOISE POLLUTION ALTERS
ECOLOGICAL SERVICES

Enhanced Pollination and Disrupted Seed Dispersal

Clinton Francis, Nathan Kleist, Catherine Ortega,
and Alexander Cruz

Ecological systems are complex and connected. This chapter shows that noise pollution can be a razor that severs links within these systems. The authors document altered bird distributions in areas with noise pollution and the consequences for seed dispersal and pollination—central ecosystem services. It is for this reason that basic ecology (from the Greek oikos, *meaning "everyone living in the house") should be pursued in tandem with applied science to provide the data necessary to best manage our parks and protected areas. Noise can reverberate through ecological systems, and to predict its path we must understand nature.*

INTRODUCTION

Human activities have altered over 75% of earth's land surface (Ellis 2011; Ellis and Ramankutty 2008). Concomitant with these surface changes is a pervasive increase in anthropogenic noise, or noise pollution, caused by expanding dendritic transportation networks, urban centers and industrial activities (Barber et al. 2010). The geographical extent of noise exposure varies by region and scale, but estimates suggest that one-fifth of the United States' land area is impacted by traffic noise directly (Forman 2000) and over 80% of some rural landscapes are exposed to increased noise levels due to energy extraction activities (Francis et al. 2011d). Despite the potentially substantial scale of noise exposure across the globe, surprisingly little is known about how these ecologically novel acoustic conditions affect natural populations and communities.

We are beginning to understand the impacts of increased noise exposure on the behaviors of individuals and the distributions of species (Bayne et al. 2008; Francis et al. 2009a, 2011a, and 2011c; Halfwerk and Slabbekoorn 2009), and several recent reviews outline potential and some known effects of noise (Barber et al. 2010; Slabbekoorn and Ripmeester 2008). Despite this recent attention given to the effects of noise, we still have limited knowledge of how these impacts scale to community and ecosystem-level processes. A few studies have shown that predators avoid noisy areas (Francis et al. 2009; Siemers and Schaub 2011), presumably because noise impairs predators' abilities to locate prey. These studies provide us with insights on how noise may directly affect predator-prey interactions, but do not provide information on whether noise may have cumulative, indirect consequences for other interactions and organisms that are not impacted by noise directly.

Our goal was to investigate whether noise pollution can reverberate through ecological communities by affecting species that provide functionally unique ecological services. We focused our efforts on ecological services provided primarily by birds because they are considered to be especially sensitive to noise pollution due to their reliance on acoustic communication (Slabbekoorn and Ripmeester 2008). However, because not all species respond uniformly to noise exposure (Bayne et al. 2008; Francis et al 2011a, 2011b, and 2011c), we can evaluate how different responses by functionally unique species impact other organisms indirectly and trigger further changes to community structure. We studied ecological services provided by *Archilochus alexandri* (black-chinned hummingbird) and *Aphelocoma californica* (western scrub-jay), which serve as mobile links for pollination and *Pinus edulis* (piñon) seed dispersal services respectively (Chambers et al. 1999; Paige and Whitham 1987; Vander Wall and Balda 1981). Because *A. alexandri* preferentially nests in noisy environments and *A. californica* avoids noisy areas (Francis et al. 2009a and 2011d), we proposed that their noise-dependent distributions could result in a higher rate of pollination for hummingbird-pollinated plants and disrupt *P. edulis* seed dispersal services in noisy areas and potentially affect seedling recruitment (Figure 3.1a).

To test these predictions, we used a unique study system that isolates the influence of noise exposure from many confounding factors common to noisy areas, such as vegetation heterogeneity, edge effects, and the presence of humans and moving vehicles (see below). We used observations, vegeta-

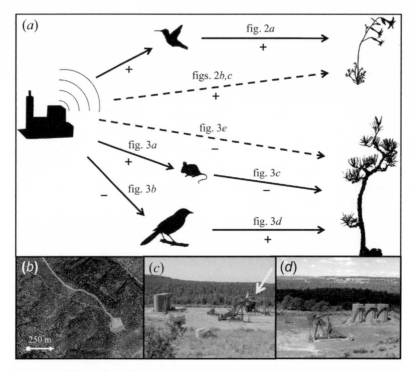

FIG 3.1: (a) Pathway by which noise alters pollination and seed dispersal services. Solid and dashed arrows denote direct and indirect interactions, respectively. Signs refer to effect direction, and support for each effect is indicated by figure number. See main text for results and citations supporting the dependence of *I. aggregata* on *A. alexandri* (arrow labeled fig. 2a) and for the functional quality of *Peromyscus* mice and *A. californica* as *P. edulis* seed dispersers (arrows labeled fig. 3c and 3d). (b) Active gas wells located at the end of access roads served as (c) noisy treatment sites due to the presence of noise-generating gas well compressors (white arrow) or (d) quiet control sites.

tion surveys, and pollen transfer and seed removal experiments on pairs of treatment and control sites to determine how ecological interactions differ in noisy and quiet areas and whether noise indirectly affects plants that depend on functionally unique avian mobile links.

MATERIALS AND METHODS

Our study took place in the Rattlesnake Canyon Habitat Management Area (RCHMA), located in northwestern New Mexico. Study area and site details can be found elsewhere (Francis et al. 2009a, 2011c, and 2011d). Briefly, RCHMA is dominated by woodland consisting of *P. edulis* and *Juniperus osteosperma* (juniper) and has a high density of natural gas wells (Figure 3.1b). Many wells are coupled with compressors that run continuously and generate noise at high amplitudes (> 95 dB[A] at a distance of 1 m), and, like most anthropogenic noise, compressor noise has substantial energy at low frequencies and diminishes towards higher frequencies (Francis et al. 2009a, 2011c, and 2011d). Additionally, human activity at wells and major vegetation features in the woodlands surrounding wells do not differ between wells with (noisy treatment sites) and without noise-generating compressors (quiet control sites, Figure 3.1 c and d) (Francis et al. 2009a), providing an opportunity to evaluate the indirect effect of noise on supporting ecological services in the absence of many confounding stimuli common to most human-altered landscapes.

Pollination Experiment

To determine whether hummingbird-pollinated flowers indirectly benefit from noise, we used a field experiment controlling for the density and spatial arrangement of hummingbird nectar resources with patches of artificial flowers that mimicked a self-incompatible, hummingbird-pollinated plant common to our study area: *Ipomopsis aggregata* (Figures 3.2 and 3.3a). In May 2010 we established seven pairs of treatment and control sites within RCHMA for the pollination experiments. Sites were paired geographically to minimize potential differences in vegetation features within each pair; however, to ensure that background noise levels were significantly different between paired sites, sites were ≥ 500 m apart and resulted in relatively quiet conditions at control sites. The resulting distance between treatment-control pairs was 767 m (± 57 s.e.m., minimum = 520 m, maximum = 954).

Artificial flower patches were established 125 m from either the wellhead or compressor on control and treatment sites respectively (Figure 3.2a). The direction of the first patch relative to the wellhead or compressor was determined randomly and the second patch was established 40 m from the first and also at 125 m from the wellhead or compressor. Prior to the experiment, at each patch we measured background noise amplitude as A-weighted decibels (dB(A)) for one minute to confirm that noise levels were significantly higher at treatment patches relative to control patches. In all cases measurements on paired treatment and control sites were completed on the same day and at approximately the same time. We measured amplitude as the equivalent continuous noise level (L_{eq}, *fast* response time) with Casella con-

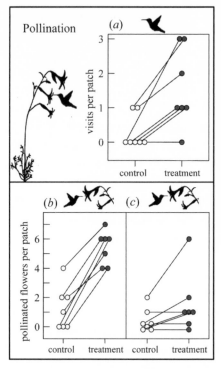

FIG 3.2: Evidence from pollination experiment. (a) *A. alexandri* visited artificial flowers on noisy treatment patches more than quiet control site patches. The values displayed reflect the sum of visits per site (14 sites total with 2 patches per site, 5 plants per patch and 3 flowers per plant). (b and c) Pollination of individual flowers was higher on treatment sites relative to control sites both (b) within and (c) between patches. Values displayed reflect the sum of pollinated flowers in both patches per site. For all panels, lines link geographically paired sites.

vertible sound dosimeter/sound pressure meters (model CEL 320 and CEL 1002 converter). We used 95 mm acoustical windscreens, and we did not take measurements when wind conditions were category three or above on the Beaufort Wind Scale (≈13–18 km/h) or when sounds other than compressor noise (i.e., bird vocalizations, aircraft noise) could bias measurements.

Artificial flowers are frequently used in pollination studies (Hurly 1996; Hurly et al. 2010) and those used in our experiment were constructed from 0.6 ml microcentrifuge tubes. This microcentrifuge tube size had been used previously in pollination experiments with *A. alexandri* (Baum and Grant 2001). To mimic the appearance of *I. aggregata*, we wrapped each microcentrifuge tube with red electrical tape. Additionally, we attached three small pieces of yellow yarn to provide a substrate for marking flowers with fluorescent dye and subsequent transfer and deposition on other flowers by pollinators. Each artificial plant consisted of three flowers attached to a 53 cm long metal rod with green electrical tape. Patches of plants were arranged in a 3 m² area with four plants marking each corner and one at the center.

Plant patches were established simultaneously or one immediately after another (≤ 30 min) on paired sites. Because *I. aggregata* nectar is 20–25%

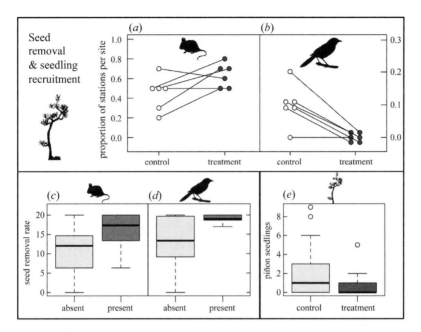

FIG 3.3. (a) *Peromyscus* mice were detected more frequently at treatment seed stations relative to control stations. (b) *A. californica* removed seeds exclusively on control sites. Values displayed in (a) and (b) reflect the proportion of seed stations per site where mice or jays were detected. (c and d) Seed removal rates (per 24 hrs) were higher when (c) *Peromyscus* mice or (d) *A. californica* were detected at a seed station. (e) *P. edulis* seedling recruitment was significantly higher on control sites relative to treatment sites. Box plots indicate the median value (solid black line), 25th and 75th percentiles (box), and whiskers denote 1.5 the interquartile range. Outliers are denoted by open circles.

sucrose (e.g., Pleasants 1983; Waser and McRobert 1998), we filled each flower with a reward of 0.4 ml 25% sucrose solution with pipettes and calibrated plastic droppers, returning each day at approximately the same time to refill the flower with the sucrose reward so that pollinators learned to use the flowers as a foraging resource. Only rarely did we encounter a single artificial flower completely depleted of the reward between visits to replenish the reward, but never all three flowers on the same plant.

We conducted observations to determine pollinator visitation rates at 11 (79%) of 14 pairs of treatment and control patches. Because the establishment of our patches took several days, prior to our observations four pairs of patches were refilled for four days, two patches were refilled for three days and five patches were refilled for two days and all observed patches had been established for > 38 hrs prior to observation. We then conducted observations at patches on pairs of control and treatment sites simultaneously or one immediately after the other. We watched flowers at focal patches for 15 min and tallied the number of visits to each plant from a distance of 5 m, using binoculars when necessary to identify arthropods visiting the flowers. All non-hummingbird pollinators were separated into their orders (Hymenoptera, Diptera, Lepidoptera) and we used Poisson generalized linear mixed models (GLMM) within the lme4 package in R (R Development Core Team 2010) to examine whether patch visitations by *A. alexandri* or other pollinators differed between treatment and control sites. Individual sites and geographically paired sites were treated as random effects.

Following focal observations, on 28 May 2010 we returned to all patches between 7:00–12:00 to refill all artificial flowers with the sucrose reward and uniquely marked one plant per patch with either yellow or red fluorescent powder (Day-Glo Color Corporation, Cleveland, Ohio) such that plants within the same site but at different patches received a unique colored powder. Use of fluorescent powder as a proxy for pollen transfer is a technique widely used in pollination studies because the transfer of powder is strongly correlated with the transfer of pollen (Dudash 1991; Kearns and Inouye 1993). Each patch was permitted 24 hrs of exposure for pollinator visits before we collected each plant for subsequent examination for powder transfer in the laboratory.

In the laboratory we used an ultraviolet lamp under dark conditions to record the presence or absence of powder on each inflorescence, noting whether the powder was from the marked plant within the same patch or the patch located at 40 m. We then used Poisson GLMMs to examine within patch and between patch pollen transfer with number of individual flowers per patch with transferred pollen as response variables. We treated each site and geographically paired treatment and control site as random effects in all models.

P. EDULIS SEED REMOVAL EXPERIMENT

We conducted *P. edulis* seed removal experiments throughout RCHMA to determine if and how seed removal rates and the community of seed predators and dispersers respond to noise exposure. *P. edulis* trees within a region typically synchronize production of large cone crops every 5–7 years (Chambers et al. 1999). As cones gradually dry and open in September, seeds not harvested by corvids from cones in the canopy fall to the ground where rodents, corvids and other bird species consume and harvest seeds for several months (Chambers et al. 1999). Monitoring rates of autumn seed removal from the ground can be problematic as seeds continue to fall from trees; therefore, we conducted our experiments in June–July when no other *P. edulis* seeds were available, similar to other studies that have examined *P. edulis* seed removal and dispersal during summer months (e.g., Vander Wall 2000).

We used six pairs of treatment and control sites that were geographically coupled. Sites met those same criteria described for the pollination experiment. The mean distance between treatment-control pairs was 821 m (± 51 s.e.m., minimum = 642 m, maximum = 1029 m). At each site we established seed stations at 10 locations within 150 m of each well or compressor. Locations were selected randomly provided that the distance between each station was ≥ 40 m and each station was located on the ground under a reproductively mature *P. edulis* tree.

Seed removal experiments lasted for 72 hrs with visits to each station every 24 hrs to quantify the daily rate of seed removal. At the beginning of each 24-hr period we simulated natural seed fall by scattering 20 *P. edulis* seeds on the ground in a 0.125 m^2 area. We then returned 24 hrs later to document the number of removed seeds, determine whether there was evidence for *in situ* seed predation by carefully searching the immediate area (≈ 2 m^2) for newly opened *P. edulis* seeds (usually conspicuous as a clumped collection of seed coat fragments from several seeds) and to again scatter 20 seeds at the station. Evidence of seed predation at a station was defined as whether recently opened (and empty) seed coats were detected during any of the three visits used to quantify seed removal rate. All seeds were collected locally within RCHMA and were handled with latex gloves so that human scent was not transferred to the seeds. During one of the four visits to each station we measured background noise amplitude following the methods described above for the pollination experiment.

To document the identity of animals removing seeds, we paired each station with a motion-triggered digital camera (Wildview Xtreme II). Cameras were mounted on a trunk or branch of an adjacent tree within 1–3 m from the seed station for a clear view, yet positioned in a relatively inconspicuous location to avoid drawing additional attention to the station. Cameras remained on each station for the entire 72-hr period and documented both diurnal and nocturnal seed removal. A positive detection of a species removing seeds was recorded only when an individual was documented removing or consuming seeds.

The number of seeds removed per 24-hr period was used to calculate a daily mean proportion of seeds removed, which we arcsine square-root transformed to meet assumptions of normality and homogeneity of variance. We used LMMs to examine whether the proportion of seeds removed differed between treatment and control seed stations or due to the presence or absence of individual species. We used binomial GLMMs to determine how the presence of individual species explained *in situ* seed predation, evidenced by the presence of newly opened *P. edulis* seed coats. For the models in which we examined the influence of individual species on seed removal or predation, we started with models containing all documented species as predictor variables and proceeded to remove each non-significant variable one at a time based on the highest *p* value until only significant effects remained. We used Poisson and binomial GLMMs to examine whether species richness of the seed removing community and detections of individual species differed at treatment and control seed stations respectively. For all models, we treated each site and geographically paired treatment and control sites as random effects. Some models evaluating detections of individual species on treatment and control sites would not converge; therefore, for these cases we used chi-squared tests to determine whether there was a difference between the total number of detections on control and treatment sites.

Seedling Recruitment Surveys

In 2007 we completed 129 random vegetation surveys on 25-m diameter vegetation plots ($\approx$ 490 m²) located on nine treatment and eight control sites, some of which were not the same sites used in the seed removal experiments, which included only six pairs of sites (12 total). Ten treatment sites were surveyed in 2007, but we excluded vegetation plots from one site from this analysis because the compressor was installed in 2006, thus confounding the acoustic conditions during which many seedlings may have been established (see below). Compressors on all other treatment sites had been in place for at least six years, but over ten years for several sites.

Because our fieldwork in previous years had documented an avoidance of noise by *A. californica* (Francis et al. 2009), in 2007 we counted all *P. edulis* seedlings per vegetation plot. We restricted counts of seedlings to those ≤ 20 cm to make sure that they had been dispersed and established relatively recently and under the same acoustic conditions that were present in 2007. We assumed seedlings ≤ 20 cm had been dispersed and established within the previous six years because one-year-old *P. edulis* seedlings were measured to have an average height of 5.3 cm (Harrington 1987) and because the closely related *Pinus cembroides* reaches a height of 1.0 m at around five years old (Segura and Snook 1992). Thus, our assumptions should be considered conservative. We analyzed seedling recruitment with the number of seedlings per plot as the response variable using Poisson GLMMs. Predictor variables included plot location on either a treatment or control site, but also plot-level features that may influence seedling establishment and recruitment, such as the number of shrubs, *P. edulis* and *J. osteosperma* trees, the amount of canopy cover, leaf litter depth, and the proportion of ground cover classified as living material, dead matter, or bare ground. Site identity was treated as a random effect. We followed the same model selection procedure described above for seed removal and seed predation. Data for seedling recruitment, plus data from the seed removal and pollination experiments have been deposited at Dryad (www.datadryad.org): doi:10.5061/dryad.6d2ps7s7.

RESULTS

Pollination

Noise amplitude values were significantly higher ($\approx$12 dB[A]) at treatment patches relative to control patches (LMM: $\chi^2 = 25.550$, d.f. $= 1, p > 0.001$) and similar to those experienced approximately 500 m from motorways (Halfwerk et al. 2011b; Summers et al. 2011). Focal observations at a subset of patches revealed that several taxa visited artificial flowers supplied with a nectar reward, yet only *A. alexandri* visits differed between treatment and control sites. *A. alexandri* visits were five times more common at treatment patches than control patches (Poisson GLMM: $\chi^2 = 6.859$, d.f. $= 1, p = 0.009$, Figure 3.2a).

Consistent with more *A. alexandri* visits to plants in noisy areas, within-patch pollen transfer occurred for 5% of control site flowers, but 18% of treatment site flowers (Poisson GLMM: $\chi^2 = 15.518$, d.f. $= 1, p < 0.001$, Figure 3.2b) and between-patch pollen transfer occurred for 1% of control site flowers and 5% of treatment site flowers (Poisson GLMM: $\chi^2 = 6.120$, d.f. $= 1, p = 0.013$, Figure 3.2c). Analyses using the presence or absence of transferred pollen at the patch level revealed the same pattern (within-patch binomial GLMM: $\chi^2 = 8.800$, $df = 1, p = 0.003$; between-patch binomial GLMM: $\chi^2 = 5.608$, d.f. $= 1, p = 0.018$).

P. Edulis Seed Removal

Noise amplitude values were consistently higher ($\approx$14 dB[A]) at treatment seed stations relative to control seed stations (LMM: $\chi^2 = 19.084$, d.f. $= 1$,

$p < 0.001$), yet neither seed removal rate (LMM: $\chi^2 = 2.209$, d.f. $= 1$, $p = 0.137$), nor documented species richness per seed station differed between sites with and without noise (Poisson GLMM: $\chi^2 = 0.461$, d.f. $= 1$, $p = 0.497$).

The majority of animals detected with motion-triggered cameras removing seeds from stations were easily identified to species; however, for two groups, *Peromyscus* mice and *Sylvilagus* rabbits, we were not always able to identify individuals to species; therefore, they were assigned to their respective genera. In total, we document eleven taxa removing seeds, 9 of which were considered seed predators. Cameras failed to detect the identity of animals that removed seeds at approximately one station per site, primarily due to battery failure. However, there was no difference in the number of camera failures between treatment and control sites that would suggest our detections were biased towards one site type over the other (binomial GLMM: $\chi^2 = 0.240$, d.f. $= 1$, $p = 0.624$); therefore, any relative differences in detections between treatment and control sites should reflect actual differences between noisy and quiet areas.

Of the nine seed predators documented removing seeds, only one, *Pipilo maculatus*, was detected more frequently on control sites relative to treatment sites (binomial GLMM: $\chi^2 = 4.133$, d.f. $= 1$, $p = 0.042$); a pattern consistent with previous findings that *P. maculatus* avoids noise in its nest placement (Francis et al. 2009). We also documented seed removal by *Peromyscus* mice and *A. californica*, considered to be primarily seed predators and important seed dispersers respectively (Chambers et al. 1999). Mice were detected at 63% of treatment seed stations and only 45% of control stations (binomial GLMM: $\chi^2 = 4.023$, d.f. $= 1$, $p = 0.045$, Figure 3.3a). In contrast, *A. californica* was detected removing seeds exclusively at control stations ($\chi^2 = 5.486$, d.f. $= 1$, $p = 0.019$, Figure 3.3b). *Peromyscus* mice and *A. californica* were also the only taxa with strong effects on seed removal and, along with *Tamias minimus*, were taxa with strong influences on patterns of seed predation at the seed station (i.e., presence of opened seed coat). Seed removal rates were approximately 30% higher at stations where *Peromyscus* mice or *A. californica* were documented removing seeds compared to stations where they were not detected (LMM: $\chi^2 = 35.775$, d.f. $= 2$, $p < 0.001$, Figure 3.3c and d). Seed predation was positively affected by the presence of *Peromyscus* mice ($\beta_{mouse} = 0.841 \pm 0.412$ SE) and *T. minimus* ($\beta_{chipmunk} = 1.199 \pm 0.544$ SE), both typically considered seed predators (Chambers et al. 1999), but negatively affected by the presence of *A. californica* ($\beta_{scrub-jay} = -2.031 \pm 1.005$ SE; binomial GLMM: $\chi^2 = 13.748$, d.f. $= 3$, $p = 0.003$). Indeed, most stations where *Peromyscus* mice (74%) and *T. minimus* (81%) were detected also had evidence of seed predation, but only 33% of stations where *A. californica* was detected were there signs of seed predation.

P. Edulis Seedling Recruitment

Consistent with the difference in animals removing seeds in noisy and quiet areas, *P. edulis* seedlings were four times more abundant on control sites relative to treatment sites ($\beta_{\text{Treatment}}$ = -1.543 ± 0.240 SE, Figure 3.3e), but number of *J. osteosperma* trees (β_{Juniper} = 0.036 ± 0.016 SE) and the proportion of dead organic ground cover (β_{Dead} = 0.023 ± 0.008 SE) had small, positive effects on seedling abundance (Poisson GLMM: χ^2 = 38.583, d.f. = 3, $p < 0.001$). However, neither of these variables, nor number of *P. edulis* trees, differed between treatment and control sites (juniper LMM: χ^2 = 0.726, d.f. = 1, p = 0.394; dead ground cover LMM: χ^2 = 0, d.f. = 1, p = 1.0; *P. edulis* LMM: χ^2 = 2.560, d.f. = 1, p = 0.110), suggesting that other habitat features can be excluded as alternative explanations for *P. edulis* seedling recruitment on treatment and control sites.

DISCUSSION

Elevated noise levels affected pollination rates by hummingbirds and *P. edulis* seed dispersal and seedling recruitment, but the direction of each effect was different. Noise exposure had an indirect positive effect on pollination by hummingbirds, but an indirect negative effect on *P. edulis* seedling establishment by altering the composition of animals preying upon or dispersing seeds. These results extend our knowledge of the consequences of noise exposure, which has primarily focused on vocal responses to noise (e.g., Francis et al. 2011a; Gross et al. 2010; Halfwerk et al. 2011b), somewhat on species distributions and reproductive success (e.g., Bayne et al. 2008; Francis et al. 2009a; Habib et al. 2007; Halfwerk et al. 2011b) and very little on species interactions (Francis et al. 2009a; Schaub et al. 2008; Siemers and Schaub 2011; Francis et al. 2012b). In an example of the latter, traffic noise negatively affects bat (*Myotis myotis*) foraging efficiency by impairing its ability to locate prey by listening to sounds generated from prey movement (Siemers and Schaub 2011). Here our data demonstrate that the frequency of species interactions can change without a direct effect of noise on the interaction itself, suggesting that noise exposure may trigger changes to numerous ecological interactions and reverberate through communities.

Increases in pollination rates were in line with our prediction based on the positive responses to noise by *A. alexandri*, both in terms of nest site selection (Francis et al. 2009a) and abundances determined from surveys (Francis et al. 2011b). Our experimental design and use of artificial flowers were advantageous because we could control for variation in density and the spatial arrangement of nectar resources that can influence pollination patterns (e.g., Feinsinger et al. 1991). However, this approach precluded us from determining whether increases in pollination in noisy areas result in greater seed and fruit production. This is probable for *I. aggregata* because it can be pollen limited throughout its range (Campbell and Halama 1993; Hainsworth et al. 1985; Juenger and Bergelson 1997) and fruit set is strongly correlated with pollinator (e.g., hummingbird) abundance (Paige and Whitham 1987). Therefore, noise-dependent increases in *A. alexandri* abundances (Francis et al. 2009a and 2011b) coupled with increases in visits to artificial flowers in this study is suggestive that *I. aggregata* plants exposed to elevated noise levels may have greater reproductive output relative to individuals in quiet areas.

Seed removal, seed predation and seedling recruitment data were consistent with one another and our expectations, suggesting that noise has the potential to indirectly affect woodland structure. It is plausible that the suite of species removing seeds may differ in June and July when we conducted our study from that found in the autumn when seeds are typically available. Yet all species documented removing seeds are year-round residents and their relative abundances are unlikely to fluctuate between treatment and control sites throughout the year. Instead, it is more likely that we underestimated the magnitude of the difference in seed dispersal quality between noisy and quiet areas for two main reasons. First, because *A. californica* typically provision young with protein-rich animal prey (Curry et al. 2002), individuals at our study area may have been foraging primarily on animal prey rather than *P. edulis* seeds during our experiments. Second, our use of seed stations on the ground did not account for seed removal from cones in the canopy by other important seed dispersers, such as *Gymnorhinus cyanocephalus* (piñon jay), a species that occurs in RCHMA, but also avoids noisy areas (Francis et al. 2009a; Ortega and Francis 2012). The degree to which these factors contribute to reduced seedling recruitment in noisy areas is unknown, but provides an interesting avenue of research for future study.

Although *A. californica* and *Peromyscus* mice had the greatest influence on seed removal rates, we were unable to track the fate of individual seeds. Nevertheless, these species influenced patterns of seed predation in a manner consistent with knowledge of how these species differ as mobile links for *P. edulis* seed dispersal and seedling establishment. Evidence of seed predation was less common at seed stations visited by *A. californica*, potentially reflecting its role as an important disperser of *P. edulis* seeds. For example, one *A. californica* individual may cache up to 6,000 *P. edulis* seeds in locations favorable for germination during a single autumn (Balda 1987). Many seeds are relocated and consumed, but many go unrecovered and germinate (Chambers et al. 1999). In contrast, although *Peromyscus* mice might function as conditional dispersers under some circumstances (Pearson and Theimer 2004; Theimer 2005), here their presence at a seed station was a strong predictor of seed predation, reflecting their primary role as seed predators (Chambers et al. 1999). Previous research using experimental enclosures to study caching behavior in the field supports our findings (Pearson and Theimer 2004; Vander Wall 2000). *Peromyscus* mice consume a large proportion ($\approx$40%) of encountered seeds and typically cache many encountered seeds that are not immediately consumed (Pearson and Theimer 2004). Yet cached seeds are often recovered and eaten ($> 80\%$) along with seeds cached by other individuals or species (Vander Wall 2000). Thus, the reduced density of seedlings in noisy areas could be explained not only by fewer seeds entering the seed bank as a result of reduced densities of important avian seed dispersers that cache many thousands of seeds, but because seeds present within the seed bank experience elevated rates of predation via cache pilfering associated with noise-dependent increases in *Peromyscus* mice.

Despite the concordance between our findings and the literature regarding the roles of *A. californica* and *Peromyscus* mice on *P. edulis* seed dispersal and predation, seedling mortality caused by key seedling predators, such as *Odocoileus hemionus* (mule deer) and *Cervus canadensis* (elk), could potentially explain the higher density of seedlings in quiet relative to noisy areas. However, ungulates such as *C. canadensis* appear to avoid areas exposed to noise from high traffic volume (Gagnon et al. 2007), suggesting that seedling mortality due to browsing ungulates should be greater in areas with less noise and leading to a pattern opposite from that which we observed. Still needed are confirmatory studies that track the fate of cached seeds and document patterns of seedling predation within noisy and quiet areas.

Despite the downstream consequences of species-specific response to noise exposure, the mechanistic reasons for species-specific responses are still not clear. *A. californica* may avoid noisy areas because noise can mask their vocal communication. Larger birds with lower-frequency vocalizations are more sensitive to noise than smaller species with higher-frequency vocalizations because their vocalizations overlap low-frequencies where noise has more acoustic energy (Francis et al. 2011a). *A. californica* is also the main nest predator in the study area (Francis et al. 2009a; Francis et al. 2012b) and it is possible that noise masks acoustic cues used to locate prey at nests (e.g., nestling and parent calls). It is also possible that these forms of acoustic interference may led to elevated stress levels that could influence patterns of habitat use (Kight and Swaddle 2011), but research on this potential link is currently lacking.

In contrast to the direct effect noise may have on *A. californica* communication and foraging, positive responses to noise by *A. alexandri* and *Peromyscus* mice likely reflect indirect responses to noise. Noisy areas may represent refugia from predators and key competitors that typically avoid noisy areas, including jays. For example, *A. alexandri* may preferentially settle in noisy areas in response to cues indicative of lower nest predation pressure from *A. californica*. Similarly, *Peromyscus* mice populations may increase in noisy areas not only because of reduced competition with *A. californica* and other jays for key foraging resources, but also in response to reduced predation by nocturnal acoustic predators that may avoid noise (e.g., Siemers and Schaub 2011), such as owls.

That noise may alter patterns of seedling recruitment adds important insights to our earlier work where we found neither *P. edulis* tree density, nor 12 other habitat features differed between treatment and control sites (Francis et al. 2009). This, however, may be slowly changing. Reduced *P. edulis* seedling recruitment in noisy areas may eventually translate into fewer mature trees, yet because *P. edulis* is slow growing and has long generation times (Floyd et al. 2003), these initial changes in stand structure could have gone undetected for decades. Such long-term changes may have important implications for the woodland community as a whole by prolonging the negative consequences of noise exposure. That is, noise may not only result in large declines in diversity during exposure by causing site abandonment or reduced densities by many species (Bayne et al. 2008; Francis et al. 2009a), but diversity may suffer long after noise sources are gone because fewer *P. edulis* trees will provide less critical habitat for the many hundreds of species that depend on them for survival (Mueller et al. 2005).

These separate experiments highlight that noise pollution is a strong environmental force that may alter key ecological processes and services. Over a decade ago Forman (2000) estimated that approximately one-fifth of the land area in the United States is affected by traffic noise, yet the actual geographic extent of noise exposure is undoubtedly greater when other sources are considered. Additionally, this spatial footprint of noise, the anthropogenic soundscape, will only increase because sources of noise pollution are growing at a faster rate than the human population (Barber et al. 2010). These data suggest that anthropogenic soundscapes have or will encompass nearly all terrestrial habitat types, potentially impacting innumerable species interactions both directly and indirectly. It is critical that we identify which other functionally unique species abandon or preferentially settle in other noisy areas around the world. Early detection of altered species distributions and the resulting disrupted or enhanced ecological services will be key to understanding the trajectory of the many populations and communities that outwardly appear to persist despite our industrial rumble.

4

ANTHROPOGENIC NOISE INCREASES FISH MORTALITY BY PREDATION

Stephen Simpson, Andrew Radford, Sophie Nedelec, Maud Ferrari, Douglas Chivers, Mark McCormick, and Mark Meekan

Noise exposure is not limited to land. Marine environments are also experiencing marked increases in sound level. The authors present data from a keystone protected area, the Great Barrier Reef in Australia, indicating that a fish predator consumes more than twice as many fish prey under exposure to motorboat noise. Using field and laboratory playback tests the authors elegantly show that these results and related stress responses are driven by noise. Clearly, understanding how noise generated by human movement in parks and protected areas, both on land and at sea, affects predator-prey interactions is critical to maintaining intact natural systems.

INTRODUCTION

Since the Industrial Revolution, anthropogenic (man-made) noise has changed the soundscape of many terrestrial and aquatic ecosystems (M. McDonald et al. 2006; Normandeau Associates, Inc. 2012; Watts et al. 2007). International legislation, such as the US National Environment Policy Act and the European Commission Marine Strategy Framework Directive, recognizes the need to assess and manage the biological impacts of noise-generating human activities. However, while recent studies have demonstrated that anthropogenic noise can detrimentally affect animal hearing thresholds, communication, movement patterns and foraging (Normandeau Associates, Inc. 2012; Shannon et al. 2015; Slabbekoorn et al. 2010), it is often difficult to translate these effects into meaningful predictions about individual fitness and population-level consequences (Morley et al. 2014; National Research Council of the National Academies 2005). This is because animals may be able to move away from noise sources, acoustic disturbance may be sporadic, and compensation by organisms may prevent long-term impacts (Bejder et al. 2006; National Research Council of the National Academies 2005; Normandeau Associates, Inc. 2012). Recent correlative evidence suggests that naval sonar may cause mortality in beaked whales (Filadelfo et al. 2009), but it is impossible to test this directly. Thus, there is a clear need for experimental studies on tractable organisms that investigate directly whether common sources of anthropogenic noise reduce survival.

In marine environments, noise pollution is derived from a variety of sources including pile-driving, seismic surveys, shipping and motorboat traffic (Normandeau Associates, Inc. 2012; Slabbekoorn et al. 2010). Much of this noise occurs in coastal regions, which are experiencing unprecedented human population growth (Small and Nicholls 2003) and thus significant rises in transportation, fishing and recreation activities that involve boating (Davenport and Davenport 2006). For example, there were more than 12.5 million registered motorboats in the USA in 2013 (National Marine Manufacturers Association 2013) and there are expected to be 0.5 million recreational motorboats using the Great Barrier Reef by 2040 (Great Barrier Reef Marine Park Authority 2014). Motorboats are therefore a prevalent and increasing source of anthropogenic noise, with emerging evidence that this noise could affect communication, orientation, and territorial behavior in fish (Whitfied and Becker 2014). Unlike industrial sources of noise such as pile-driving and commercial shipping, it is relatively straightforward to design studies that use motorboats in controlled experiments to test impacts of noise on marine organisms.

Here, we examine the effect of motorboat noise on predator-prey dynamics and survivorship using a model coral reef system that lends itself to manipulation, observation and replication: the Ambon damselfish *Pomacentrus amboinensis* and its predator, the dusky dottyback *Pseudochromis fuscus*. Damselfishes share life-history traits with the majority of benthic and coastal fishes and invertebrates, typified by demersal, site-attached adults that produce pelagic larvae that develop in open water before settling to suitable habitat where they will live as juveniles and adults (Leis and McCormick 2002). Upon settlement to reef habitat, young naïve fish encounter a suite of novel predators and suffer high rates of mortality that make the first few days post-settlement a critical population bottleneck (Almany and Webster 2006). We tested the impact of motorboat noise on the post-settlement survival, physiology and performance of *P. amboinensis* when exposed to the predator *P. fuscus*, thus providing a direct assessment of the fitness consequences of anthropogenic noise. We found that both motorboat noise and direct disturbance by motorboats elevated stress and reduced anti-predator responses, more than doubling mortality by predation.

RESULTS

Survival on Patch Reefs

Pomacentrus amboinensis suffers natural mortality rates of ~50% in the first 5 days after settlement to reef habitat (McCormick and Hoey 2004). We tested whether motorboat noise affected the likelihood of mortality at this time by placing settlement-stage individuals on small isolated experimental patch reefs on sandflats (Munday et al. 2010), with underwater sound systems broadcasting either recordings of ambient habitat noise or ambient noise with the addition of noise from motorboats passing nearby (10–200 m away). Addition of boat noise in the playback recordings had a significant negative effect on survival of *P. amboinensis* (Cox's $F = 4.30$, n_{amb}, $n_{boat} = 39$ fish on individual reefs, $p < 0.001$): 79% of recruits survived the 72 h observation period in the control treatment, but only 27% survived in the boat-noise treatment (Figure 4.1). It is possible that some of this additional mortality was directly caused by noise-induced physiological changes (Lagardere 1982). However, given that mortality at this life-history stage is driven predominantly by predation (Munday et al. 2010), we used a series of further experiments to examine the factors driving increased predation and to assess the consequences of boat noise on predator-prey interactions and prey fitness.

Metabolic Rate

Recent laboratory work on fish has established the possibility that anti-predator behavior in European eels (*Anguilla anguilla*) could be compromised by a noise-induced elevation in stress, as indicated by an increase in active metabolic rate of individuals in sealed tubes exposed to different playbacks of noise (Simpson et al. 2015). We used the same approach and found that settlement-stage *P. amboinensis* used 20% more oxygen over 30 min when experiencing playback of motorboat noise compared to playback of ambient noise (independent samples t-test: $t = 9.04$, df $= 57$, $n_{amb} = 30$, $n_{boat} = 29$, $p < 0.001$; Figure 4.2A). Since playback of field recordings in tanks does not fully replicate the acoustic conditions in open water with real noise sources, we repeated this study *in situ* with motorboats. We

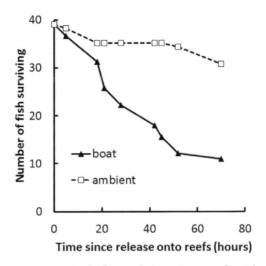

FIG 4.1: Survival of *P. amboinensis* on reefs with and without playback of boat noise. Field-based survival of *P. amboinensis* during 72 h following release onto experimental patch reefs with playback of ambient or boat-noise recordings using underwater speakers.

found a similar result using the same experimental design: settlement-stage *P. amboinensis* used significantly more (33%) oxygen when exposed to motorboats passing compared to ambient conditions ($t = 6.29$, n_{amb}, $n_{boat} = 18$, $p < 0.001$; Figure 4.2B).

Anti-Predator Behavior

Noise-induced stress could reduce the likelihood that prey detect the approach of predators, and thus do not react with an appropriately rapid startle response (A. Chan and Blumstein 2011; A. Wright et al. 2007). We designed a looming-stimulus experiment (Fuiman and Cowan 2003) for use with motorboats in open water to test the effect of boat noise on the anti-predator behavior of settlement-stage *P. amboinensis*. We found that when motorboats were passing, *P. amboinensis* were six times less likely to startle to a simulated predator attack compared to fish tested in ambient conditions (Fisher's exact test: n_{amb}, $n_{boat} = 30$, $p = 0.005$; Figure 4.2C). Of those fish that startled, the response time after the release of the stimulus was 22% slower (independent samples *t*-test: $t = 4.28$, $n_{amb} = 28$, $n_{boat} = 18$, $p < 0.001$; Figure 4.2D)

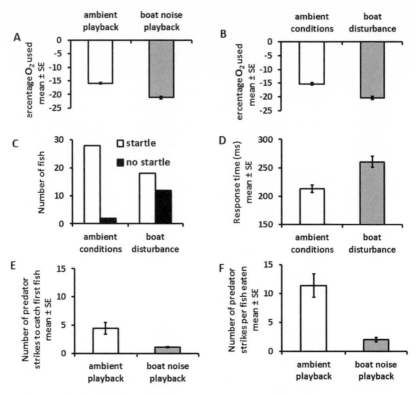

FIG 4.2: Impacts of boat noise on active metabolic rate and anti-predator response of *Pomacentrus amboinensis*, and predatory success on *P. amboinensis* by its natural predator *Pseudochromis fuscus*. (A) Mean ± SE oxygen depletion for *P. amboinensis* exposed to playback of ambient or boat-noise recordings in tanks ($n = 29$ for each treatment). (B) Mean ± SE oxygen depletion for *P. amboinensis* exposed to ambient conditions or boats motoring nearby ($n = 19$ for each treatment). (C) Number of *P. amboinensis* exhibiting a startle response to a looming stimulus with or without boats motoring nearby ($n = 30$ for each treatment). (D) Mean ± SE time taken to startle to a looming stimulus by those individuals in (C) that exhibited a startle response (ambient conditions: $n = 28$; boat motoring nearby: $n = 18$). (E) Mean ± SE number of strikes made by *P. fuscus* to catch the first *P. amboinensis* while exposed to playback of ambient or boat-noise recordings in tanks ($n = 18$ for each treatment). (F) Mean ± SE number of strikes per prey item in the same experiment.

with motorboats passing compared to ambient conditions. Consequently, the "predator" was 31% closer to fish that startled during exposure to boat noise (mean ± SE: 4.17 ± 0.26 cm) than in ambient conditions (2.56 ± 0.28 cm; $t = 4.11, p < 0.001$).

Strike-Success Rate of Predators

The extent to which such a reduction in anti-predator responses could influence the survival of prey also depends on the impact of boat noise on the performance of the predator. Therefore, we observed interactions between *Pseudochromis fuscus* and *P. amboinensis*, an established predator-prey model system (Ferrari et al. 2011), in predation trials conducted with and without motorboat noise. To allow detailed observation, we initially conducted the experiment in tanks (Ferrari et al. 2011), using playback of field recordings. Predation by *P. fuscus* was more successful in trials with boat-noise playback than in trials with ambient-noise playback, with the predators needing 74% fewer strikes to capture their first prey (Mann-Whitney U: $U = 34.5$, $n_{amb}, n_{boat} = 18, p < 0.001$; Figure 4.2E) and 82% fewer strikes per prey capture overall (Mann-Whitney U: $U = 27.5, n_{amb}, n_{boat} = 18, p < 0.001$; Figure 4.2F).

Mortality Due to Predation

To examine the consequences of these noise-related effects for the fitness of *P. amboinensis*, we investigated survival likelihood in 15 min predator-prey trials. In the tank-based experiment, there was a significant effect of boat-noise playback (Mann-Whitney U: $U = 52, n_{amb}, n_{boat} = 18, p < 0.001$), with 2.9 times as many *P. amboinensis* consumed compared to when there was playback of ambient noise (Figure 4.3A). Importantly, in a similar experiment in open water, 2.4 times as many *P. amboinensis* were consumed by *P. fuscus* when motorboats were passing compared with ambient conditions ($U = 99, n_{amb}, n_{boat} = 22, p < 0.001$; Figure 4.3B).

DISCUSSION

Our study demonstrates a direct impact of anthropogenic noise on predator-prey dynamics and quantifies, for the first time, the negative consequences for prey survival. Combining laboratory and field experiments, utilizing playbacks and real noise sources, we provide strong evidence that motorboat noise can have detrimental effects on anti-predator behavior, potentially as a

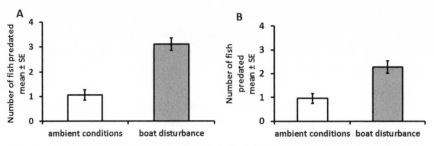

FIG 4.3: Direct impact of boat noise on predation of *P. amboinensis* by its natural predator *Pseudochromis fuscus*. (A) Mean ± SE number of *P. amboinensis* eaten of 10 individuals during 15 min trials in controlled tank conditions, with and without playback of ambient or boat-noise recordings. (B) Mean ± SE number of *P. amboinensis* eaten from 5 individuals during 15 min trials in open water conditions, with and without boats motoring nearby.

result of increased stress. In our model system, boat noise favored the predator, with the prey suffering reduced fitness; the winners and losers in other predator-prey interactions will depend on the relative hearing sensitivities and noise tolerances of the species involved, as well as the particular noise source. For instance, previous tank-based playback studies have shown that additional noise can reduce the foraging success of fish and crabs (Purser and Radford 2011; Wale et al. 2013), although whether this translates into consequences for fitness is unknown. It remains to be determined whether repeated exposure to boat noise would result in increased tolerance by *P. amboinensis*, and thus a lessened impact over time (Bejder et al. 2006). However, tolerance can only develop, and compensation is only possible, in animals that survive predatory attacks.

Elevated stress in response to noise was indicated by the increased active metabolic rate of *P. amboinensis* (Barton 2002); previous work has also suggested that noise can cause stress in fish (Simpson et al. 2015; Wysocki et al. 2006). A general allostatic stress response could result in decreased locomotor activity (Metcalfe et al. 1987) or affect attention (A. Chan and Blumstein 2011). Both are possible explanations for the reduced likelihood of startle responses by prey to predatory attacks, also seen in laboratory work on eels and crabs utilizing playback and simulated predator assays (Simpson et al. 2015; Wale et al. 2013). Reduced performance of prey in response to a predator could also arise from noise-induced distraction (A. Chan et al. 2010a) or from the masking of acoustic cues indicating the approach of a predator (Brumm and Slabbekoorn 2005).

In this study, we focused on the critical life-history period immediately following settlement of larval reef fish to benthic habitat. Our experiments suggest that motorboat noise could increase mortality at this transitional stage; future experiments are needed to test the spatial scale of impact for a range of species at key life-history stages. In those coral reef environments where motorboat noise is a frequent event, for example the Great Barrier Reef, this level of disturbance could affect the demography of impacted populations. A range of stressors increasingly threatens coral reefs (Burke et al. 2011; Ferrari et al. 2015), yet reefs generate important revenue for many countries through tourism, and provide food and livelihoods through fisheries to 500 million people (Burke et al. 2011). If sufficient resilience is to be retained for reef ecosystems to survive predicted global climate change, managing current, local environmental stressors has been proposed as an essential goal. Our work highlights the need for anthropogenic noise to be included in environmental management plans and, in general, the importance of assessing the direct fitness consequences of anthropogenic noise.

METHODS

Permits and Ethical Approval

This study was conducted during October–November in 2012 and 2014 at Lizard Island Research Station (14°40'S 145°28'E), Great Barrier Reef, Australia, with permission and ethical approval from: Lizard Island Research Station, Great Barrier Reef Marine Park Authority, James Cook University (A2081), Australian Institute of Marine Science and University of Exeter (2013/247).

Acoustic Stimuli and Playback Conditions

Daytime ambient recordings were made in the bay in front of Lizard Island Research Station (where all field trials were conducted) at five different inshore sandy-bottom locations, at 3–5 m depth and always >100 m from reefs. At each location, a recording was also made with one of five of the research station boats (5 m long aluminum hulls with 30 hp Suzuki outboard motors) motoring at various speeds 10–200 m from the hydrophone and accelerometer, replicating the kinds of boat operations common in coral reef environments. Much of the Great Barrier Reef lagoon is less than 40 m deep and fishermen, divers and tourists typically operate motorboats in shallow water (<10 m deep) over and through reefs. At popular sites on the Great Barrier Reef, many boats may pass over a reef each hour. Recordings were taken from a kayak moored using an anchor without chain to avoid unwanted noise (e.g., waves on the hull of a boat), and were made 1 m above the seabed for 5 min. Acoustic pressure was measured using a calibrated omnidirectional hydrophone (HiTech HTI-96-MIN with inbuilt preamplifier, manufacturer-calibrated sensitivity -164.3 dB re 1V/μPa; frequency range 0.02–30 kHz; calibrated by manufacturers; High Tech Inc., Gulfport, MS) and a digital recorder (PCM-M10, 48 kHz sampling rate, Sony Corporation, Tokyo, Japan). Particle acceleration was measured using a calibrated triaxial accelerometer (M20L; sensitivity following a curve over the frequency range 0–3 kHz; calibrated by manufacturers; Geospectrum Technologies, Dartmouth, Canada) and a digital 4-track recorder (Boss BR-800, 44.1 kHz sampling rate, Roland Corporation, Los Angeles, CA). Recording levels used with each setup were calibrated using pure sine wave signals from a function generator with a measured voltage recorded in line on an oscilloscope.

For the playback experiments, a unique compilation of three of the five ambient recordings (made using Audacity v2.0.2, http://audacity.sourceforge .net) was used for each control track, and compilations of three of the five boat-noise recordings made at the same five locations were used for the boat-noise tracks. The sound systems used for playback of ambient and boat-noise recordings consisted of a battery (12V 7.2 Ah sealed lead-acid), WAV/ MP3 player (GoGEAR Vibe, frequency response 0.04–20 kHz; Philips, Netherlands), amplifier (M033N, 18W, frequency response 0.04–20 kHz; Kemo Electronic GmbH, Germany) and speaker (University Sound UW-30; maximal output 156 dB re 1 μPa at 1 m, frequency response 0.1–10 kHz; Lubell Labs, Columbus, OH).

Ambient and boat-noise recordings, and recordings of their playback in open water and in experimental tanks (descriptions below), were analyzed using MATLAB v2010a: Fast-Fourier Transformation was used to calculate power spectral density for comparison of sound levels for each treatment across the frequency range 0–3000 Hz (Figure 4.4). Playback using speakers in both natural settings and in tanks alters the characteristics of the original recordings. However, analysis of spectral content and sound levels showed that characteristics of the original recordings were at least partially retained in playback and that these characteristics differed between playback of ambient and motorboat noise in both the field and in tanks. As boat-noise playbacks in the tank were too loud to characterize fully using the accelerometer due to instrument sensitivity ("clipping" was observed), we could not describe the full extent of particle acceleration levels, although it is clear that they were louder than the ambient playback.

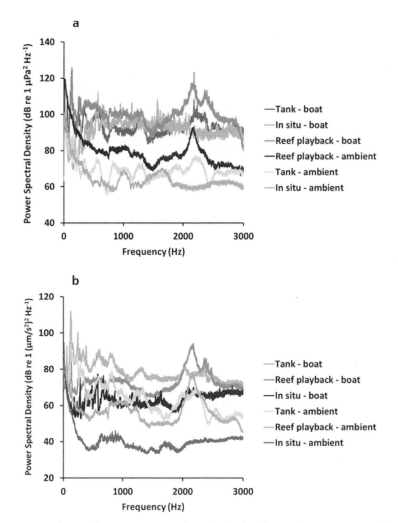

FIG 4.4: Analysis of acoustic stimuli and playback conditions. Spectral content shown in both (a) acoustic pressure and (b) particle acceleration domains for original field recordings of boat-noise and ambient conditions, playback of ambient and boat-noise recordings in the field, and playback of ambient and boat-noise recordings in experimental tanks. Mean power spectral density of 1 min of each sound condition is shown. In both the pressure and particle acceleration domains, there is a clear difference between boat and ambient conditions, whether using playback or real noise, and in tanks or *in situ*. However, received levels of particle acceleration at the fish were higher in playback treatments than with motorboats. Sounds were analyzed in MATLAB v2010a, fft length = sampling frequency (48 kHz for sound pressure and 44.1 kHz for particle acceleration, both result in 1 Hz bands).

Impact of Boat Noise on Survival on Patch Reefs

Following a ~3 week pelagic larval stage, young *P. amboinensis* settle onto patch reefs and continuous reefs in shallow waters 1–15 m deep. In this novel habitat, juveniles are exposed to a diverse range of predators that use both ambush (lizardfish *Synodus dermatogenys* and small cods *Cephalopholis microprion*) and pursuit (dottybacks *Pseudochromis fuscus* and wrasse *Thalassoma lunare*) tactics. These species are regularly observed capturing juvenile reef fish that venture away from shelter (T. Holmes and McCormick 2006), and all species were seen around the experimental reefs during this experiment. We tested the effect of boat-noise playback on the mortality rate of settlement-stage *P. amboinensis* released onto experimental patch reefs.

Our experimental reefs comprised pieces of healthy and dead bushy hard coral *Pocillopora damicornis* (~18 × 15 × 18 cm) placed on sandflats. Ten such reefs were spaced 3 m apart in a circle at four different shallow-water (3–5 m) sites that were separated by >400 m on the backreef of the fringing reef. At each site, a sound system (details above) was moored so that the speaker was suspended in the center of the reefs. Patches were cleared of any fish or large invertebrates using hand nets prior to release of experimental fish. Sound systems played either tracks of ambient noise or alternated between tracks of 5 min ambient and 5 min boat noise (details above).

Settlement-stage fishes were collected overnight using light traps (Meekan et al. 2001) moored ~500 m offshore in open water (~10–20 m depth) around Lizard Island, transported in 60 L tubs back to the research station where they were sorted into species and *P. amboinensis* were transferred to 30 L aquaria supplied with a continuous flow of aerated seawater. Fish were held for four days (27–30° C, 14:10 h natural light:dark cycle) and fed twice daily *ad libitum* with newly hatched *Artemia* sp. On the day of the release, *P. amboinensis* were placed into individually labelled 1 L plastic bags of seawater and kept in a water bath of flowing seawater until deployment in the field. To allow the identification of experimental fish from any other fish that might naturally recruit to our experimental reefs, fish were tagged subcutaneously with a red fluorescent elastomer tattoo using a 27-gauge hypodermic needle (Burke et al. 2011). This left a visible 1.5–2 mm long stripe of color on the flank of the fish. Tagging with a single elastomer tattoo has been found to have no influence on the mortality or growth of this species (Hoey and McCormick 2006), but has demonstrated that loss of individuals from reefs is not due to post-settlement migrations (Hoey and McCormick 2004). Fish were transported in bags to the field site in a shaded 60 L water bath to maintain a stable temperature and offer conditions of diffused light to minimize stress of transfer.

A single *P. amboinensis* was selected at random and placed on each of the 10 reefs at the four sites, and mortality was then monitored twice daily for 3 days. The experiment was repeated to control for any site effects, with the sound treatments (i.e., boat-noise playback with ambient reef sound, or playback of ambient reef sound) at each site reversed for the second block. Survival (up to 72 h) of *P. amboinensis* in the two acoustic treatments was compared using multiple-sample Survival Analysis, which uses a Cox's proportional hazard model (Statistica 9.0). Survival curves for fish within each treatment were calculated and plotted using the Kaplan-Meier product-limit method, which is a non-parametric estimator of survival that incorporates incomplete (censored) observations, such as those cases where fish had not died by the end of the census period. The difference in survival of fish between the boat noise and ambient noise treatments was compared using the Cox-Mantel test with a Cox's F statistic.

Impact of Boat Noise on Metabolic Rate

The methods for assessing metabolic rate from oxygen depletion followed those of Simpson et al. 2015. Two experiments were conducted: the first, in tanks, examined the impact of playback of boat noise; the second, in the field, examined the impact of motorboats. For both experiments, *P. amboinensis* were collected in light traps (as above), and were starved for 20 h prior to the experiment to avoid any inter-individual effects of metabolic demands due to digestion (Killen et al. 2014). At the start of the experiment, individual fish were randomly allocated to each treatment, to avoid any biases arising from preferential capture. Each fish was transferred by a scoop and sealed in a weighted 120 ml opaque plastic tube (12 cm long, estimated to be 90–95% acoustically transparent based on typical acoustic impedance of polystyrene vs. water). The tube was filled with fully aerated seawater (89–91% O_2 saturation, 28–29°C), with the top sealed underwater to avoid air bubbles. Dissolved O_2 content of the water was measured (Dissolved Oxygen and Temperature Meter HI 9164, Hanna Instruments Inc., Woonsocket, RI) before and after 30 min of exposure to sound. After each trial, fish were weighed and measured to test for bias in the size of fish allocated to different treatments.

In the first experiment, four exposure tanks were used; these were placed on separate benches to avoid the transfer of sound. Each consisted of a smaller experimental plastic tank (40 × 30 × 30 cm, 2 mm walls, 25 cm water depth) inside a larger plastic tub (70 × 50 × 60 cm, 3 mm walls, 25 cm water depth), with the speaker suspended in the larger tank to avoid contact with the sides. Two randomly allocated tanks received two different ambient playback tracks (details above) while the other two received different boat-noise tracks. Fifty-eight fish were evenly split in an independent-measures design between the two sound treatments, with each weighted tube placed in the allocated exposure tank for the 30 min trial period. No significant differences were found in the weight (mean ± SE, ambient: 40.7 ± 2.8 mg, boat: 42.9 ± 0.8 mg, t-test: $t = 0.78$, $p = 0.437$) or size (mean ± SE, ambient: 11.7 ± 0.1 mm, boat: 11.9 ± 0.1, t-test: $t = 1.63$, $p = 0.108$) of fish allocated to the two treatments.

In the second experiment, a weighted plastic crate was placed on sand in 2 m water. P. amboinensis were transported to the test site as for the patch reef experiment (details above), before sealing in tubes that were then placed in the crate. In two blocks of trials (treatment order reversed on the second day of testing), fish were randomly allocated and exposed to either one of two boats driving at 10–200 m distance from the crate, or to ambient conditions. Thirty-six fish were evenly split in an independent-measures design between the two sound treatments, with each weighted tube placed in the crate for the 30 min trial period. No significant differences existed in the weight (mean ± SE, ambient: 40.5 ± 1.4 mg, boat: 38.7 ± 1.2 mg, t-test: $t = 0.94$, $p = 0.351$) or size (mean ± SE, ambient: 12.5 ± 0.1 mm, boat: 12.5 ± 0.1, t-test: $t = 0.47$, $p = 0.640$) of fish allocated to the two treatments.

Impact of Boat Noise on Anti-Predator Behavior

We adapted the simulated "ambush predator" looming-stimulus experiment (Munday et al. 2010), which isolates the visual component of a predatory strike, for use in open-water conditions. The stimulus consisted of a 60 cm length of 22 mm PVC pipe with a black end cap that emerged from a larger pipe secured to a concrete block. The stimulus was remotely released and powered by a speargun rubber so that it travelled at high speed for 350 ms toward a 250 ml plastic holding pot. The stimulus, which appeared as a black disk increasing in size as it moved toward the fish, was prevented from hitting the holding pot by a lanyard.

P. amboinensis were collected, housed and transferred to the test site as described above. Individual fish were transferred by scoop into holding pots with fresh aerated seawater 15 min before the experiment to minimize stress. During this time, all fish returned to normal swimming behavior and ventilation rates. For each trial, a pot with a fish was attached in position on a second concrete block for 1 min to acclimatize to the experimental arena on the seabed prior to release of the stimulus. An observer snorkeled at the surface (to avoid the noise of scuba) and was hidden from the fish behind a large plastic tub, upon which an underwater video camera (HDR-XR520VE, 25 fps, Sony Corporation, Tokyo, Japan) was positioned to film the experiment. After 1 min acclimatization, the stimulus was released.

In two blocks of trials (treatment order reversed on the second day of testing), fish were randomly allocated to be exposed to one of two different boats driving continually at 10–200 m distance from the experiment or to ambient conditions. Limited visibility (<10 m) meant that, when present, the boat could not be seen by the fish at the test site. Thirty fish were evenly split between the two treatments and blocks in an independent-measures design. The videos were analyzed without sound (thus "blind" to the acoustic treatment), to determine whether P. amboinensis startled (a rapid shift in position or obvious directional change in swimming trajectory between consecutive frames [Simpson et al. 2015]) in response to the looming stimulus. When the fish did startle, the time taken to startle (from initiation of looming stimulus release) and the distance between the stimulus and the fish at the point of startle, were also calculated.

Impact of Boat Noise on Strike-Success Rate of Predators

We tested the effect of boat-noise playback on the interaction between settlement-stage P. amboinensis and the predator Pseudochromis fuscus. Like many dottybacks, P. fuscus is resident in a small territory on coral heads, and hunts for newly settled damselfishes that typically shelter in the branches of a single coral head because of the risk of further relocation. Adult P. fuscus were collected from reefs >2.5 km to the east of our study site by divers using a dilute clove oil solution and hand nets, and kept separately in a flow-through aquarium system (subsurface inflow of water, no bubblers) for a minimum of 3 days prior to use in the experiment (Ferrari et al. 2011). Each day, P. fuscus were fed two live P. amboinensis that had been collected the previous night using light traps. The day prior to testing, P. fuscus were not fed; predators naturally feed episodically in the wild, and not feeding for 24 h ensured consistency between trials. P. fuscus were transferred by scoop (to minimize stress from handling) into six 40 × 30 × 30 cm plastic tanks (2 mm walls, 25 cm water depth; subsurface inflow of water, no bubbler) inside larger plastic tubs (70 × 50 × 60 cm, 3 mm walls, 25 cm water depth) on separate benches to avoid sound transfer. Each tank contained a live Po-

cillopora damicornis coral head (˜15 × 15 × 15 cm) collected locally. *P. fuscus* were given 24 h to acclimatize to their experimental habitat.

Settlement-stage *P. amboinensis* were collected using light traps and kept in holding tanks as above. Since they were collected prior to settlement and contact with reef habitat these fish were naïve to *P. fuscus*. Before testing, 10 *P. amboinensis* were transferred by scoop into a 1 L jug and allowed to recover for 15 min, during which time they resumed normal swimming behavior and ventilation rates. An experimental arena was constructed from a smaller experimental plastic tank (40 × 30 × 30 cm, 2 mm walls, 25 cm water depth, acoustically transparent) inside a larger plastic tub (70 × 50 × 60 cm, 3 mm walls, 25 cm water depth). A speaker (details above) playing either an ambient or a boat track was suspended in the larger tank to avoid contact with the sides. The allocation of ambient or boat playback treatments was alternated between the experimental benches each day, and the order of the trials was counterbalanced to avoid any effects of time of day on behavior.

At the start of each trial, 10 *P. amboinensis* were released into a section of the experimental tank separated from the *P. fuscus* by a transparent Perspex screen. After 30 s, the screen was lifted and the activities of predator and prey observed from behind a hide for 15 min. Every "strike" (targeted lunge) at a prey fish and "capture" (prey fish caught and eaten) were recorded. After 15 min, the *P. fuscus* was removed and placed in a polythene bag for measurement of size. No significant difference was found in the length (mean ± SE, ambient: 75.9 ± 1.1 mm, boat: 76.0 ± 1.2, *t*-test: $t = 1.23$, $p = 0.221$) of *P. fuscus* allocated to the two treatments. Thirty-six trials (six blocks of three trials per treatment) were conducted, with *P. fuscus* and *P. amboinensis* randomly allocated between treatments; predator and prey fish were only ever used once in the experiment.

Impact of Boat Noise on Mortality Due to Predation

Data on survival of prey during interactions with predators were collected from two experiments. The first was the tank-based experiment described above, in which the number of *P. amboinensis* remaining at the end of the 15 min trial was counted in addition to data on strike success of the predator. The second modified the tank experiment to test the impact of boat noise in the field. Here, trials were conducted in upturned 30 L plastic aquaria, each containing a live *Pocillopora damicornis* coral head (˜15 × 15 × 15 cm), placed on sheets of Perspex on the sand in 2–3 m of water; separate aquaria were at least 10 m apart.

Adult *P. fuscus* used in this experiment were collected and housed as described above. On the day of the experiment, *P. fuscus* that had not been fed for 24 h were carried to a shaded location on the shore in individual holding pots held in 60 L bins. One *P. fuscus* was placed into each arena and given 15 min to acclimatize to its surroundings; during this time, either one of two different boats was driven continually at 10–200 m distance from the arena (boat treatment) or the fish was exposed to ambient conditions. Five settlement-stage *P. amboinensis* (collected and housed as above) were transferred by scoop into a 500 ml plastic pot and given 15 min to recover from capture. They were then taken by snorkelers to the experimental arena and released. The snorkelers left the arenas immediately after release. After a treatment time of 15 min, the number of *P. amboinensis* remaining in the aquaria was counted by snorkelers. All fish were then removed and the size of each *P. fuscus* measured. The order of trials was alternated between the two days of experiments to avoid any effects of time of day on experimental outcomes. No significant differences were found in size (mean $\pm$ SE, ambient: 75.2 ± 1.1 mm, boat: 75.1 ± 1.5, t-test: $t = 0.07$, $p = 0.942$) of *P. fuscus* allocated to the two treatments. Forty-four trials (two blocks of 11 trials per treatment; order of treatments alternated between blocks) were conducted, with *P. fuscus* and *P. amboinensis* randomly allocated between treatments; predator and prey fish were only ever used once.

5

LIGHT AND NOISE POLLUTION INTERACT TO DISRUPT INTERSPECIFIC INTERACTIONS

Taegan McMahon, Jason Rohr, and Ximena Bernal

The loss of quiet is often concomitant with a deficit in darkness. Noise and light pollution have strong potential to interact and synergistically drive changes in animal behavior, distributions, and reproductive success. This chapter provides evidence of these interactions. The authors reveal that frogs in a protected area in Panama are affected more strongly by parasitic frog-biting midges in quieter and darker sensory environments. Importantly, at low noise levels, increased light intensity lowered midge abundance, and at high noise levels no midges were present regardless of light level. Intact host-parasite interactions are one example of the inner workings of natural systems that reflect their intricacy and resilience. These fine-scale interactions are not why people visit parks and protected areas. Nevertheless, "to keep every cog and wheel is the first precaution of intelligent tinkering" (Leopold 1949, 190).

INTRODUCTION

Humans have modified most of the earth's surface, rapidly increasing the rate and scale of urbanization (Grimm et al. 2008; Ellis 2011). This process is complex and has rapidly altered habitat structure by modifying the type and amount of vegetation present while also changing patterns of abiotic factors resulting in novel light, noise, and temperature landscapes (McKinney 2002). For example, cities are predictable islands of heat in a backdrop of otherwise cooler environments (Shochat et al. 2006; Gaston et al. 2012). Similarly, artificial nighttime lighting forms a grid that expands across the globe (Longcore and Rich 2004; Elvidge et al. 2014). Elevated levels of low frequency noise produced by traffic and industry are pervasive (Barber et al. 2011; Ortega 2012) and characterize human-dominated ecosystems (P. Warren et al. 2006).

Heat, light, and noise pollution are common in urban environments (Douglas 1983), and affect the abundance, behavior and distribution of many species (Hoelker et al. 2010b; Gomes et al. 2016). Urban heat islands, for instance, can facilitate colonization by warm-tolerant species, which can have adverse impacts on native species (Shochat et al. 2006). Similarly, anthropogenic light can affect species abundance and distribution. Some species avoid light-polluted areas while others may alter their activity patterns; for example some species of urban songbirds start singing earlier in the morning in light polluted areas (Dominoni et al. 2013). More subtle effects can also occur. Streetlights negatively affect moth defensive behaviors and disrupt moth flight, navigation, vision, and feeding (Acharya and Fenton 1999). While increased light levels often have negative fitness consequences for species, positive effects are also possible (see Acharya and Fenton 1999).

For species that use acoustic signals, such as many insects, birds and frogs, anthropogenic noise may generate acoustic interference affecting their communication systems (Slabbekoorn et al. 2010; Halfwerk et al. 2011a; Brumm 2013; Haven 2015; Kleist et al. 2016) and ultimately alter their intraspecific interactions (Slabbekoorn and Ripmeester 2008; Francis et al. 2009a; Ortega 2012; Kleist et al. 2016). Artificial noise, light, and heat can select for signaling strategies that affect the behavior and physiology of organisms (Slabbekoorn and Peet 2003; Patricelli and Blickley 2006), modulate habitat preferences (Parris et al. 2009; Hoelker et al. 2010b) and ultimately influence the fitness of individuals living in urban environments (Fernández-Juricic et al. 2005; Barber et al. 2009; Francis et al. 2009a and 2009b; Parris et al. 2009; Laiolo 2010; Halfwerk et al. 2011b). Consequently, anthropogenic heat, lighting, and noise represent novel and important evolutionary challenges to many organisms.

Although urbanization concurrently modifies temperature, light, and noise levels (Ellis 2011), studies focusing on urbanization often only consider the effect of one of these abiotic factors (Halfwerk and Slabbekoorn 2015). In cases where more than one of these stressors are studied simultaneously, they are often examined independently. If these factors interact antagonistically, synergistically, or additively, then studies considering them in isolation could misestimate their impacts on biodiversity and species interactions.

Much like focusing on the effects of increased light, noise, and temperature independently, studies of urbanization often investigate individual species, paying less attention to potential consequences on species interactions (see Gaston et al. 2013). The handful of studies that have considered the effects of urbanization on species interactions suggest that the effects can be profound and complicated. For example, light pollution extends the foraging times of some crepuscular organisms increasing their temporal overlap and thus competition with diurnal species (Hoelker et al. 2010a; Francis et al. 2012a; Rich and Longcore 2013). Additionally, anthropogenic noise reduces the amount of time male frogs spend chorusing and impacts mixed-species breeding aggregations (Kaiser et al. 2011) with potential consequences on reproductive success, including reduced access to females. To truly understand the impacts of urbanization on communities, we need to study how species interactions are altered by multiple ecologically relevant pollutants (Halfwerk and Slabbekoorn 2015).

Given that parasitism is a very common consumer strategy and can impact food web stability, energy flow, and the health of humans, wildlife, and ecosystems (Lafferty et al. 2008), we need to understand how urbanization impacts this important and often overlooked interaction in communities. The effects of heat and chemical pollution on host-parasite interactions have been examined extensively (McMahon et al. 2013; Raffel et al. 2013 and 2015; Rohr et al. 2013). However, the effects of urban light and noise pollution on host-parasite interactions have not been studied and thus are poorly understood (Bradley and Altizer 2007).

In an effort to obtain a more complete understanding of the intricate ways in which anthropogenic environments affect natural systems, we investigated the combined effects of urban heat, noise, and light pollution on the abundance and behavior of nocturnal, frog-biting midges (*Corethrella* spp.) and their nocturnal host, túngara frogs (*Engystomops pustulosus*). Corethrellid midges are attracted in great numbers to their hosts using the mating calls of the frogs (McKeever 1977; Bernal et al. 2006; Borkent 2008). The midges depend on the mating calls produced by the frogs to locate and successfully feed on their host (Bernal and de Silva 2015). In túngara frogs, once they reach the calling male, the midges walk to the nostrils of the frog, where they obtain a blood meal (de Silva et al. 2014). As in other species of hematophagous insects, female midges use this blood meal for egg production and mating success (Borkent 2008) so finding a host frog is thus, a critical component of their mating success.

We first conducted a field study in urban and rural areas to test for associations among temperature, light, and noise on the abundance of both túngara frogs and midges. We then conducted two field experiments to partition the main and interactive effects of noise and light pollution on the abundance and frog-finding behaviors of the midges. Given the importance of nighttime conditions and the intricacies of frog calls to the mating success of both túngara frogs and frog-biting midges, we hypothesized that noise and light pollution would interfere with these communication networks reducing the abundance of both species. Because frog-biting midges are active under laboratory conditions only at low light intensities (de Silva and Bernal 2013), and not present during the day in the wild (Bernal and McMahon pers. obs), we predicted that light pollution would be detrimental to them. However, because the midges find the vicinity of their host using mostly the call of the frog, we predicted that there would be a noise-by-light interaction and that noise pollution might be even more detrimental than light pollution. Although both noise and light levels affect communication in túngara frogs (Rand et al. 1997; A. Baugh and Ryan 2010; Halfwerk et al. 2016), these frogs occur across a wide range of light conditions and thus we expected that they would be less susceptible to light pollution than frog-biting midges. In contrast to our predictions for light and noise pollution, because of the generally stable warm nocturnal temperature conditions in the lowland tropics where these two species occur, we did not predict a strong effect of urban heat pollution on the distribution of either species.

MATERIAL AND METHODS

Urban vs. Rural Survey

To quantify the relationships among temperature, light, and noise levels and the abundance of túngara frogs (*Engystomops pustulosus*) and their midge (*Corethrella* spp.) parasites, we surveyed 49 túngara frog calling sites in urban and rural areas (surveyed between 9/17–10/3/2012; each night the survey started half an hour after sunset, was conducted before moonrise, and did not continue past midnight; during this survey period the moon was waning). The urban sites were found within a 10 km transect, which started at Ancon Hill (Smithsonian Tropical Research Institute (STRI), Tupper-Tivoli Complex; N 08°57.742' W 079°32.649') in Panama City and continued along Omar Torrijos H. Avenue (urban sites were all within the city limits). The rural sites were found along Omar Torrijos H. Avenue near or within a 10 km transect of Gamboa (N 09°06.780' W 079°64.884'), which is a small town surrounded by the mature rainforest of the Soberanía National Park (rural sites were all outside of the city and surrounded by forested areas). We conducted an auditory and visual encounter survey for frog calling sites. When a calling site was detected, we searched for additional calling sites within 15 m in all directions. Most sites were identified and located auditorily. At each site, we recorded abundance of túngara frogs counting all frogs present at the site which included both male and female frogs. We also counted all frog-biting midges on or flying above the frogs, the number of túngara frog foam nests (egg masses), and the presence of other flying insects to rule out the potential effect of variation in insecticide concentrations across breeding sites as insecticides could impact abundance of insects in general. For each site, we calculated the average of three measurements of light and noise intensity and substrate temperature. All measuring instruments were held 1 m above the center of the calling site. Light intensity was measured using a digital light meter (Lux/FC, Sper Scientific), noise intensity was measured with a digital-display sound-level meter (RadioShack # 33–2055; measurements included natural and anthropogenic sounds), and temperature was measured with a hand-held non-contact infrared thermometer (MT6 Raytek MiniTemp). We conducted this survey over six nights at the same time each night, and used the date of data collection as a blocking factor in the analysis to account for temporal and environmental variation.

Midge Attraction Experiment

Our field survey indicated that temperature was not a significant factor driving the distribution of frogs or midges, and thus we focused on light and noise pollution in our experiments. To test whether light and noise levels affect the ability of frog-biting midges to locate their hosts, we applied three fully crossed levels of light (0.1, 0.1,5 and 0.30 lx) and noise pollution (45, 60, and 75 dB sound pressure level [SPL; re. 20 mPa at 1m]; referred to as frog call only, frog call and low city noise, and frog call and high city noise treatments, respectively) to sound traps. The sound traps used were modified versions of Center for Disease Control (CDC) miniature light traps (McKeever and Hartberg 1980). Each trap broadcast the same recording of a túngara frog call at 80 dB SPL (re. 20 mPa at 1m, a natural calling decibel) from a speaker and MP3 player (isound 1603 and Sylvania SMP2200, respectively; the speaker simultaneously played the frog call and respective city noise treatment), representing the calling intensity characteristic of this species (Ryan 1985).

To mimic conditions equivalent to noise pollution to the traps, traffic noise was recorded from Panama City, Panama, and broadcast at a trap at 4, 60, or 75 dB SPL. These noise pollution levels were chosen to match the range of noise intensity that we found in Panama City while sampling frog populations (see Figure 5.1); we used this range because it was the range frogs

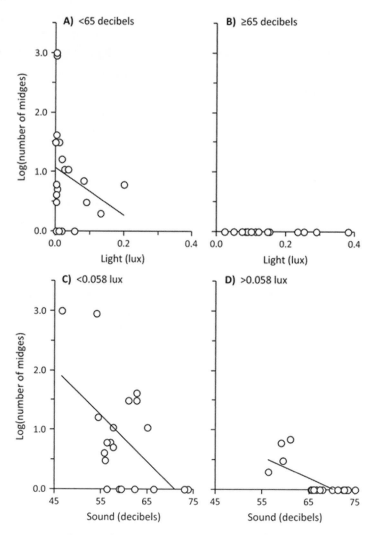

FIG 5.1: Interaction ($\chi^2_1 = 3.30$, $p < 0.001$) between light and noise levels on the number ($\log_{10} +1$) of frog-biting midges (*Core-thrella* spp.) counted per *Engystomops pustulosus* frog calling site at 49 breeding sites distributed from Panama City (urban) to Gamboa (rural) in the Republic of Panama. Panels A) and B) show the effects of light levels on midge captures above and below median sound levels (65 decibels), respectively, whereas panels C) and D) show the effects of sound levels on midge captures above and below the median light level (0.058 lux), respectively. Also shown are best-fit lines.

were experiencing in the urban setting. To achieve the three desired light intensity levels, we placed artificial white lights at each trap that were covered with an opaque plastic shield to partially obscure the light source. The light was placed 1 m away from each trap to avoid an increase in temperature at the traps due to lighting (we verified there was no change in temperature at the end of each sampling period).

To conduct this experiment, we positioned three sound traps equidistant from a stream and forest edge in Gamboa, Panama (9°07.0' N, 79°41.9'W). The traps were separated by over 20 m so the sound from the treatments was not detectable by a human standing at a different trap (McMahon, pers. obs.) and thus increased the chances that the treatments would be independent. Each trap was assigned one of the nine light-by-noise pollution treatments. We captured midges for 10 minutes at each trap and then the midges were euthanized and counted. There was a 20-minute break between trials, the traps were assigned a new treatment, and the midges were captured for 10 minutes and then counted. We conducted three trials per night; all nine treatments were tested each night in random order and location to reduce potential spatial and temporal biases. This procedure was repeated over 10 nights (surveyed between 9/19–9/29/2012) for a total of 10 replicates or temporal blocks of each of the nine treatments.

Manipulative Feeding Experiment

The *Midge Attraction Experiment* described above examined the ability of frog-biting midges to find their host under different levels of noise and light pollution, but because there were no frogs present, it could not evaluate whether midges that successfully found the host would have successfully fed. To address this question, we conducted another field experiment where we could observe the behavior of the midges in response to the túngara frogs, when exposed to the same light and noise treatments as in the *Midge Attraction Experiment*. Nine male túngara frogs were collected from Gamboa and placed individually inside nine circular containers (4 × 10 cm; height/diameter) covered with mesh with openings (2 mm) large enough for the midges to access the frogs. This container was placed on top of a speaker that broadcast the same noise treatments (frog call and city noise) used in the previous experiment. We did not want the frogs themselves to call in this experiment because frog-biting midges are sensitive to variation in túngara frog call properties (Bernal et al. 2006; Aihara et al. 2015). To prevent the frogs from calling, we did not provide standing water, which túngara frogs require to call. Both the speaker and frog container were placed inside a larger container (an opaque plastic container 13 × 24 × 35 cm, height/width/length), containing 20 frog-biting midges (collected with a hand-operated insect aspirator) that did not have a fresh blood meal (recognizable because midges have expanded red abdomens after blood ingestion). The container was open on top and was covered by fine white mesh that prevented the midges from escaping while allowing observations. The containers received the same light treatments used in the *Midge Attraction Experiment* and were placed outside in the field approximately 20 m apart from each other to reduce treatment interference given that the stimuli did not carry that far (McMahon, pers. obs.).

Over a two-hour period, six 1-minute observations were completed per treatment, recording number of animal movements (midge: walks or flights; frog: jumps). At the end of the 2-hour trial, we recorded the number of midges 1) on the container holding the frog, 2) on the frog, and 3) with a blood meal. These procedures were repeated across 10 nights (conducted between 9/17–9/29/2012; each time we started half an hour after sunset before moon rise) resulting in 10 replicates or temporal blocks of each treatment. The location of each treatment was randomized within and across the temporal blocks and 180 midges and nine frogs were collected and released each night so that different individual midges and frogs were used in each trial.

Statistical Analysis

Data were analyzed with R statistical software. Significance was attributed when $p < 0.05$. Frog-biting midges were considered a guild and thus their numbers were pooled together following previous studies interested in interspecific interactions between túngara frogs and frog-biting midges (Bernal et al. 2006; Trillo et al. 2016). We conducted multimodel inference analyses (package: glmer, function: glmer, family: poisson) considering all possible models using the dredge function in the MuMIn package. For these analyses we report model-averaged, weighted coefficients and associated p-values from models with delta AICc < 4).

Urban vs. Rural Survey A generalized linear model (package: glm, function: glm, family: Gaussian) was used to determine whether urbanization had an effect on temperature, light, and noise levels. For noise levels, the number of calling frogs was included as a covariate in the analysis. We used a generalized linear model (package: glm, function: glm, family: Poisson) to determine if there was an effect of rural or urban location on midge abundance. We used the same package and function to test whether light and noise levels correlated with one another across the sampling sites. We also used a negative binomial generalized linear model (package: glm, function: glm.nb) to determine if there was a difference in the number of frogs and frog foam egg nests in urban and rural sites; for the frog foam egg nests we used number of frogs at each site as a covariate. To determine the effects of temperature, light, noise, night, canopy cover (open/closed) and number of frogs on the abundance of frog-biting midges at each site, and to determine the effects of those same factors on the abundance of túngara frogs at each site, we conducted multimodel inference analyses (package: glmmADMB, function: glmmadmb, zeroinflated, family: nbinom) considering all possible models using the dredge function in the MuMIn package. For these analyses we report model-averaged, weighted coefficients and associated p-values from models with delta AICC < 4.

Midge Attraction Experiment We analyzed the effect of light and noise levels and their interaction on the number of frog-biting midges collected in acoustic traps using a general linear model (package: glmmADMB, function: glmmadmb, family: nbinom) with night and treatment order as blocking factors.

Manipulative Feeding Experiment We used a general linear mixed model (package: nlme, function: glmer, family: poisson) to test how light and noise levels, and their interaction affected the movement of the midges and frog, the proportion of midges on the frog container or the proportion of midges that obtained a blood meal, treating night as a blocking factor and the container as a random effect.

RESULTS

Urban vs. Rural Survey

Urban sites had significantly higher light, noise, and temperature than rural sites ($F_{1,39} = 29.73$, $F_{1,39} = 30.51$ and $F_{1,39} = 140.70$, respectively, and $p < 0.0001$ for all analyses; Table 5.1). Light and noise intensity were positively correlated with one another ($\chi^2_1 = 19.73$, $p < 0.0001$). There was no significant difference in túngara frog abundance between rural and urban sites (Table 5.1; $\chi^2_1 = 1.53$, $p = 0.13$) but there were more foam nests in rural than urban sites (Table 5.1; $\chi^2_1 = 2.06$, $p = 0.04$). Neither, light, noise, tempera-

TABLE 5.1: Abiotic factors related with anthropogenic changes, host (*Engystomops pustulosus*) and parasite (*Corethrella* spp) abundance at 49 frog breeding sites found between Panama City (urban) and Gamboa (rural) in the Republic of Panama. Mean ± SEM are shown.

Factor	Urban	Rural
Light intensity	0.16 ± 0.02 lx	0.11 ± 0.02 lx
Noise intensity	69.0 ± 0.80 dB	59.2 ± 1.00 dB
Temperature	27.6 ± 0.09 °C	25.9 ± 0.04 °C
Túngara frog abundance	6.09 ± 2.63 frogs/site	4.05 ± 1.11 frogs/site
Foam nest abundance	2.06 ± 0.74 nests/site	0.24 ± 0.23 nests/site
Frog-biting midge abundance site	0.00 ± 0.00 midges/site	67.75 ± 43.27 midges/

ture, nor the interaction between light and noise were significant predictors of frog abundance (light: $\chi^2_1 = 1.56$, $p = 0.12$; noise: $\chi^2_1 = 1.27$, $p = 0.21$; temperature: $\chi^2_1 = 0.36$, $p = 0.72$; light*noise: $\chi^2_1 = 0.36$, $p = 0.72$; Table 5.1 for model selection information).

In contrast to the abundance of the frogs, there were more frog-biting midges per calling site in rural than urban sites, where there were no frog-biting midges found (Table 5.1; $\chi^2_1 = 2214.9$, $p < 0.001$). Average temperature did not significantly affect midge abundance ($\chi^2_1 = 0.24$, $p = 0.08$) and thus could not account for the absence of midges in the urban sites. However, midge abundance was associated positively with the number of frogs ($\chi^2_1 = 5.88$, $p < 0.0001$) and negatively with light ($\chi^2_1 = 3.24$, $p = 0.001$) and noise levels ($\chi^2_1 = 2.94$, $p = 0.003$). Additionally, there was an interaction between light and noise levels ($\chi^2_1 = 3.30$, $p < 0.001$); at low levels of noise, light intensity was negatively associated with midge abundance, but at high levels of noise, there were no midges regardless of light level (Figure 5.1). Flying insects other than frog-biting midges, such as mosquitos, were found at all of the sites.

To reduce the likelihood that a third variable that differed between urban and rural sites was the true factor causing the decline in midges across the rural to urban gradient, we tested whether variation in light and noise levels in the rural sites only was also associated negatively with midge abundance. Within rural sites, midge abundance was again not significantly associated with temperature ($\chi^2_1 = 0.82$, $p = 0.41$), but was positively associated with frog abundance ($\chi^2_1 = 24.63$, $p = 0.001$). Additionally, light and noise intensity remained the most important negative predictors of midge abundance

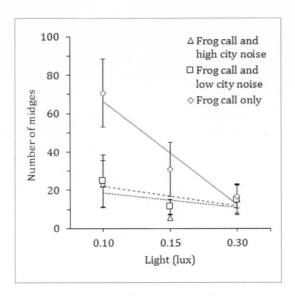

FIG 5.2: Number of frog-biting midges (*Corethrella* spp.) collected in acoustic traps (Gamboa, Panama) using a fully crossed 3x3 design: no, low, or high artificial light treatments (0.10, 0.15, 0.30 lux, respectively) crossed with *Engystomops pustulosus* frog call plus no (solid trendline), low (dashed trendline), or high (dotted trendline) city noise (45, 60, and 75 decibels, respectively). Shown are means ± 1 SE.

($\chi^2_1 = 8.75$, $p = 0.03$ and $\chi^2_1 = 5.01$, $p = 0.02$, respectively). At low noise levels, light intensity had a significant negative effect on midge abundance but at high noise levels, light did not matter because of the strong negative effects of high noise intensity (interaction: $\chi^2_1 = 66.96$, $p < 0.0001$).

Midge Attraction Experiment

There was no significant effect of order of presentation of the treatments ($\chi^2_1 = 0.09$, $p = 0.77$), but there was a significant effect of sampling night on the number of midges collected ($\chi^2_1 = 44.25$, $p < 0.0001$). Despite fluctuations mediated by environmental conditions, we found a significant interaction between light and noise pollution ($\chi^2_1 = 18.88$, $p = 0.0008$; Figure 5.2). As with the field survey, at low light intensity, the number of midges was significantly reduced by artificial noise, whereas at high light, few midges were collected regardless of noise level.

Manipulative Feeding Experiment

We examined the effect of light and noise pollution on midge behavior once they are in close proximity to their host and found that light and noise intensity reduced the ability of midges to find and feed on their frog hosts. There was an effect of light intensity on midge activity but no effect of noise intensity or an interaction between these factors ($\chi^2_1 = 9.26$, $p = 0.002$, $\chi^2_1 = 3.09$, $p = 0.08$; light and noise respectively; Figure 5.3A). In terms of the number of midges that found the frog container, there was an interaction between light and noise intensity ($\chi^2_1 = 14.4$, $p = 0.006$). At low light levels, there was a negative effect of noise intensity on the number midges that reached the frog container, but at high light, few midges reached the frog container regardless of noise treatment (Figure 5.3B). On the other hand, noise and light pollution significantly increased the movement of the frogs (noise: $\chi^2_1 = 57.70$, $p < 0.001$; light: $\chi^2_1 = 57.60$, $p < 0.001$; Figure 5.3C), but there was no interaction between noise and light on frog activity ($\chi^2_1 = 0.31$, $p = 0.57$). Midges did not feed in treatments with artificial noise or light (0% had successful feeding events), whereas 40% of the replicates with no artificial noise or light had successful feeding events (noise and light: $\chi^2_1 = 5.46$, $p = 0.03$).

DISCUSSION

Results from our field survey and two field experiments demonstrate that anthropogenic noise and light pollution associated with urbanization disrupt the túngara frog and frog-biting midge, host-parasite, interaction. Most previous work on the effects of light and noise pollution on the responses of organisms has considered the effects of these abiotic factors independently (Halfwerk and Slabbekoorn 2015), often also investigating only one species at a time. Our results show that the effects of artificial light and noise depend on the level of the other factor and highlight that the interaction between light and noise pollution can have important consequences on species interactions that would be missed if these factors were studied in isolation.

Although the abundances of hosts and parasites are typically correlated within their range (Hudson and Dobson 1995), parasitic frog-biting midges are absent in urbanized areas where their túngara frog hosts are present. Given that other flying insects, such as mosquitos, are found throughout all areas surveyed, the lack of frog-biting midges at the frogs' breeding sites in the city is unlikely to be caused by higher insecticide use in urban than rural populations. Additionally, although urban sites had slightly warmer

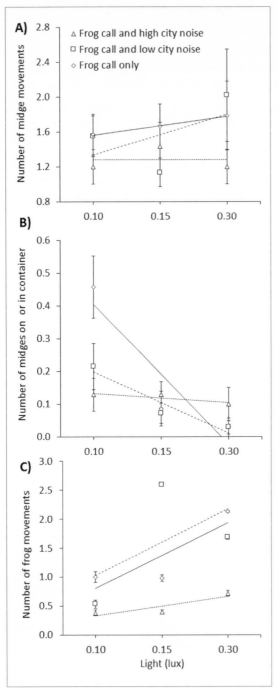

FIG 5.3: Effect of artificial light and noise on the number of A) frog-biting midge (*Corethrella* spp.) movements, B) midges that successfully located the container holding the frog (*Engystomops pustulosus*; please note, the data points for the frog call only and frog call plus low city noise at 0.3 lux, are the same values and therefore overlap one another) and C) frog movements within the container with a fully crossed 3x3 design with no, low, or high artificial light treatments (0.10, 0.15, 0.30 lux, respectively) crossed with frog call plus no (solid trendline), low (dashed trendline), or high (dotted trendline) city noise (45, 60, and 75 decibels, respectively). Shown are means ± 1 SE.

conditions, midge abundance was not affected by temperature in the range documented in this study.

Variation in noise and light pollution, both across the urban and rural sites and within the rural sites, were the main predictors of midge abundance. In general, higher levels of artificial noise and light resulted in decreased abundance, and even absence, of frog-biting midges. This pattern suggests that even low levels of light or noise (tested here: 0.15 lux, 60 dB) can affect midge abundance and disrupt this host-parasite interaction. Our two experiments demonstrated that these field patterns were indeed causal. Noise intensity was an important negative predictor of midge abundance in the field survey and artificial noise and light levels decreased the ability of the midges to find the frogs. In fact, in both experiments we found that noise over approximately 60 dB affected parasite behaviors, such as the ability of the midges to localize their host. This reduced performance is probably caused by acoustic interference that results from the presence of anthropogenic noise. Given the frequency overlap of anthropogenic noise and the calls of túngara frogs, this novel source of noise could reduce the ability of the midges to detect and localize the frogs based on their calls. Some species of frogs, for example, increase the frequency of their mating calls in traffic noise, which may alter both inter- and intra-specific interactions (Parris et al. 2009). In addition to using acoustic signals to locate hosts for blood meals, frog-biting midges also use acoustic signals for intraspecific communication. Specifically, the midges use the wingbeats of conspecifics for courting purposes and to deter rival males (de Silva et al. 2015). The frequency bandwidth of the mating signal of the midges also overlaps with the spectral properties of anthropogenic noise and efficient communication during mating is thus also likely reduced with noise pollution. Therefore, acoustic interference elicited by anthropogenic noise could compromise midge fitness by affecting reproductive potential in two ways, 1) interfering with host detection and location and thus blood meal acquisition that will, in turn, reduce egg production, and 2) diminishing the chances of mating. These factors combined may have reduced the ability of the midges to have sustainable populations in urban areas where anthropogenic noise is high.

While it is clear that frog-biting midges depend on the acoustic signals produced by their hosts to acquire blood meals, little is known about how these midges hear. It is possible that frog-biting midges rely on antennal hearing to detect and respond to the calls of their anuran host. Frog-biting midges, however, respond to frog calls at distances greater than those expected to be within the range of antennal hearing suggesting a more elaborate, tympanic organ may be involved in hearing (Page et al. 2014). Current research is investigating the hearing mechanisms in this midge; regardless of the specific organ(s) involved, given the spectral overlap of traffic noise and the calls of túngara frogs, a reduction in the ability of the midges to hear and respond to the frog calls is to be expected.

Our results also revealed an effect of anthropogenic light on midge abundance and their ability to successfully locate and bite their host. In the *Midge Attraction Experiment*, increased light intensity reduced midge abundance, and in the *Manipulative Feeding Experiment*, midges reduced their movement when exposed to any level of artificial light, indicating that light is also an important stressor affecting the behavior of the midges. The reduced activity of the midges under high light levels made them less likely to find a frog host and is consistent with their lack of activity during the day in captivity and the wild. While it is expected that their host seeking behavior is timed with those hours in which the frogs are actively calling at night, the midges concentrate all their activities to the night hours becoming almost immobile when light intensity increases (de Silva and Bernal 2013). Ultimately, reduced activity at high light levels might be a strategy to minimize predation risk. In fact, when túngara frogs were presented with frog-biting midges in daylight hours, frogs readily consumed them (McMahon and Bernal, pers. obs.). Additionally, artificial light and noise increased the movements of the frogs, which could result in increased risk for the midges. Host movement and defensive behaviors can be a significant source of mortality for hematophagous insects (Walker and Edman 1985; Edman and Scott 1987), so augmented frog activity in high light conditions could further deter approaches by midges.

We found that túngara frogs in urban habitats were free of their frog-biting midge parasites. At this point, we do not have a good understanding of how the absence of this parasite may affect the fitness of urban túngara frogs. Given that the midges can take substantial amounts of blood from túngara frogs (~10% of blood volume in a night of active calling, Bernal unpub data), losing this parasite may result in higher reproductive success. Male frogs in areas that are not attacked by midges may be able to invest more energy into their signaling strategy, such as, increasing call rate, hours of activity each night or chorus attendance which would result in increased attractiveness to female frogs. The absence of this host-parasite relationship in urban areas could also potentially influence other organisms and processes in the community. For example, túngara frogs are preyed upon by frog-eating bats who also depend on the mating call of the frogs to detect, localize them and finally eat them (Tuttle and Ryan 1981). Given that these bats may be able to compensate for the effect of noise through acoustic interference by using other cues (Gomes et al. 2016), frogs in midge-free breeding areas may encounter yet another source of antagonistic selection limiting their signaling strategies. Urban frogs, however, by being in areas with light and noise pollution may have also escaped a different natural enemy. Frog-biting midges can transmit blood parasites, such as *Trypanosoma tungarae*, to túngara frogs (Bernal and Pinto 2016), so the absence of midges could have impacts on infectious disease dynamics in this community.

We found that both light and noise pollution disrupt host-parasite interactions. These findings highlight that more research is needed on the effects of urbanization on species interactions, parasitism in particular, and the importance of considering interactions among types of pollutants. This work is necessary to accurately assess the impacts of urbanization and altered interactions on ecological communities and ecosystems.

THE EXPERIENCE
OF NATURAL QUIET

Part 2 of this book addresses the impacts of noise pollution on the quality of the visitor experience in parks and protected areas. The opportunity to experience natural quiet can be a vital part of visiting many national parks, and many park visitors report that hearing the sounds of nature is an important reason why they visit national parks. Moreover, natural quiet can have restorative effects on human health and well-being. The five studies described in this part of the book suggest that human-caused noise can substantially diminish the potential experiential benefits of natural quiet. Study authors point out that there is often a strong relationship between the impacts of noise pollution on the natural environment and the experience of visiting the national parks. For example, alterations in animal distributions and behavior caused by noise pollution can diminish the opportunity for visitors to see and hear iconic wildlife.

6

HUMAN RESPONSES TO SIMULATED MOTORIZED NOISE IN NATIONAL PARKS

David Weinzimmer, Peter Newman, Derrick Taff,
Jacob Benfield, Emma Lynch, and Paul Bell

The Organic Act of 1916 that created the National Park Service charged the agency with protecting the remarkable ecological and experiential values of the national parks. Aesthetic values were at the heart of protecting national park landscapes from the beginning and remain central to national park management. This chapter introduces the idea that anthropogenic noise can have important impacts on aesthetic quality and demonstrates the relationship between aural and visual perceptions of national park environments.

INTRODUCTION

Recreationists' motivations for desired psychological outcomes (whether it be solitude, family bonding, or connecting with nature) are determinants in their participation in, and, ultimately, satisfaction with, their recreation experiences. Managers should aim to provide a range of activity types and setting attributes (resource, social, and managerial) to address the diversity of experiences and psychological outcomes desired by visitors (Driver, Brown, Stankey, and Gregoire 1987). Thus, if people visit a protected area motivated by the expectation to experience natural sounds and to recuperate from a stressful week, then the presence of excessive, unnatural noise will likely influence their enjoyment and acceptability of resource conditions (Marin, Newman, Manning, Vaske, and Stack 2011). Research shows that many visitors share these motivations to escape the noise and commotion of their everyday lives in their outdoor recreation experiences (Driver, Nash, and Haas 1987).

BENEFITS OF NATURAL SOUNDS

Of all of the benefits offered by protected areas, one of the most frequently cited and sought after by visitors is the opportunity to experience natural quiet and the sounds of nature. In one study of national parks, more than 90% of visitors mentioned the enjoyment of natural quiet and the sounds of nature as primary motivators for their visit (McDonald, Baumgartner, and Iachan 1995). The typical visitor wants to hear birds singing, wind rustling leaves, and water cascading down a creek—not people talking on cell phones, radios, dogs barking, or loud children (Kariel 1980; Kariel 1990; Pilcher, Newman, and Manning 2009). The desire of the general public to escape the stress of everyday life by visiting a protected area is a driving force behind the growing realization by natural resource managers that natural soundscapes (the human perception of acoustical environments) need to be protected like any other valuable natural or cultural resource (Newman, Manning, and Trevino 2010). There is clearly a need for more research targeted to the psychological and social impacts of motorized transportation noise in protected areas. Such research would allow protected area managers to undertake more informed cost-benefit analyses for potentially implementing messaging strategies or regulations to control these sources of noise in order to protect natural soundscapes.

DANGERS OF EXCESSIVE ENVIRONMENTAL NOISE

Excessive environmental noise poses distinct threats to human well-being. There are general dangers in both physical and psychological dimensions, which are often highly interrelated. High levels and durations of noise events have been associated with cardiovascular problems, elevated levels of stress, decreased immune function, and impairments of cognition. Aydin and Kaltenbach (2007) found an increase in objective circulatory parameters (e.g., hypertension) and subjective factors (e.g., annoyance) in areas exposed to high levels of airport noise. Babisch, Fromme, Beyer, and Ising (2001) studied traffic noise exposure and stress hormones and reported increased endocrine response to high levels of environmental noise. Ising and Krupp (2004) present a similar model, with greater attention to the mediating parameters that influence the physical and psychological outcomes of noise exposure.

Babisch et al. (2001) also implicated coping mechanisms and control over the stimulus as moderating factors in physiological responses to noise. This is one reason why children may be more vulnerable to the detrimental effects of noise, as they tend to lack control over their surroundings and the ability to escape their environment if noise is excessive (S. Cohen, Evans, Krantz, and Stokols 1980). This combination of psychological risk factors is also implicated in the development of feelings of helplessness, which in turn can lead to illness and depression (Seligman 1975).

G. Evans, Bullinger, and Hygge (1998) combined longitudinal physiological and quality of life measurements to demonstrate a detrimental and enduring effect of airport noise on health and well-being. Staples (1996) identified a list of cognitive variables that are sensitive to disruption by chronic levels of environmental noise, including decreased school performance and learning ability in children, decreased tolerance for frustration, and fewer instances of helping behavior. Staples (1996) emphasized the importance of subjective responses to noise as more predictive of adverse reactions to noise than the physical acoustic properties alone. She also proposed the application of environmental stress theory to noise exposure. Environmental stress theory identifies two critical factors in predicting adverse reactions to noise: appraisal of the event as threatening important personal needs or goals and subjective lack of control over one's exposure. Individual reactions to noise will be influenced by these factors, leading to varying degrees of psychological distress, physical symptoms, and annoyance.

PROTECTION OF PARK SOUNDSCAPES

The National Parks Overflights Act of 1987 was one of the first federal efforts to systematically protect natural soundscapes on public lands. This law also motivated new psychological research on the impacts of noise in protected areas (Gramann, 1999). Several important policies on soundscape protection followed. In response to the fact that its parks were rapidly becoming louder, in 2000 the U.S. National Park Service (NPS) issued Director's Order #47: Soundscape Preservation and Noise Management "to articulate National Park Service operational policies that will require, to the fullest extent practicable, the protection, maintenance, or restoration of the natural soundscape resource in a condition unimpaired by inappropriate or excessive noise sources" (National Park Service 2000). The order established guidelines for monitoring and planning to preserve park soundscapes. The National Parks Air Tour Management Act of 2000 required the Federal Aviation Administration (FAA) to work with the NPS to address and mitigate adverse impacts of commercial air tours on NPS resources, including natural soundscapes. And in 2006, the NPS Management Policies included several directives related to soundscapes, including the affirmation that "The Service will preserve, to the greatest extent possible, the natural soundscapes of parks" (National Park Service 2006). Controversies surrounding snowmobile use in Yellowstone, scenic air tours over the Grand Canyon, and personal watercraft use in multiple NPS units continue to highlight the growing threat of human-caused noise to the park experience, wildlife, and natural and cultural resources, as well as the potential for conflict between stakeholders (N. Miller 2008).

RESEARCH GAPS

Despite the need to provide opportunities for natural quiet and the established detrimental impacts of anthropogenic noise in protected areas, very little research has compared the effects of different common sources of noise on visitor experiences. One of the most prevalent sources of noise in many NPS units is motorized transportation noise. In addition to primitive forms of travel, such as backpacking, stock use, or cross-country skiing, many parks offer trails for snowmobiling and scenic air tours by propeller planes. National parks are also popular destinations for motorcycle groups, who take advantage of the many amenities surrounding national parks and their gateway communities.

Previous scientific research regarding motorcycle use and behavior typically concerns safety (e.g., helmet use or conspicuity) but rarely human dimensions. Several studies have investigated the impact of aircraft and snowmobiles on visitor experiences and resource conditions. Mace, Bell, and Loomis (1999) used the landscape assessment paradigm to investigate impacts from helicopter tour noise at 40 dB(A) and 80 dB(A) on evaluations of simulated scenic overviews in Grand Canyon National Park. Self-reported changes in affective state in response to helicopter noise and landscape scenes were also assessed. Negative aesthetic and affective impacts were associated with helicopter noise at both sound pressure levels (though more markedly at the higher decibel level), relative to natural sounds. The role of source attribution in laboratory participants' responses to helicopter noise was presented in a study by Mace, Bell, Loomis, and Haas (2003). This study tested the same 60 dB(A) auditory stressor but attributed its purpose to different sources: scenic overflights for tourists, backcountry maintenance activities by park management, or search and rescue operations. Findings suggested that 60 dB(A) helicopter noise was equally disruptive to the experiences of potential park visitors, regardless of whether the noise was attributed to tourist overflights or official park management activities.

Tarrant, Haas, and Manfredo (1995) surveyed visitors to four wilderness areas in Wyoming about their feelings regarding aircraft overflights, including reported levels of annoyance and other dimensions of wilderness experience. The authors looked at the effects of visitor characteristics like motivations and attitudes, as well as factors related to noise exposure (e.g., number of overflight events, sound pressure level). This study affirmed the complexity in evaluating the psychological effects of aircraft overflights on wilderness visitors since multiple visitor characteristics and exposure variables affected visitor evaluations across a range of dimensions. M. Davenport and Borrie (2005) conducted qualitative research with snowmobile users in Yellowstone National Park to determine meanings of snowmobiling experiences for those who engaged in the controversial activity. The study addressed the question of how appropriate the activity is for the setting and explored the ways in which snowmobile users related to the park.

Despite this attention by researchers to aircraft and snowmobiles, a lack of social science research (especially related to soundscapes) has focused specifically on motorcycles. This is somewhat surprising, given the ubiquity of motorcycles in national parks, especially in the most heavily visited areas. A recent NPS report found that visitors to the Blue Ridge Parkway were exposed to noise from an average of 25 different motorcycles over a given 20-minute time period at a popular scenic overlook in the park (Rochat 2011). Some research suggests that engine noise, in general, is evaluated as particularly annoying due to the fact that it is irregular and unpredictable. In addition, rhythmic sounds are rated more negatively than continuous sounds (Kryter 1985). Similarly, Kariel (1990) summarizes research findings about the influence of the physical characteristics of sound on annoyance, implicating noises characterized by high frequency and intensity, rhythmic, and intermittent sounds as increasing the risk of causing annoyance relative to sounds with other physical properties. It is important to note that motorcycle engines generate noise that is both rhythmic and irregular, which is thus likely to cause greater annoyance from visitors to natural settings who are exposed to the noise.

Recent research has begun to look for differences in human responses between common noise sources in national parks and other natural settings. Benfield, Bell, Troup, and Soderstrom (2010a) compared the effects of natural sounds with aircraft sounds, ground traffic sounds, and human voices. The authors found that all three types of anthropogenic sounds had negative impacts on participant ratings of environmental quality (whereas natural sounds alone had no substantial effects on ratings). Mace, Corser, Zitting, and Denison (2013) investigated differential effects between three types of aircraft noise on human laboratory respondents. Results indicated that all three types of aircraft noise had negative consequences on several psychological dimensions, relative to natural sounds conditions. Helicopter noise was evaluated as the most detrimental type of aircraft noise on simulated national park experiences. Propeller plane noise was perceived to be nearly as disruptive as helicopter noise, while the jet condition produced the smallest effects of the three aircraft noise conditions.

CURRENT STUDY

Managers from different NPS units have identified the noise produced by propeller planes, snowmobiles, and motorcycles as significantly impacting the social and biological resources of their parks (National Park Service n.d.a). Based on previous research, it would not be surprising to find important distinctions among these sources in how they impact visitors to protected areas, whether due to the physical characteristics of the noise, attitudes about the sources, themselves, or expectations for a particular park (Kariel 1990; Tarrant et al. 1995; Benfield et al. 2010a; Mace et al. 2013). Such a finding would suggest that "motorized noise" cannot be treated as a single, homogeneous category for soundscape research. Negative impacts to visitor experiences (whether feelings of increased annoyance, lack of tranquility, diminished naturalness, or some other detrimental experiential effect) may depend on the specific source and visual context of the motorized noise present.

This experiment employed a laboratory simulation with both visual and auditory stimuli to determine the effects of different sound sources on participant evaluations of landscape conditions. The study design was based on an established landscape assessment procedure using common aesthetic and experiential indicators of quality (Mace et al. 1999; Benfield et al. 2010a). It is hypothesized that assessments of landscape quality will be more negative when motorized sounds are present than when natural sounds are present. Furthermore, it is expected that there will be differences in landscape evaluations between the three motorized noise sources that were simulated in this study. Improved knowledge about the specific impacts of different recreational motorized noise sources would allow park managers to develop better educational and regulatory interventions to protect high-quality visitor experiences in parks and other protected areas.

VALIDITY OF LABORATORY SIMULATIONS

Previous research suggests that laboratory simulations of natural settings can be more cost-effective and internally valid than data collected from the field. Careful experimental control of potential confounds allows simulations to isolate the factors that influence the experiences of park visitors. Furthermore, results from well-designed studies permit generalization of results beyond laboratory participants to the broader population of park visitors. Daniel (1990) demonstrated strong correlations (R ≥ 0.8) between ratings of landscapes based on color photo slides and those obtained directly onsite. C. Anderson, Lindsay, and Bushman (1999) found that when experimental variables represent core psychological processes, results obtained from student samples generalize to more diverse populations. The authors found similar effect sizes across a range of variables when comparing results between laboratory and field participants. L. Anderson, Mulligan, Goodman, and Regen (1983) also found strong correspondence between results obtained in the field, in the lab, and by questionnaire. In this study testing evaluations of different categories of sounds, the authors found that the specific methodology used explained far less variance in the data than did the setting (natural or built) and the source (natural or mechanical) of the sound. In all cases, respondents rated natural sounds more positively, human and animal sounds neutrally, and mechanical sounds more negatively, again reinforcing the idea that basic psychological processes, including human responses to sound, are similar regardless of how they are measured.

METHODS

Subjects and Procedure

This study was conducted in a psychology laboratory at Colorado State University. The overall sample size was N = 75, consisting of undergraduate and graduate students enrolled at Colorado State University who volunteered in exchange for course credit. All responses were recorded on iPad 2nd Generation computers (Apple Inc., Cupertino, CA) programmed with iSURVEY software (Contact Software Limited, Wellington, New Zealand). The study employed a repeated measures, within-subjects design that has been successfully utilized in previous soundscape experiments (Mace et al. 1999; Benfield et al. 2010a; Mace et al. 2013). The landscape assessment paradigm was closely adapted from the scales and procedures first described in

Mace et al. (1999). Each session lasted about one hour. All participants completed landscape assessments for six different conditions. There were three conditions that featured only natural sounds (i.e., birds, wind, and water sounds)—one each recorded from Denali, Glacier, and Yellowstone National Parks. There were three conditions that included the natural sounds plus an overlaid motorized sound—propeller plane, motorcycle, or snowmobile sounds. Each participant experienced all six conditions, in one of six possible pseudo-randomized orders. Different orders of conditions were used to test for the possibility of order effects. Half of the participants rated each natural condition before the corresponding motorized condition; the other half rated the motorized conditions first.

Auditory Stimuli

Six sound clips were prepared with assistance from the NPS's Natural Sounds and Night Skies Division. All clips were extracted from actual acoustic recordings from the specific parks (in the appropriate season) represented in the laboratory simulations. Raw data files were trimmed to 45 seconds, selecting for the best window from the raw data, based on sound quality and included sound events. Seven-second fade-in and fade-out effects were added to the clips to simulate movement of the sound sources and to create a realistic noise event for an observer on foot. The natural and motorized sounds were audible for nearly the entire 45-second duration of the sound clips, although the sounds were faint at the start and end of the clips as they faded in and out. Across all trials, natural sounds alone were audible approximately 50% of the time (three of six sound conditions), while motorcycle, snowmobile, and propeller plane sounds were each audible for approximately 17% of the time (one condition each). These conditions are representative of the soundscape conditions that could be encountered in these parks, especially during the high-use visitor seasons.

Data files were normalized so that all natural clips had the same standardized sound energy levels, and all motorized clips had equivalent sound energy levels. The clips were then embedded into a PowerPoint presentation and tested using a Larson Davis 824 sound level meter calibrated to 94 dB at 1000 Hz, paired with a GRAS artificial ear headphone system (½ inch microphone type 40AG) and Quiet Comfort 15 Bose headphones. The loudness of the auditory stimuli was measured by sound pressure level using the decibel scale. While dB is flat weighted (no corrections applied), dB(A) applies a correction to better represent the sensitivity of the human ear to certain frequencies. Very high (6.3 kHz and above) and very low (below 1 kHz) frequency sounds are discounted, and mid-range frequencies are emphasized with dB(A), which was the metric employed in this study. Normalized sound files were calibrated so that the headphones would deliver ˜45 dB(A) natural clips and ˜60 dB(A) motorized clips. All auditory stimuli were selected to represent existing sound sources from the actual parks in order to maximize the utility of the findings for park managers. The propeller plane sound clip consisted of a single general aviation fixed-wing aircraft; the motorcycle clip contained sounds from a pair of cruiser motorcycles [type "C" vehicles from Rochat's (2013) classification system] passing the observer; and the snowmobile clip included a single machine with a four-stroke engine approaching and passing.

Visual Stimuli

Landscape photographs were obtained from park staff, and settings were selected to represent typical views from scenic overlooks, without evidence of obvious anthropogenic influence. Winter scenes were selected for Yellowstone to match the snowmobile sounds. Summer scenes were selected for Denali and Glacier to pair with propeller plane and motorcycle sounds, respectively. The motorized sound sources were not visible in the scenes. Participants were instructed to imagine taking a hike in each landscape and to "try to place yourself into the scene." Participants viewed each scene for 45 seconds. After 25 seconds, a message automatically appeared on each slide informing participants to complete the landscape assessments on the iPad. Two practice scenes were included prior to the actual experiment. Following the practice scenes, very few participants had difficulty completing the ratings in the allotted 20 seconds. Stimuli were blocked so that participants viewed four slides (with the same repeated sound clip) for each condition. Consequently, four sets of landscape assessments (eight dimensions each) were obtained for each condition. In total, participants rated 24 landscape scenes (six conditions with four scenes per condition), as displayed in Figure 6.1.

RATING SCALES

Landscape scenes were rated according to the following dimensions: naturalness, freedom, preference, annoyance, solitude, scenic beauty, tranquility, and acceptability. Specific dimensions were selected to be consistent with previous work using the landscape assessment paradigm (Mace et al. 1999; Benfield et al. 2010a). Participants reported their ratings of the dimensions

Glacier National Park | Yellowstone National Park | Denali National Park & Preserve
(4 scenes/condition) | (4 scenes/condition) | (4 scenes/condition)

Condition 1: | Condition 2: | Condition 3: | Condition 4: | Condition 5: | Condition 6:
Natural | Motorcycle | Natural | Snowmobile | Natural | Propeller

FIG 6.1: Example of study design for one possible order of conditions.

on a 10-point visual analog scale ranging from "Very low" (coded 1) to "Very high" (coded 10). The annoyance dimension was reverse coded prior to analysis. Repeated measures analysis of variance (ANOVA) tests and paired-samples t-tests were conducted on the landscape assessments to investigate whether there were significant differences between sound sources across participants.

RESULTS

Ratings from the four slides in each block were averaged to obtain a mean score for each dimension in each condition. The mean scores were then combined for the eight dimensions to get an overall rating for each condition. Tables 6.1 and 6.2 show the reliability analyses for the composite scores in the natural sounds conditions (Table 6.1) and the motorized sounds conditions (Table 6.2). The resulting Cronbach's α values for each condition were sufficient to justify combining the individual dimensions into a single rating (Cronbach's α ranged from .85 to .90 for the six conditions). For several of the composite variables, the value for Cronbach's α could be slightly improved by removing a single item from the scale. However, the pattern was not consistent across variables; in order to maintain consistency, all items were retained in the scales.

An initial analysis tested for the potential confounding effects of stimulus presentation order on the composite variables, depending on which of the six possible orders of scenes and sound conditions the participants received. Six separate one-way ANOVAS were performed to look for significant differences in composite scores based on order (Table 6.3). Of the six composite variables, five showed no significant differences in order [Glacier natural: $F(5, 69) = 0.77$, p = .574; Yellowstone natural: $F(5, 69) = 2.16$, p = .069;

TABLE 6.1. Reliability Analysis for Composite Natural Variables

Variable	Item total correlation	Alpha if item deleted	Cronbach alpha
NATURAL DENALI CONDITION			.87
Acceptability	.76	.84	
Freedom	.72	.84	
Naturalness	.78	.85	
Preference	.71	.84	
Scenic beauty	.62	.86	
Solitude	.62	.85	
Tranquility	.54	.87	
Annoyance[1]	.50	.86	
NATURAL GLACIER CONDITION			.89
Acceptability	.80	.86	
Freedom	.83	.86	
Naturalness	.79	.87	
Preference	.75	.87	
Scenic beauty	.56	.89	
Solitude	.46	.92	
Tranquility	.79	.86	
Annoyance[1]	.65	.88	
NATURAL YELLOWSTONE CONDITION			.90
Acceptability	.78	.87	
Freedom	.80	.88	
Naturalness	.72	.89	
Preference	.70	.89	
Scenic beauty	.68	.88	
Solitude	.51	.90	
Tranquility	.76	.88	
Annoyance[1]	.65	.89	

1. Annoyance variable reverse coded.

propeller plane: $F_{(5, 69)} = 1.08$, $p = .381$; motorcycle: $F_{(5, 69)} = 0.42$, $p = .832$; snowmobile: $F_{(5, 69)} = 0.67$, $p = .648$]. Only the Denali natural sound condition produced a significant difference between orders, $F_{(5, 69)} = 2.60$, $p = .033$. However, Bonferroni post-hoc tests revealed no significant differences in the pair-wise comparisons between the various orders of slide presentations and sound conditions. Taken together, these results suggest that order was not an important factor in participants' ratings of the landscapes. This finding further supports the creation of composite variables for the sound conditions, combined across different slide and sound orders.

TABLE 6.2. Reliability Analysis for Composite Motorized Variables

Variable	Item total correlation	Alpha if item deleted	Cronbach alpha
MOTORCYCLE CONDITION			.85
Acceptability	.62	.83	
Freedom	.75	.81	
Naturalness	.68	.82	
Preference	.69	.82	
Scenic beauty	.42	.85	
Solitude	.68	.82	
Tranquility	.75	.81	
Annoyance[1]	.25	.88	
PROPELLER CONDITION			.89
Acceptability	.76	.87	
Freedom	.74	.87	
Naturalness	.65	.88	
Preference	.75	.87	
Scenic beauty	.50	.89	
Solitude	.75	.87	
Tranquility	.81	.86	
Annoyance[1]	.46	.90	
SNOWMOBILE CONDITION			.86
Acceptability	.68	.83	
Freedom	.80	.81	
Naturalness	.64	.83	
Preference	.62	.84	
Scenic beauty	.45	.85	
Solitude	.63	.84	
Tranquility	.77	.82	
Annoyance*	.31	.88	

1. Annoyance variable reverse coded.

The descriptive statistics for the composite variables for each condition are displayed in Table 6.4. The mean values across participants for the natural sound conditions were "very high" (Denali: M = 9.24, SD = 0.64; Glacier: M = 9.27, SD = 0.65; Yellowstone: M = 9.05, SD = 0.79). The mean values for the motorized sound conditions, however, were more neutral with higher variance (propeller plane: M = 6.21, SD = 1.51; motorcycle: M = 5.73, SD = 1.35; snowmobile: M = 5.89, SD = 1.44). Difference scores were computed by subtracting the motorized source from the corresponding natural condition (Denali natural -propeller plane: M = 3.03, SD = 1.47; Glacier

TABLE 6.3. Comparison of Sound Condition Order
Effects Using One-Way ANOVA

Variable	df	F	p
Natural (Denali)	5, 69	2.60	.033*
Natural (Glacier)	5, 69	0.77	.574
Natural (Yellowstone)	5, 69	2.16	.069
Propeller Plane	5, 69	1.08	.381
Motorcycle	5, 69	0.42	.832
Snowmobile	5, 69	0.67	.648

*One-way ANOVA significant at $p < .05$. Bonferroni post-hoc tests revealed no significant differences ($p > .05$) in pair-wise comparisons between orders.

TABLE 6.4. Descriptive Statistics for Composite Variables

Variable[1]	Mean[2]	SD
Natural (Denali)	9.24	0.64
Natural (Glacier)	9.27	0.65
Natural (Yellowstone)	9.05	0.79
Propeller Plane	6.21	1.51
Motorcycle	5.73	1.35
Snowmobile	5.89	1.44

1. Computed by averaging 8 dimensions.
2. Coded 1 = "very low" to 10 = "very high."

natural -motorcycle: M = 3.55, SD = 1.31; Yellowstone natural -snowmobile: M = 3.16, SD = 1.33). These scores indicate how much participants' ratings changed in the motorized conditions relative to the natural conditions and represent an internal baseline that controls for differences between parks. Thus, a higher mean score signifies a larger change from the natural sound baseline condition. This is important given that each motorized source was tested in only a single park (where it has the highest management concern). The largest percent reduction in ratings was observed in the motorcycle condition (38%), followed by the snowmobile (35%) and propeller plane (33%) conditions.

Before proceeding with statistical analysis between motorized sources, one-way ANOVA was performed to investigate potential differences between the composite scores for the three natural conditions. Due to the lack of a significant difference between conditions [$F(2, 222) = 2.33$, $p = .100$], the ratings from Denali, Glacier, and Yellowstone natural sound conditions were combined for subsequent analyses. In order to test for global differences between natural and motorized conditions, the three combined natural conditions (M = 9.19, SD = 0.60) were compared to the three combined motorized conditions (M = 5.94, SD = 1.31) by a paired-samples t-test. The results indicate that there was indeed a significant difference in landscape assessments between natural and motorized conditions overall, $t_{(74)}$ = 23.36, p < .001. This finding supports the expectation that landscape assessments would be lower (more negative) in motorized noise conditions than in natural sounds conditions, across aesthetic dimensions. Lower evaluations of landscape quality and experiential conditions in the presence of motorized noise sources, generally, are consistent with the results described by Mace et al. (1999), Benfield et al. (2010a), and Mace et al. (2013).

To determine whether landscape assessments differed between the three motorized noise sources (i.e., the impact of motorcycle, snowmobile, and propeller plane noise had differential effects on visitor experiences), repeated measures ANOVA was utilized to test for within-subjects differences between motorized sound conditions. Repeated measures ANOVA, which accounts for individual variability in participants' ratings, was conducted for the composite variables, as well as for the eight individual dimensions (Table 6.5). Mauchly's Test of Sphericity was used to test the assumption that the variance of the difference of all variables was a constant. Where Mauchly's Test indicated that the assumption was violated (for the freedom and naturalness dimensions), the Greenhouse-Geisser correction was used to adjust the F-value and degrees of freedom obtained from the repeated measures ANOVA. The difference scores for the three motorized sources of noise differed in six of eight dimensions, as well as for the combined variables [overall: $F_{(2)}$ = 8.50, p < .001, partial η^2 = .10; acceptability: $F_{(2)}$ = 5.97, p = .003, partial ^{2} = .08; freedom: $F_{(1.8)}$ = 4.24, p = .019, partial η^2 = .05; naturalness: $F_{(1.8)}$ = 6.09, p = .004, partial η^2 = .08; preference: $F_{(2)}$ = 8.38, p < .001, partial η^2 = .10; solitude: $F_{(2)}$ = 7.81, p = .001, partial η^2 = .10; tranquility: $F_{(2)}$ = 3.28, p = .040, partial η^2 = .04].

The difference scores were not significantly different between motorized sources for the annoyance and scenic beauty dimensions [annoyance: $F_{(2)}$ = 1.65, p = .196, partial η^2 = .02; scenic beauty: $F_{(2)}$ = 0.86, p = .426, partial η^2 = .01]. Overall, motorized noise had a large impact on ratings of annoyance, but there were no significant differences between the motorized conditions in the magnitude of the effect observed. By contrast, motorized noise had little effect on participants' assessments of scenic beauty for the landscapes. In all six sound conditions (including the motorized noise con-

TABLE 6.5. Tests of Within-Subjects Effects between Three
Motorized Sources from Repeated Measures ANOVA

Variable	df	F	p	partial η^2
OVERALL	2	8.50	<.001*	.10
Acceptability	2	5.97	.003*	.08
Annoyance	2	1.65	.196	.02
Freedom	1.8[1]	4.24	.019*	.05
Naturalness	1.8[1]	6.09	.004*	.08
Preference	2	8.38	<.001*	.10
Scenic Beauty	2	0.86	.426	.01
Solitude	2	7.81	.001*	.10
Tranquility	2	3.28	.040*	.04

1. Sphericity cannot be assumed; Greenhouse-Geisser correction used to adjust degrees of freedom.

*Significant at $p < .05$.

ditions) scenic beauty was rated very highly, suggesting that participants were separating the visual and auditory components of the total experience.

For the six individual dimensions and the combined variable with a significant F-value from the repeated measures ANOVA, pair-wise comparisons were conducted to determine between which conditions the significant differences lie. Paired-samples t-tests were run for each comparison (three comparisons per variable), and the Bonferroni correction was used to adjust the alpha level for multiple comparisons (Table 6.6). For five of the six individual aesthetic dimensions affected by noise (acceptability, freedom, naturalness, preference, and solitude) and the combined variable, motorcycle noise had a significantly stronger negative impact on landscape assessments relative to the natural sounds condition than propeller plane or snowmobile noise. For the annoyance, scenic beauty, and tranquility dimensions, there were no significant differences in the ratings between motorized sound sources at the Bonferroni-corrected alpha level. There were no significant differences in the decline in ratings from the natural baseline between the propeller plane and snowmobile noise conditions for any of the aesthetic dimensions.

DISCUSSION

Previous research has shown that natural sounds contribute to high-quality experiences for visitors to protected areas. The present study aimed to investigate differences between motorized noise sources in their effects on poten-

TABLE 6.6. Means, Standard Deviations, and Percent Reductions for Difference Scores from Three Motorized Sound Conditions

| | MOTORCYCLE | | PROPELLER | | SNOWMOBILE | |
Variable	Mean (SD)	% Diff[1]	Mean (SD)	% Diff	Mean (SD)	% Diff
OVERALL	3.55[a] (1.31)	38	3.03[b] (1.47)	33	3.16[b] (1.33)	35
Acceptability	4.24[a] (2.04)	46	3.65[b] (2.19)	40	3.74[b] (2.25)	42
Annoyance	5.58 (2.43)	58	5.15 (2.80)	54	5.25 (2.71)	56
Freedom	2.98[a] (2.11)	32	2.44[b] (1.85)	26	2.66[ab] (2.08)	29
Naturalness	2.56[a] (1.80)	27	2.00[b] (1.72)	21	2.13[ab] (1.60)	23
Preference	3.27[a] (1.92)	36	2.54[b] (2.00)	29	2.62[b] (1.98)	30
Scenic Beauty	0.80 (1.05)	8	0.72 (0.95)	8	0.88 (1.02)	10
Solitude	4.33[a] (2.20)	43	3.58[b] (2.22)	38	3.77[b] (2.18)	41
Tranquility	4.61 (2.04)	50	4.18 (2.38)	43	4.24 (2.03)	48

1. "% Diff" refers to percentage reduction from natural condition.

Note. Means with different letter superscripts differ significantly at $p < .017$ (Bonferroni-corrected for three comparisons per variable).

tial park visitors. Our laboratory simulation demonstrated the detrimental aesthetic and psychological effects of three common sources of recreational motorized noise on participants' evaluations of landscape scenes, which is consistent with previous laboratory research that found negative evaluations of landscape quality and experiential conditions in the presence of motorized noise sources, generally (Mace et al. 1999; Benfield et al. 2010a; Mace et al. 2013). Landscape assessments along eight aesthetic and experiential dimensions were evaluated very highly in natural sounds conditions (i.e., birds, wind, and water) across three national park settings. When motorized vehicle noise was added to the natural sounds, however, landscape ratings were negatively impacted. The presence of recreational motorized noise yielded evaluations that were 33–38% lower, on average, than when only natural sounds were audible.

Furthermore, the extent of the decline in ratings differed between the motorized sound sources on most of the dimensions. In all cases where there was a significant decline (for the acceptability, freedom, naturalness, preference, and solitude variables), motorcycle noise was most detrimental to participants' reported experiences. Note that although there was a significant difference in tranquility between the motorized conditions from the repeated measures ANOVA, the change was no longer statistically significant in the pair-wise comparisons after Bonferroni correction for multiple comparisons. The annoyance dimension did not produce significant differences between motorized sources, as all three sources yielded very large increases in reported levels of annoyance (58% for motorcycle, 56% for snowmobile, and 54% for propeller plane). Similarly, there were no differences between motorized sources in their impacts on ratings of scenic beauty, which were evaluated very highly in all sound conditions, both natural and motorized (snowmobile noise produced the largest reduction in ratings of scenic beauty at 10%). The absence of differences between sound conditions for evaluations of scenic beauty is consistent with the findings reported by Mace et al. (2013). While the motorcycle sound clip produced the largest reductions in evaluations of landscape quality for several variables, propeller planes and snowmobiles were not significantly different from each other in terms of the size of the observed decrease in ratings for any of the dimensions that were tested (although, importantly, both sources were significantly different from natural sounds).

STUDY LIMITATIONS AND FUTURE RESEARCH

A potential limitation of the present study related to the experimental design is the confounding relationship between the park and source variables. In the interest of minimizing participant burden, the number of conditions each participant received was intentionally restricted. Thus, each sound condition was limited to one park only—the actual park in which the audio file was recorded and where there is management interest in that particular sound source. For instance, the motorcycle audio files were only matched with the slides from Glacier (where park managers are concerned about the level of motorcycle noise currently experienced by visitors to the park). Participants never responded to the motorcycle sound condition while viewing scenes from either Denali or Yellowstone. As a result, the source of the sound and the park scenes present a potential confound in the design of the study.

With the current design, it was not possible to separate the effects of the sound source from the visual context in which they were experienced. If the scenes from one park were more appealing than those from another park, participants' responses may have been influenced in a manner independent of the sound condition. For example, Benfield et al. (2010a) showed that the effect of sound was more detrimental for landscape scenes rated higher in scenic beauty. L. Anderson et al. (1983) demonstrated the importance of the interaction between visual and auditory components in evaluations of natural settings. The authors found that the appropriateness of sounds for a setting affected evaluations of that setting. More specifically, if there was incongruence between auditory (i.e., motorized) and visual (i.e., natural) information in an environment, then ratings of aesthetic quality could be negatively impacted. It is possible that the effects of visual-auditory congruence may partially explain the results obtained in the present study. However, it was expected that by looking at differences within each park—by comparing the motorized noise sources to their own natural baseline conditions with difference scores and percent changes—the impact of this design feature should be minimized. This particular design was selected to obtain as much information as possible from various common sources of noise in protected areas, while also minimizing participant burden. Results indicated that participants' assessments differed only minimally between parks during the natural sounds baseline conditions.

Participants never saw the source of the sounds, nor were they informed that they would be hearing motorcycles, propeller planes, and snowmobiles, specifically. Some participants may have recognized the sound of the motorcycles, for instance, while others may not. Thus, the impact of the noise in the motorcycle condition may be underestimated for participants who have negative attitudes about motorcycles in protected areas, but who may not have recognized the source of the sound that they experienced. The same is true for the snowmobile condition, which may be less familiar to many participants, and the propeller plane condition, in which participants may not have attributed the noise to a recreational air tour (as opposed to, for example, a commercial aircraft). Feedback from several participants during debriefing suggests that this latter situation may have indeed been the case.

It may be useful in future studies using this experimental paradigm to instruct participants that the sounds that they hear will all be forms of motorized recreation. However, as described above, Mace et al. (2003) found that noise source attributions for helicopters (whether tourist overflights, rescue missions, or park maintenance operations) did not prevent negative evaluations by laboratory participants. Future research should also attempt to clarify whether it is the motorcycle itself that is adversely impacting participants' assessments, or if the effect can be attributed to the fact that motorcycles may symbolize the presence of nearby roads and development (more than propeller planes or snowmobiles). It is possible that participants in the current study assessed the landscapes more negatively in the motorcycle noise condition in the acceptability, freedom, naturalness, preference, and solitude dimensions for this reason.

Finally, the relationship between noise exposure and human response involves more factors than can be accounted for with a single laboratory study. The source of the noise is just one of many variables influencing the experiences of potential national park visitors. Park experiences and land-scape evaluations are also affected by individual differences in motivations, expectations, and personal preferences for soundscape conditions (Marin et al. 2011; Taff et al. 2014; Tarrant et al. 1995). Moreover, group characteristics and social norms can also have important impacts on the responses of visi-tors to acoustic conditions encountered in the field (N. Miller 1995–97; Frei-mund, Vaske, Donnelly, and Miller 2002). Future research can address some of these potential covariates. For example, it would be possible to control for individual differences in visual preferences for different types of scenes by pretesting the landscape photographs to determine whether some scenes (e.g., summer scenes, water features, or alpine settings) are inherently more appealing to participants than others, irrespective of soundscape condition.

In addition to these suggestions for future research that would address some of the limitations of the current study, two directions for further study are specifically recommended. First, more work should be done to discern the moderating factors that influence the observed effects of motorized noise at the individual level, such as attitudes toward NPS management of sound-scapes and motorized recreation in national park settings. Second, a better understanding of the restorative aspects of natural soundscapes on visitor health and well-being would further support management policy aimed at protecting opportunities for natural quiet and the sounds of nature.

MANAGEMENT IMPLICATIONS

This study validates the wisdom of preserving highly-natural park sound-scapes, just like other biophysical and social resources, as NPS policy re-quires. Managers should be aware of the anthropogenic noises common in their parks and how those noises are evaluated by their visitors. Some parks may have issues with excessive motorcycle noise; others may have conflicts between user groups due to backcountry use of snowmobiles in winter; still others may receive visitor complaints due to the high number of scenic air tours that disturb wilderness values. Current findings suggest that, at the very least, motorcycle noise (and perhaps noise from road traffic, generally) should be a top priority for management attention. Managers should take steps to protect the experiences and opportunities for which visitors come to their parks—whether by visitor education, public outreach, or by park policy and management objectives. Just like visitor crowding, ecological impacts, or unhealthy air quality, park soundscapes characterized by high levels of anthropogenic noise can undermine potential benefits for the public.

The findings from the present study underscore for park managers the importance of protecting natural soundscapes, especially from excessive motorized noise, in order to preserve high-quality visitor experiences. Providing opportunities to experience natural sounds is a central factor contributing to visitors' evaluations of acceptable resource and experiential conditions. In order to conserve the multi-dimensional values that the NPS is mandated to protect, excessive anthropogenic noise sources must be identified and managed to mitigate their social impacts. Participants in the current laboratory simulation clearly perceived and evaluated the landscapes more favorably on a variety of aesthetic and experiential dimensions when they were exposed to natural, rather than motorized, sounds.

7

NATURAL SOUND FACILITATES MOOD RECOVERY FROM STRESS

Jacob Benfield, Derrick Taff, Peter Newman, and Joshua Smyth

The last chapter explored the relationship between aural and visual perceptions of national parks. In this chapter, the authors show the power of soundscapes in providing restorative benefits independent of those produced by visual quality. This relationship between soundscapes and stress reduction in humans is an important link and strengthens the argument that healthy parks lead to healthy people.

AFFECTIVE RECOVERY FROM STRESS USING SOUND: EVIDENCE OF NATURAL SOUND RESTORATION

Research has consistently demonstrated that visual access to natural environments facilitates attention restoration, improves mood, and can enhance physiological stress recovery and health more generally (e.g., Berman, Jonides, and Kaplan 2008; Devlin and Arneill 2003). For example, Tennessen and Cimprich (1995) showed that college students who viewed natural window scenes from their residence hall room scored higher on multiple measures of directed attention and reported having greater focus. Likewise, Benfield, Rainbolt, Bell, and Donovan (2015) showed that students in course sections with a natural window view performed better over the length of a semester than students in the same course but with a concrete wall window view.

Related to health, stress, and physiological recovery, seminal work by Ulrich (1984) showed that natural views could speed recovery from gall bladder surgery while also reducing pain medication usage and negative interactions with hospital staff members. Similar health promoting and/or stress reducing effects have been shown for laboratory participants (Laumann, Gärling, and Stormark 2003), prisoners (E. Moore 1981), dental patients (Heerwagon 1990), office workers (Kaplan 1993; Shin 2007), and those living within a 3-km radius of green spaces in cities (van den Berg, Maas, Verheij, and Groenewegen 2010).

Conversely, research has shown that urban, or manmade, environments typically do not produce these same restorative effects. For instance, A. Taylor, Kuo, and Sullivan (2002) showed that inner-city girls who had an exclusively urban view from their residence exhibited lower levels of concentration, impulse inhibition, and lessened ability to delay gratification compared to those with an urban view containing green spaces. Hartig and colleagues (2003) showed that urban walks did not result in the reduced blood pressure observed in response to walking in natural settings. This research also showed that participants in the urban condition showed decreased performance on attention tasks as well as worsened affect. Participants in the urban conditions reported lessened positive affect and greater anger when comparing pre-urban to post-urban exposure (Hartig, Evans, Jamner, Davis, and Garling 2003). Additionally, much of the research cited previously in support of nature views (e.g., Berman et al. 2008; Laumann et al. 2003; van den Berg et al. 2010) utilized an urban environment as a comparison group when quantifying the restorative benefits of nature. In sum, evidence suggests that looking at natural elements is typically physically and mentally beneficial; in contrast, looking at manmade or urban elements is often not beneficial and sometimes even detrimental.

THE AUDITORY ENVIRONMENT: SOUNDSCAPES

Although humanity relies heavily on vision when it comes to information gathering and processing, emerging research related to auditory environments, or "soundscapes," has shown that auditory experiences of the environment are also important. For example, researchers have shown that the presence of human-caused or urban sounds from automobiles, aircraft, or voices can have deleterious effects on memory (Benfield, Bell, Troup, and Soderstrom 2010b), scenic evaluations (Weinzimmer, Newman, Taff, Benfield, Lynch, and Bell 2014), affective state (Mace, Bell, and Loomis 1999), and responses to historical park tours (Rainbolt, Benfield, and Bell 2012).

Unfortunately, very little research on natural soundscapes has evaluated the potential positive effects of natural soundscape exposure. Some research suggests that these sounds are, at the very least, perceived to be beneficial. For example, Ratcliffe, Gatersleben, and Sowden (2012) conducted semi-structured interviews with adults regarding their perceptions of what makes an environment restorative. From those interviews, the sound of birds singing was isolated as the most salient component of natural sounds leading to stress reduction and attention recovery. Along similar lines, research by Payne (2008 and 2013) conducted in urban parks shows that individuals in urban parks classified natural sounds as being perceived as more restorative and rated different soundscapes as more or less restorative based largely on the amount of audible natural sounds. In all of these cases, the actual restorative potential of the sounds was left untested and the results instead speak to perception of sounds as either restorative or not. Although important and intriguing, it is important that this work be extended to include outcomes that better capture actual experiences; we now turn to such evidence and remaining gaps in the knowledge base.

Jahncke, Hygge, Halin, Green, and Dimberg (2011) explored the negative effects of office noise, but also included a natural sound restoration condition; sound exposure was followed by physiological and cognitive measures. Although they showed some detrimental effects of office noise exposure related to cognitive abilities (memory), and then some evidence of improved restoration via self-report for those in the natural sound condition when compared to the office noise restoration condition, the results were inconclusive regarding the restorative effect of sound. Specifically, results showed that cortisol levels (a naturally occurring hormone in the body that increases in response to stress) decreased during the course of the office noise exposure period (despite it being intended as a stressful situation). No restoration effect was shown after that period for either cognitive or physiological outcomes; this may indicate that there is no benefit for natural sounds or, alternatively, that the participants were not adequately depleted or stressed prior to the recovery period. Given the results of the cortisol assays (suggesting decreasing, not increasing, physiological stress), the latter interpretation seems the more likely conclusion. As such, the potential restorative effects of sounds is still not clearly understood.

This relative dearth of evidence regarding the restorative abilities of natural sound is particularly troubling when recreation and leisure research often shows that outdoor recreators are seeking solace from the sounds of urban living (Driver, Nash, and Haas 1987) or actively seeking natural quiet (Haas and Wakefield 1998; Marin, Newman, Manning, Vaske, and Stack 2011). Similarly, with legislative actions being taken in both the United States and across Europe to better manage noise in both natural and urban settings (e.g., Jensen and Thompson 2004) and promote natural recreation areas as a source of health (e.g., Rochat 2011), a lack of evidence regarding the benefits of natural sounds, or the detriments of manmade sounds, fails to provide an evidence base upon which to make decisions and may expose such legislation and management plans to criticisms or legal actions.

Thus, this project examined the restorative potential of natural versus urban auditory stimuli in hopes of connecting it with prior work on restorative visual environments while also providing empirical evidence in the context of soundscape research and the benefits of nature more generally. This was done in a laboratory-based experiment utilizing a sample of urban participants from the Philadelphia metropolitan area.

METHOD

Participants

A convenience sample of 133 participants were recruited from introductory level courses at a college campus servicing the greater Philadelphia area. Participants were racially diverse (54.1% white) and of traditional college age (M = 19.09, SD = 2.12).

Materials and Procedure

Following informed consent, participants completed demographic (age, sex, race) and psychological (e.g., HEXACO personality) questionnaires before providing baseline measurements of affective state using the 16-item Brief Mood Introspection Scale (BMIS; Mayer and Gaschke 1988). This measure consists of single emotional words (e.g., lively, tired) that are rated along a 4-point range from "Definitely do NOT feel" to "Definitely feel." These adjectives were combined to create aggregates for Pleasant-Unpleasant (16 items; $\alpha = .84$), Positive-tired (7 items; $\alpha = .74$), and Negative-relaxed (6 items; $\alpha = .74$). The BMIS can also be scored to include an Arousal-Calm mood scale, but the data collected for that subscale showed unacceptable internal consistency ($\alpha = .51$), and the subscale was therefore not included in any subsequent analyses.

Following baseline measurement, all participants were exposed to a stress-inducing video similar to that utilized in prior research (Bosch et al. 2001; Takai et al. 2004). Specifically, participants watched a 3-minute video of a hand tendon replacement surgery that included close-up footage of a human hand being cut open to expose the tendon, the tendon being relocated in the hand using medical equipment, and the tendon inside the hand being sutured using a needle and thread. This video was used to create an initial state of discomfort and negative mood in participants to address the concern that prior research on sound restoration (e.g., Jahncke et al. 2011) may have failed to create a sufficiently stressful preliminary task from which participants needed to recover.

After the stress-inducing task, participants were again assessed using the BMIS and then randomly assigned to one of four soundscape conditions—natural sounds, natural sounds with voices, natural sounds with motorized noise, or a control condition. The natural sound condition was created from recordings taken by the U.S. National Park Service (NPS) and contained bird-song and leaves rustling in the wind. The voices and motorized noise conditions were created by adding NPS recordings of voices or motorized craft to the original natural sound file; the control condition contained no sound clips. All sound conditions were presented through noise cancelling head-phones which were worn during the entire experimental procedure. Sound levels were set at 45–50 dB(A) and sound exposure was limited to 3 minutes to assess rapidly occurring changes in affective state. No visual information was provided to accompany the sound exposure treatments with participants instructed to focus attention on a "+" in the middle of a computer screen during the three minute exposure task. Thus any effects observed following the restoration period would be primarily, if not entirely, based on auditory stimuli rather than visual stimuli (and, further, all visual stimuli were identical across experimental conditions).

A third and final BMIS measurement was reported immediately following the soundscape exposure. Participants were then debriefed and given research credit toward a course requirement.

RESULTS

Initial statistical tests were conducted to verify both the homogeneity of groups and the effect of the manipulation. Results showed that the four experimental groups did not differ on baseline ratings of BMIS subscales (all p's > .32) or on BMIS ratings following the stressor task (all p's > .30). Additionally, results showed that all four groups' BMIS scores were negatively affected by the stress-inducing video (all p's < .001) showing that the stressor task was effective at producing dysphoric mood.

To test the restorative potential of the sounds, a mixed factorial Repeated Measures Analysis of Variance (R-ANOVA) was conducted with sound condition (control, natural, voice, motorized) being predictive of changes in BMIS scores from post-stressor (pre-sound exposure) to post-sound exposure. Results were supportive of the hypotheses (see Table 7.1). Significant BMIS

TABLE 7.1. Results from Repeated Measure ANOVAs
for Each of the 3 BMIS Subscales

		F	Sig.	Partial η²2
Pleasant-Unpleasant	**Mood**	11.58	.001	.082
	Condition	0.36	.779	.008
	Mood X Condition	7.60	.000	.150
Positive-Tired	**Mood**	3.74	.055	.028
	Condition	0.09	.965	.002
	Mood X Condition	6.22	.001	.126
Negative-Relaxed	**Mood**	46.64	.000	.266
	Condition	1.06	.369	.024
	Mood X Condition	2.18	.093	.048

Note: Bold type numbers indicate statistically significant effect.

change by sound condition interactions were shown for both the Pleasant-Unpleasant subscale ($F = 7.60$, $p < .001$, partial $\eta^2 = .150$) and the Positive-Tired subscale ($F = 6.22$, $p = .001$, partial $\eta^2 = .126$); a marginal interaction between sound condition and affective restoration was also shown for the Negative-Relaxed subscale ($F = 2.18$, $p = .093$, partial $\eta^2 = .048$).

Follow-up analyses showed that the natural sound condition showed greater recovery from the stressor compared to both the control and anthropogenic sound conditions. For the Pleasant-Unpleasant score, participants in the natural condition were the only ones to show improved affect from post-stressor ($M = 2.43$, SD = 0.51) to post-recovery ($M = 2.77$, SD = 0.43); the other three conditions showed no significant changes from pre-recovery levels (see Figure 7.1). The Positive-Tired scale showed a similar trend with natural sound condition scores improving from pre- to post-recovery ($M = 2.31$ vs. 2.47) and scores in other conditions either staying the same (voices) or decreasing (control and aircraft). Figure 7.2 displays the mean change for each condition on the Positive-Tired subscale of the BMIS.

Because the interaction between sound condition and mood recovery did not reach conventional levels of statistical significance for the Negative-Tired subscale, those simple effects were not reported in the text. However, the trend seen in the other two subscales was replicated; the natural sound condition resulted in greater lowering of Negative-Tired ratings when compared to the other three conditions (fjM = -0.48 vs. -0.25, -0.18, and -0.28 for control, aircraft, and voices, respectively). These simple effects suggest that the marginal result for the interaction was largely due to all sounds showing

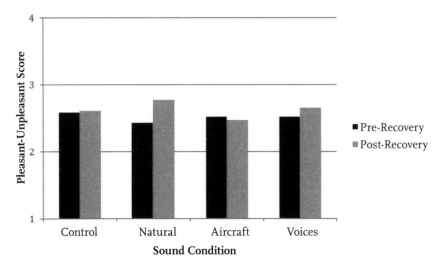

FIG 7.1: Change in pleasure-unpleasant score from pre- to post-recovery for each sound condition.

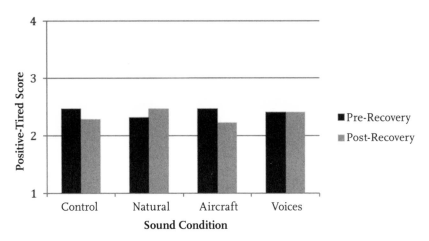

FIG 7.2: Change in positive-tired score from pre- to post-recovery for each sound condition.

recovery in the same direction, but with natural sounds doing so more effectively. As such, the effect is likely harder to detect, and thus this analysis was likely statistically underpowered.

DISCUSSION

Based on research showing restorative effects for natural scenes and visual stimuli, the current study hypothesized that natural sounds would similarly show a positive effect on individuals' affective recovery following a stressful event (independent from any benefit from visual stimuli). Consistent with that hypothesis, the data show that individuals experiencing natural sounds (in the absence of any relevant visual cues) showed greater enhanced mood recovery than those hearing no sounds or those hearing anthropogenic sounds (aircraft or voices). This provides strong preliminary evidence for the restorative aspects of soundscapes on mood, and has a number of implications for research and practice.

These findings directly connect to prior work on visually restorative natural environments and thus may potentially benefit from existing theory and research in the area. For example, given this broader prior literature on the effects of visual stimuli, the effects of natural acoustic sounds on restoration may be similarly broad; that is, soundscapes may influence not only mood but could also have benefits on attention, cognition, or other health and performance outcomes. Although speculative at this point, this line of reasoning leads directly to a number of intriguing research possibilities. For example, the research by Ulrich (1984) found that natural views can aid surgery recovery and pain management; it may be that exposure to natural sounds has similar effects while also being easier to implement as a restorative soundscape could be provided in the absence of natural window views (or even windows themselves). Similarly, research on visual restoration shows that the view from a window can improve attention (Tennessen and Cimprich 1995) or school performance (Benfield et al. 2014); natural soundscape exposure may have the same effect in these domains as well. Unfortunately, the current study focused only on reported mood recovery from an acute stressor and broader health and cognitive outcomes were not assessed. Such predictions, although based on prior research, still need to be tested in future research.

The laboratory-based nature of the study instills a high degree of experimental validity regarding the presence of the positive effect, but is also limited in its direct application to outdoor recreation settings (i.e., ecological validity). Future research focused on stress recovery and other potential benefits will need to test the real world limits and generalizability of this restorative acoustic effect. Field experiments with urban park users or visitors to national parks would provide a more ecologically valid test of stress recovery as would cross-sectional studies of stressful situations taking place in different, naturally occurring soundscapes, some nature dominated and others less so.

Finally, although this study utilized a powerful manipulation of emotional state, as well as measures sensitive to changes in affective state, study outcomes were still reliant on self-report. More rigorous testing of the restorative properties and potential of natural sounds should be conducting making use of biomarkers indicative of stress and subsequent recovery. Changes in heart rate variability, stress hormone levels, and/or skin conductivity would be strong indicators of not only stress recovery but also provide a plausible biological mechanism linking these processes to manifest disease outcomes (e.g., hypertension, diabetes, and cardiovascular disease). Additionally, research examining the potential moderating role of listener restorative expectations, motivations, or attitude toward noise would also be valuable next steps in quantifying the restorative potential of natural sounds. Overall, this work provides important initial experimental evidence on the restorative properties of natural sound and suggests many additional lines of future research.

8

A PROGRAM OF RESEARCH TO SUPPORT MANAGEMENT OF VISITOR-CAUSED NOISE AT MUIR WOODS NATIONAL MONUMENT

Robert Manning, Peter Newman, Kurt Fristrup,
David Stack, and Ericka Pilcher

In the introduction to this book, we described a management-by-objectives framework for managing park resources and experiences. This is an adaptive process, with key components that include the formulation of indicators and standards of quality, monitoring, and adaptive management action as needed to maintain desired conditions. This chapter describes a program of research at Muir Woods National Monument designed to help implement a management-by-objectives framework to monitor and manage visitor-caused noise in the park.

MUIR WOODS NATIONAL MONUMENT

Muir Woods, a unit of Golden Gate National Recreation Area, lies just north of San Francisco and is a popular visitor attraction, accommodating more than three-quarters of a million visits. The park is known for its 560-acre (227 ha) grove of old-growth redwoods. Most visitors experience the park by walking the main trail, which extends about a mile (1.6 km) from the park entrance and follows Redwood Creek.

Human-caused noise has been a management issue in the park for more than two decades. Initial attention was focused on protection of the threatened northern spotted owl (*Strix occidentalis caurina*) during its breeding season (Monroe et al. 2007). More recently, this has expanded to include consideration of the impacts of human-caused noise on the quality of the visitor experience (Manning et al. 2005; Pilcher et al. 2009). This work has been guided by the Visitor Experience and Resource Protection framework, an example of an adaptive, management-by-objectives framework, and supported by a program of research.

INDICATORS AND STANDARDS OF QUALITY

Initial phases of research at Muir Woods focused on identifying indicators and standards of quality for the visitor experience. The first phase was exploratory, collecting baseline data about visitors and visitor use patterns and probing for issues that generally affect the quality of the visitor experience (Manning et al. 2005). A survey of a representative sample of visitors was conducted and a 55% response rate was attained, yielding 406 completed questionnaires. Using a series of open- and close-ended questions, "peacefulness," "quiet," and "the sounds of nature" were found to have a positive influence on the quality of the visitor experience, and "noisy visitors," "loud talking," and related issues were found to substantially detract from the quality of the visitor experience.

Given the apparent importance of soundscape-related issues in the park, the second phase of research was designed to focus more specifically on soundscape-related indicators (Pilcher et al. 2009). Visitors to the park were asked to participate in a "listening exercise." This exercise was conducted at three locations in the park (three points along the park's main trail), and visitors were asked to engage in the exercise as they passed each of the three points. A total of 280 visitors participated in the exercise, which consisted of listening to and identifying the sounds heard in the park and rating the extent to which each type of sound was "pleasing" or "annoying." An "importance/performance" analysis of resulting data (Figure 8.1) suggests potential soundscape-related indicators of quality (Hollenhorst and Gardner 1994; Manning 2007). This analysis suggests that natural sounds such as water flowing in Redwood Creek, birds calling, and wind blowing in the trees are good indicators of quality that contribute to the visitor experience, and visitor-caused noise, such as visitors talking and boisterous behavior, is a good indicator of quality that detracts from the visitor experience. The former sounds are heard by large percentages of visitors and are rated as very pleasing, while the latter sounds are also heard by large percentages of visitors but are rated as very annoying.

The third phase of research was designed to help formulate standards of quality for visitor-caused noise in the park (Newman et al. 2007). Five 30-second audio clips were prepared that included a range of natural and

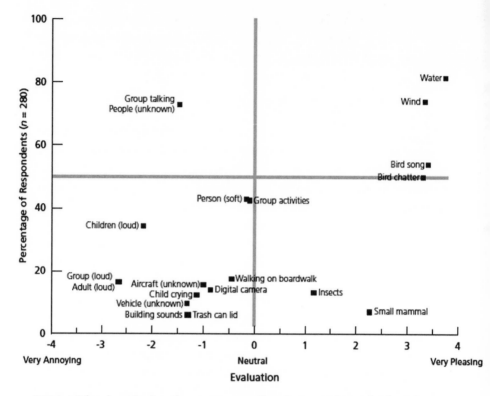

FIG 8.1: The chart depicts the percentage of study respondents that heard various types of sounds by their mean rating on a scale of very annoying to very pleasing.

visitor-caused sounds. All these sounds were recorded in Muir Woods, and the resulting audio clips were created by the National Park Service's Natural Sounds and Night Skies Division. These sound clips were ordered by increasing decibel levels, with visitor-caused sounds increasingly masking the park's natural sounds and ranging from 31 to 48 decibels. In other words, the sound clips started with a relatively quiet natural setting with wind, birds, and flowing water, and became increasingly saturated with human sounds in each subsequent sound clip. The audio clips were incorporated into a survey administered to a representative sample of visitors; a response rate of 53% was attained, yielding 286 completed questionnaires. After listening to each sound clip, respondents were asked to rate the acceptability of the sound on a scale that ranged from -4 ("very unacceptable") to +4 ("very acceptable"). In addition, respondents were asked to indicate which audio clip was most like the soundscape conditions they had experienced in the park.

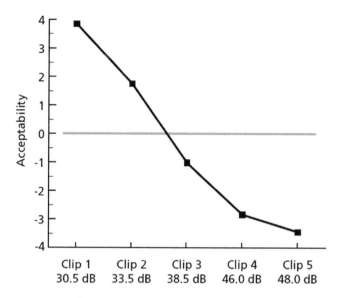

FIG 8.2: This social norm curve depicts average acceptability of visitor-caused noise on the Cathedral Grove
Trail at Muir Woods at various volumes. The sound clips
comprise human-caused noise recorded on the trail and
played back for survey respondents at varying loudness.
Respondents find greater levels of visitor-caused noise
(and decreasing levels of natural sounds) to be increasingly unacceptable.

Respondent acceptability ratings for each of the five audio clips were averaged, and mean ratings were plotted to construct a social norm curve (Figure
8.2) (Manning 2007). This curve indicates that respondents find greater
levels of visitor-caused noise (and decreasing levels of natural sounds) to be
increasingly unacceptable. The point at which aggregate ratings fall out of
the acceptable range and into the unacceptable range (i.e., the point at which
the social norm curve crosses the neutral point on the acceptability scale)
is between audio clips 2 and 3, or 36.7 decibels. Respondents reported the
audio clip that best represented the soundscape conditions they experienced
in the park on the day they participated in the visitor survey. Most visitors
(42.8%) reported that audio clip 2 was most representative, 40.9% reported
that audio clip 1 was representative, 12.9% thought audio clip 3 was representative, and 3.4% considered audio clip 4 representative. This means that
more than 15% of respondents are hearing visitor-caused noise that is louder
than the social norm.

MONITORING

To measure the sound levels in the park, researchers installed a camouflaged acoustic monitoring system approximately 2 yards (1.8 m) off the main trail in the Cathedral Grove portion of the park. This device recorded A-weighted decibel levels (dB[A]) every second. This decibel level is a metric that is an aggregate of sound levels across the range of audible frequencies, weighted to express typical human sensitivities to each frequency band (Fahy 2000). The system used at Muir Woods is certified to measure sound levels accurate to 1 dB(A) and measures sound levels in 31 one-third-octave bands. As noted in the previous section, sound was also monitored by means of a visitor survey that asked respondents which of five sound clips was most representative of the conditions they experienced in the park.

MANAGEMENT

As noted, more than 15% of visitors to Muir Woods reported hearing more visitor-caused noise than the social acceptability norm as defined in Figure 8.2. Moreover, if visitor use continues to rise, violation of noise-related standards of quality is likely to increase, suggesting that management actions are needed to help ensure that noise-related standards of quality are maintained. But which actions might be effective and acceptable to visitors?

The professional literature on parks and outdoor recreation suggests that a range of management actions can be taken to address the impacts of visitor use (Manning 2011; Manning et al. 2017). For example, visitor use levels might be limited or visitor behavior might be altered through educational programs. Generally, educational programs are preferred to visitor use limits because they do not restrict public access to parks and related areas (Peterson and Lime 1979; McCool and Christensen 1996). However, little research has been conducted to test the effectiveness and acceptability of educational programs to address excessive visitor-caused noise.

A program designed to sensitize visitors to human-caused noise at Muir Woods and to encourage them to reduce the noise they generate was applied experimentally at Cathedral Grove ("Quiet Zone") and throughout the park ("Quiet Day") on selected days. During these "treatments," signs asking visitors to turn off cell phones, to encourage children to walk quietly, and to talk in a lower voice were strategically placed around the park. Visitor-caused noise was monitored during these periods as well as during a "control" period in which neither treatment was applied. A visitor survey was adminis-

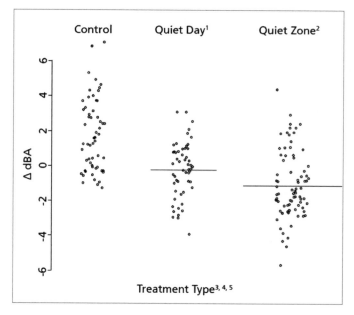

FIG 8.3: A-weighted decibel (dB[A]) levels during control and treatments.

Notes:

1. The difference between the control and quiet day sound levels is 1.96 dB(A).

2. The difference between the control and quiet zone sound levels is 2.84 dB(A).

3. Each circle represents a mean dB(A) level for one hour.

4. Chart data were measured from 10 a.m. to 6 p.m.

5. The difference in mean dB(A) among the control and two treatments is significant at the p = 0.01 level.

tered during the treatment and control periods to assess how the educational program affected visitor behavior and how acceptable visitors found it to be.

A-weighted decibel readings were significantly lower on treatment days than on control days (Figure 8.3). During the Quiet Zone treatment, sound levels dropped an average of 2.84 dB(A), which translates into a near doubling of "listening area" (Fahy 2000). This means that visitors during the Quiet Zone treatment had a substantially greater opportunity to hear the natural sounds of Cathedral Grove. The reduction in sound level of 1.96 dB(A) during the Quiet Day treatment was not as dramatic, but was still statistically significant. Findings from the visitor survey indicated strong support for both parts of the educational program (Table 8.1).

TABLE 8.1. Support for the Use of Educational Programs
to Reduce Visitor-Caused Noise at Muir Woods

Management action	Percentage*
QUIET ZONE	
I strongly support the implementation of a "quiet zone."	71.6
I support the implementation of a "quiet zone."	26.4
I oppose the implementation of a "quiet zone."	1.2
I strongly oppose the implementation of a "quiet zone."	0.8
QUIET DAY	
I strongly support the implementation of a "quiet day."	72.0
I support the implementation of a "quiet day."	23.3
I oppose the implementation of a "quiet day."	4.3
I strongly oppose the implementation of a "quiet day."	0.4

Note: Data were derived from a survey of visitors to Muir Woods in summer 2007.

*$x^2 = 5.19$, $p = 0.158$, and Cramer's $V = 0.101$.

CONCLUSION

Soundscapes are an issue of increasing importance in national parks, and visitor-caused noise is in turn a potentially important component of this issue. At Muir Woods, the visitor experience is enhanced by the sounds of nature—water flowing in Redwood Creek, wind blowing through the old-growth forest, animals calling—but visitor-caused noise can mask these sounds and otherwise detract from the quality of the park experience. Findings from the program of research are being considered as part of the new management plan that is being developed by Golden Gate National Recreation Area (including Muir Woods), and Muir Woods has implemented a permanent quiet zone at Cathedral Grove.

As with other types of visitor-caused impacts in parks, the issue of visitor-caused noise can be analyzed and managed through application of a management-by-objectives framework by (1) formulating indicators and standards of quality for visitor-caused noise, (2) monitoring indicator variables, and (3) taking management actions to help ensure that standards of quality are maintained. Moreover, this management approach can be supported by a program of research that provides an important empirical foundation for this work.

9

THE ROLE OF MESSAGING ON ACCEPTABILITY OF MILITARY AIRCRAFT SOUNDS IN SEQUOIA NATIONAL PARK

Derrick Taff, Peter Newman, Steven Lawson, Alan Bright,
Lelaina Marin, Adam Gibson, and Tim Archer

To protect natural quiet and soundscapes in national parks, managers must think about approaches such as zoning, as outlined in the previous chapter. However, visitor expectations can play a key role in defining the quality of visiting national parks. This chapter introduces the ideas that messaging is an important part of managing parks for high-quality visitor experiences, and that this may have important implications for managing natural quiet.

INTRODUCTION

Mandates such as the U.S. 1972 Noise Control Act, the 1987 National Parks Overflights Act, and recent National Park Service (NPS) policy directives require protection of the acoustic environment as a resource, similar to that of the flora and fauna present in our national parks, and specifies that parks should integrate monitoring and management efforts to protect park soundscapes (Newman et al. 2010). Accordingly, the NPS Natural Sounds and Night Skies Division, which is dedicated to the protection of the acoustic environment, has begun to improve monitoring and planning efforts in many parks, such as Grand Canyon National Park, Muir Woods National Monument, and Sequoia and Kings Canyon National Parks (Keizer 2008).

Located in south-central California, Sequoia was established in 1890, long before the preservation of park soundscapes was of high concern. However, annual visitation now exceeds one million visitors, many of whom escape to this iconic park to experience the sounds of nature (Marin 2011). Due to its proximity to military installations, military aircraft are prevalent above Sequoia National Park, and exposure to sounds produced by aircraft has been found to detract from visitor experiences (Krog and Engdahl 2005; Mace et al. 1999; Mace et al. 2004; Mace et al. 2003; N. Miller 1999; N. Miller et al. 1999; Tarrant et al. 1995). Sequoia staff have instituted multi-day "Wilderness Orientation Overflight Pack Trips" in which they take military officials into the backcountry to increase understanding of the effects of aircraft noise on the park resources and its visitors, leading to improved cooperation between federal entities (Keizer 2008). Despite these advances, military overflights and associated acoustic impacts are likely to continue given Sequoia's proximity to military installations, and in turn, visitor experiences may be depreciated. This provides reasoning and need for effective management strategies that mitigate adverse effects of military aircraft sounds in Sequoia. Indirect management such as educational information can effectively reshape visitor attitudes, perceptions and expectations so that they are more supportive of management actions and policies (Manning 2003). Could educational messaging alter visitor expectations and perceptions of military aircraft sounds in Sequoia? Would informing visitors that they may hear or see military aircraft while recreating in Sequoia increase or decrease acceptability of this anthropogenic sound intrusion? The purpose of this study was to (1) assess whether indirect management actions in the form of educational messaging can significantly affect visitor acceptability of military aircraft sounds, (2) enhance understanding of the strengths and/or limitations of educational messaging as it pertains to soundscape management, and (3) suggest potential educational messaging strategies that may be applied in Sequoia and other NPS units to improve soundscape management.

SOUNDSCAPE MANAGEMENT–INDICATORS AND STANDARDS

The majority of Americans consider opportunities to experience the sounds of nature as an important reason for protecting national parks (Haas and Wakefield 1998). Research suggests that visitors often retreat to parks to experience the sounds of nature, such as wind, water, and natural quiet (Mace et al. 2003; Haas and Wakefield 1998; Driver et al. 1991; C. McDonald et al. 1995). Anthropogenic sounds, such as loud voices, vehicles, and aircraft have been found to detract from visitor experiences by masking the sounds of nature (Krog and Engdahl 2005; Mace et al. 2003; N. Miller 1999; N. Miller et al. 1999; Tarrant et al. 1995; Bell et al. 2009; Benfield et al. 2010b; Pilcher et al. 2009).

Anthropogenic sounds can cause resource and social impacts, and mandates require the NPS to preserve the natural soundscape as a resource (Newman et al. 2010; Ambrose and Burson 2004; Jensen and Thompson 2004), therefore requiring managers to determine how much change should be allowed within the environmental resources, recreation experiences, and the resulting management actions. This requires that descriptive (unbiased data) and evaluative (subjective measure) components be addressed, so that management objectives (desired conditions) and ensuing indicators and standards of quality can be established (Manning 2011). Indicators are "quantifiable proxies or measures of management objectives," while standards "define the minimum acceptable condition of indicator variables" (Manning 2011, 23).

Recent research has helped inform the NPS concerning effective sound-related indicators and standards of quality that help managers protect, maintain, and restore the natural acoustic environment. Pilcher et al. (2009) conducted a two-phase study in Muir Woods National Monument in which suggested sound-related social indicators and standards of quality were established. Phase one focused on descriptive evaluations, by asking respondents to listen to the surrounding environment, and to determine the degree to which sounds heard were pleasing or annoying. The results of phase one suggested that visitor-caused sounds, such as groups talking, were frequently heard and rated as annoying, and therefore would be a good indicator of quality. Phase two focused upon the evaluative component by specifically addressing varying levels of visitor-caused sounds to determine respondents' threshold, or the point at which sounds become unacceptable, and subsequently established a suggested standard of quality. A series of soundclips was created from the recordings of the area, each containing varying levels of visitor-caused sounds. Respondents were asked to listen to each soundclip and indicate how acceptable or unacceptable the sounds of other visitors they heard in the soundclips were to them. On average, visitors reported that sounds stemming from visitors talking at a level of 38 A-weighted decibels dB(A) or greater were unacceptable. Acoustic monitoring data for the area suggested that visitor standards (i.e., sound levels no greater than 38 dB[A]) were being violated at least a portion of the time, potentially degrading the quality of the visitor experience (Pilcher et al. 2009). Potential management actions were suggested, such as indirect messaging, which could be implemented in the study area to alter visitor behaviors and decrease visitor-caused noise.

SOUNDSCAPE MANAGEMENT–EDUCATIONAL MESSAGING

A subsequent experimental study in Muir Woods National Monument addressed strategies for managing visitor-caused sounds by implementing simple signage denoting either a "quiet zone" or a "quiet day" (Manning et al. 2010; Stack 2008; Stack et al. 2011). Results of the experimental study suggested that educational messaging in the study area designating a "quiet day" led to a sound-level decrease equivalent to 793 visitors, while designating "quiet zone" signage led to a sound-level decrease equivalent to 1,150 visitors (Pilcher et al. 2009; Stack 2008). Implementation of "quiet zone" messaging decreased visitor noise by 3 dB(A), essentially doubling a visitor's listening area, or opportunity to hear sounds within the area (Stack et al. 2011). The results of this study demonstrate the positive influence indirect management, such as educational messaging, can have on visitor behaviors and preservation of park soundscapes. However, there has been limited research that evaluates the role of messaging in modifying visitor perceptions and evaluations of sounds.

One study of the effects of messaging on perceptions and evaluations of anthropogenic sounds in natural settings (Mace et al. 2003) used a college psychology laboratory to examine the effects of contextual messaging on evaluations of helicopter noise within park settings. Using simple messages notifying participants that the helicopter sounds that they were evaluating could be attributed to "tourist overflights," "backcountry maintenance operations," and the "rescue of a backcountry hiker," the researchers determined whether contextual factors affected evaluative judgment of the noise. Findings indicated that regardless of the reason attributed to the sound, amplified helicopter noise resulted in lower evaluations of the park setting and greater levels of annoyance, suggesting that park management–related noise disturbances are just as annoying as other aircraft noise sources. This study advanced understanding of how messaging may or may not influence perceptions and evaluations of sounds in parks; however, the study was conducted solely within a laboratory setting with college student respondents who evaluated only helicopter noise. Despite previous research suggesting that lab- and field-based evaluations are similar (Malm et al. 1981; Stamps 1990), the messaging applied in this study may not have induced elaboration, or contemplative thought, about the message content among the participants for at least two reasons. First, the messages may have lacked relevance, given that the respondents were not visitors in the evaluated parks. Second, the messages were simplistic (i.e., "tourist overflights," "backcountry maintenance operations," and the "rescue of a backcountry hiker") and may not have contained enough information to influence respondent attitudes. N. Miller et al. (1999) evaluated messaging concerning military aircraft and associated noise impacts in a park setting to determine if messaging could alter expectations and perceptions. This cooperative study between the US Air Force and the NPS at White Sands National Monument evaluated whether informing visitors that they may hear or see aircraft would reduce adverse effects of military aircraft on park visitors (N. Miller et al. 1999). Approximately half of the visitors sampled were exposed to an NPS-formatted sign with a neutral message stating, "Military aircraft can regularly be seen or heard on this trail" (N. Miller et al. 1999, 6). Only 40% of respondents who could have seen the sign remembered seeing the message, but of that subset, results suggested that information decreased respondent annoyance due to aircraft noise (N. Miller et al. 1999). The message applied in the study was not based on theoretical communication frameworks, but instead, was created with neutrality in mind, so as not to provide a subjective evaluation of the presence of military aircraft (Nick Miller, personal communication, 11/30/11). The findings suggested that informative messaging could affect perceptions and evaluations of aircraft, even by using a non–theoretically based, neutral message. These results could potentially be limited due to the location of White Sands National Monument—it is surrounded by White Sands Missile Range and Holloman Air Force Base. This location requires that visitors travel through the missile range in order to reach the park entrance, suggesting that some visitors to the park may have already been aware of the presence of military and associated sounds. This potential limitation does not negate the effectiveness of educational messaging, but warrants further investigation within a park setting in which the presence of military aircraft would not be as obvious.

The results of these studies suggest that educational messaging can be applied as an effective management strategy to decrease anthropogenic noise, and potentially alter perceptions of anthropogenic sounds depending upon the context and environment in which sounds are heard. These studies have advanced understanding of the role of messaging, but have applied little theoretical basis to message design. Furthermore, applying educational messaging in a park in which visitors may not be as readily cognizant of the presence of military aircraft, may result in different acceptability of associated sounds. This study builds upon previous research by designing an informative message based upon theory, and determining if that message has the potential to alter acceptability of military aircraft sounds.

THEORETICAL BASIS: ELABORATION LIKELIHOOD MODEL

The Elaboration Likelihood Model (ELM) (Petty and Cacioppo 1981; Petty and Cacioppo 1996) is one of the most prominent theoretical approaches applied to the design of persuasive messages for visitors in parks and protected areas (Absher and Bright 2004). This model postulates that there are two routes to persuasion: (1) the central route, which occurs through thoughtful, motivated consideration of information; and (2) the peripheral route, which induces change without deep reflection on information (Petty and Cacioppo 1996). The ELM focuses upon the processes by which message features influence attitudes (Booth-Butterfield and Welbourne 2002) by better understanding the level of elaboration (i.e., extent to which a message is scrutinized) that a particular communication strategy has upon an individual (Petty and Cacioppo 1986). Perhaps most importantly, central route attitude change demonstrates "greater resistance to counter-persuasion than attitude changes that result mostly from peripheral cues" (Petty and Cacioppo 1986, 21).

Educational communication strategies in parks and protected areas often rely on central route processing (Marion and Reid 2007), but situational and personal variables like motivation, message relevancy, potential distractions, ability, previous experiences, and knowledge, all affect the level of elaboration, and determine whether central or peripheral processes occur (Petty and Cacioppo 1986; Booth-Butterfield and Welbourne 2002; Perloff 2003). It is unrealistic to motivate central processing within every visitor because it is "inevitable that people will rely on mental shortcuts" and instead process information through a peripheral route (Perloff 2003, 129). Therefore, effective messaging design requires consideration of variables that are thought to enhance and motivate understanding such as personal relevance, personal responsibility, the number of messages, and message sources (Petty and Cacioppo 1986), while also considering factors that may inhibit attitude change. For example, Manfredo and Bright (1991) found that elaboration was affected by source credibility (i.e., information from a government agency) and also determined that respondents' prior knowledge had a strong effect on elaboration and acquisition of new beliefs. Wiles and Hall 2005 evaluated the role of personal relevance and need for cognition with regard to knowledge about and attitudes toward wildland fire, and found that those factors were extraneous to change in attitude or knowledge.

As noted above, effective messaging design requires consideration of variables that are thought to enhance and motivate understanding such as personal relevance, personal responsibility, the number of messages, and message sources (Petty and Cacioppo 1986). While interpretive strategists cannot always reach visitors due to situational and personal variables, developing messages that are strong and impactful, by making them relevant to the visitor (Ham 2009; Ham et al. 2009), may lead to more central route processing, which in turn results in more enduring behavior change.

ARGUMENT STRENGTH

Strong messages, or messages that contain substantial argument strength, can stimulate and enhance elaboration (Petty and Cacioppo 1986). Strong messages provide relevant, reasonable, quality information that can be used to influence attitudes. Alternatively, weak messages lack argument strength and therefore are not as effective in triggering elaboration (Petty and Cacioppo 1986; Petty and Wegener 2008). Attitudes that align or match with presented information are thought to be reinforced with strong arguments, while recipient attitudes that mismatch may not change if the message does not have the strength to stimulate elaboration (Petty and Wegener 2008; Lavine and Snyder 1996; W. Wood 2000). Furthermore, framing arguments to trigger recipient values or goals increases elaboration potential and likelihood of attitude change (W. Wood 2000).

Argument strength and framing can be tested through elicitation studies, in which a small sample of respondents evaluate a series of potentially useable statements, to determine which are perceived as containing quality, relevant, stimulating information (Petty and Cacioppo 1986; Petty and Wegener 2008). Those messages that exhibit the most effect, are typically the strongest, and have the most significant power to influence attitudes (Petty and Cacioppo 1986; Petty and Wegener 2008). Furthermore, impactful messages have qualities that will increase the prospect of elicitation, potentially altering mismatching attitudes and increasing attitudes that already align with the concepts presented (Petty and Cacioppo 1986; Petty and Wegener 2008; Lavine and Snyder 1996; W. Wood 2000; Ziegler et al. 2007). For example, Tarrant et al. (1997) used an elicitation study to formulate belief statements and, subsequently, persuasive arguments for use in a study examining communication techniques and attitudes toward ecosystem management. Schroeder et al. (2012) used an elicitation study to first determine hunters' beliefs affecting attitudes toward controlling utilization of lead shot, prior to testing communication strategies for increasing support of restrictions on lead shot use.

METHODS

This research originates from findings from a two-phase study to help inform social indicators and standards of quality pertaining to sounds in Sequoia National Park. During phase one, a visitor survey was conducted at Sequoia in the summer of 2009 to determine what visitors were hearing (Marin 2011). Approximately 50% of respondents heard aircraft, and approximately 72% of those found the associated sounds to be unacceptable (Newman et al. 2012). These findings led to phase two, this study, which applied a theoretically-based educational message and military aircraft soundclips, to evaluate visitor standards related to aircraft sounds in Sequoia, as described below.

Elicitation Study

An elicitation study was used to determine which informative message should be applied during phase two. To evaluate message strength, several messages were tested during the spring of 2010 using a paper survey instrument. Thirty-eight undergraduate natural resources students at Colorado State University were asked to evaluate how hearing or seeing aircraft flying overhead during a visit to Sequoia would affect their experience. The students were then informed that they would be presented with three messages intended to provide information to park visitors about potential reasons for hearing and/or seeing aircraft while in the park. The message that resulted in the most robust argument strength, and therefore effect on respondents' acceptability was "Military aircraft are allowed to conduct training flights over Sequoia National Park in an effort to help keep the United States of America safe. Consequently, visitors hiking in this area of the park can sometimes hear and/or see military aircraft flying overhead." This message was then applied at Sequoia during phase two, to determine the effect of messaging on respondent acceptability and standards of quality pertaining to military aircraft noise in the park.

Survey Questionnaire and Military Aircraft Soundclips

The questionnaire developed for this study was designed to determine the acceptability of sounds recorded in the park, which included natural, and military aircraft overflight sounds. Two questionnaire versions were used in the study in order to test the effect of messaging on acceptance of military aircraft sounds. Prior to respondents' rating the acceptability of the same soundclips, the "primed" survey provided the message that was established through elicitation methods, informing visitors about military aircraft. This was followed by instructions asking visitors to indicate how acceptable it would be to hear the following sounds while hiking in this area of the park. The "unprimed" survey only asked respondents to indicate how acceptable it would be to hear the following sounds while hiking in this area of the park, without any mention of military aircraft. The acceptability of the aircraft soundclips was rated on a 9-point scale (-4 = Very Unacceptable; 0 = Neutral; 4 = Very Acceptable).

The soundclips evaluated during this study were extracted from recordings at Sequoia taken by the National Park Service Natural Sounds and Night Skies Division acoustical monitoring equipment during July and August of 2009. Recordings were collected using an Edirol R05 recorder, a Larson–Davis 831 sound level meter with a pre-polarized ½ inch condenser PCB 377B20 model microphone affixed to a tripod approximately five feet from the ground. The monitoring equipment recorded "33 one-third-octave band level measurements every second" concurrently with an "A-weighted, summary of aggregate sound level" which measured both one-third-octave and A-weighted sound levels as decibels, respective to "10–12 W/ m²" standard sound intensity (Fristrup et al. 2010, 3). Sound events were analyzed and extracted into MP3 format using the NPS Sound Pressure Level Annotation Tool spectrograms and Adobe Audition 1.0 with the assistance of NPS staff. Soundclips were chosen to typify both natural ambient and military aircraft overflight episodes from several days and times, so as to represent various potential visitor experiences at Sequoia. Ultimately, five, forty-second, A-weighted soundclips, ranging in their average of 1 s Leq sound pressure levels were chosen for field application (Ambrose and Burson 2004; Stack et al. 2011; Fristrup et al. 2010; Fahy 2000, Fristrup 2009). One recording clip contained natural ambient sounds from the park, consisting predominantly of wind, birds, and water, which were at maximum level 28 dB(A). The additional four recordings contained natural ambient sounds masked by military aircraft, which resulted in varying Leq measurements ranging from 66 dB(A) down to 33 dB(A). The soundclips were played beginning with the natural ambient recording, followed by the 66 dB(A), the 53 dB(A), the 46 dB(A), and 33 dB(A) military aircraft recordings for both the "primed" and "unprimed" samples.

STUDY AREA

Sequoia is in close proximity to several military installations in California, including Lemoore Naval Air Station, China Lake Naval Air Weapons Center, and Fort Irwin National Training Center for the us military. In addition, there are nearby military installations in Nevada, including Nellis Air Force Base and the military test ranges associated with Area 51. Despite collaborative efforts between Sequoia and military officials (Keizer 2008), military overflights and associated acoustic impacts are likely to continue. Furthermore, previous research suggests that Sequoia visitors are hearing aircraft, and the majority find associated sounds unacceptable (Newman et al. 2012), providing greater rationale for this study.

SURVEY ADMINISTRATION

The visitor survey was administered to hikers exiting Sequoia's Crescent Meadow and Wolverton trailheads during the summer of 2011, yielding a total n = 146 ("primed" n = 74; "unprimed" n = 72) and a response rate of 88%. Willing respondents were asked to complete an on-site paper survey instrument after listening to the soundsclips described above. Camp chairs placed beside the trailheads were provided for respondents so that they could listen to the soundclips and subsequently complete the questionnaire. The soundclips were played for the respondents through Bose Quiet Comfort 15 noise-cancelling headphones to eliminate extraneous noise.

DATA ANALYSES

Independent samples t-tests were used to determine if messaging statistically affected acceptance of military aircraft sounds. Statistical and practical significance were examined through consideration of p-values, Eta values, and the magnitude of differences in the mean acceptability ratings of military aircraft sound levels from the "primed" and "unprimed" samples. Practical significance was addressed by evaluating Eta values (η) categorized as having either a "minimal" (η = rv.10), a "typical" (η = rv.30), or a "substantial" effect measures (η = rv.50) (J. Cohen 1988; Vaske 2008; Vaske et al. 2002).

RESULTS

Primed vs. Unprimed

Recording one, which contained natural ambient sounds from Sequoia, but no military aircraft, resulted in similar, non-statistically different mean

TABLE 9.1. Comparison of primed (Respondents notified of military aircraft presence through messaging) and unprimed (Respondents not informed of military aircraft presence) visitors at Sequoia National Park.

Soundclips	Sample	N	Mean	SD	t-value	p-value	Eta
Recording 1 – peak 28 dBA	Primed	74	3.59	1.0	-.170	.719	.014
natural ambient wind, water	Unprimed	72	3.63	1.2			
and bird							
Recording 2 – peak 66 dBA	Primed	74	-.08	2.6	3.40	.001	.272
natural ambient masked by	Unprimed	72	-1.42	2.1			
military aircraft							
Recording 3 – peak 53 dBA	Primed	74	-.31	2.5	3.60	<.001	.284
natural ambient masked by	Unprimed	72	-1.64	2.0			
military aircraft							
Recording 4 – peak 46 dBA	Primed	74	.12	2.6	2.90	<.001	.230
natural ambient masked by	Unprimed	72	-.97	2.0			
military aircraft							
Recording 5 – peak 33 dBA	Primed	74	.18	2.6	2.10	.005	.169
natural ambient masked by	Unprimed	72	-.65	2.2			
military aircraft							

Note: Variables coded on a 9-point scale (-4 = Very Unacceptable; 0 = Neutral; 4 = Very Acceptable).

values between the "primed" and "unprimed" samples (primed M = 3.59, unprimed M = 3.63) (Table 9.1). Recording two, which contained military aircraft peaking in sound pressure at 66 dB(A) resulted in statistically-different mean values between respondents notified of the presence of military aircraft and those that were not (primed M = -.08, unprimed M = -1.42, p = .001, η = .272). Recording three, which contained military aircraft peaking at 53 dB(A) also resulted in statistically-different mean values between "primed" and "unprimed" respondents (primed M = -.31, unprimed M = -1.64, p = <.001, η = .284). Recording four, which consisted of military aircraft sounds peaking in sound pressure at 46 dB(A) also resulted in statistically-different mean values between samples (primed M = .12, unprimed M = -.97, p = <.001, η = .230). Recording five, which contained the lowest level of military aircraft sound pressure peaking at 33 dB(A), also resulted in statistically-significant differences between "primed" and "unprimed" respondents (primed M = .18, unprimed M = -.65, p = .005, η = .169) although the effect size suggests a minimal relationship. Three of the four soundclips that contained military aircraft resulted in statistically-significant mean differences with typical effect sizes between samples.

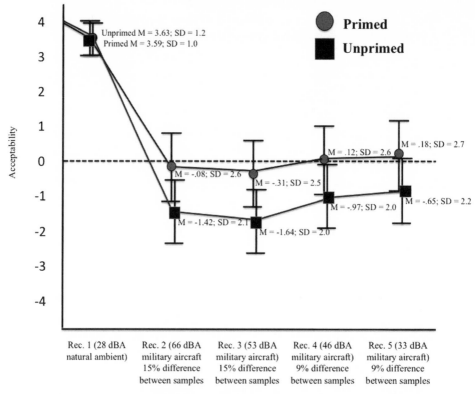

FIG 9.1: Plotted trend comparing primed and unprimed respondent acceptability of military aircraft soundclips.

Soundclip Acceptability

Results suggest that, on average, both "primed" and "unprimed" respondents found the natural ambient soundclip to be very acceptable (Figure 9.1), but upon hearing soundclips two (aircraft peaking at 66 dB(A)) and three (aircraft peaking at 53 dB(A)), mean values dropped below acceptability for both samples. Evaluation of soundclips four and five resulted in mean values that were acceptable for "primed" respondents, but unacceptable for "unprimed" respondents. "Primed" respondents' mean acceptability of the military aircraft soundclips was approximately 15% more acceptable than "unprimed" respondents for recordings two and three, and 9% more acceptable for recordings four and five.

DISCUSSION

The purpose of this study was to determine if indirect management actions in the form of educational messaging could significantly affect visitor acceptability of military aircraft sounds. Subsequently, our goal was to increase understanding of the strength of educational messaging and to discuss how it may be applied to soundscape management in Sequoia and other units. This study demonstrated that a theoretically-based and tested message could be applied in a park to effectively alter visitor attitudes, perceptions, expectations, and therefore, average thresholds of acceptability for military aircraft sounds. Although policies, technology, in-flight behaviors, and improved relations between military and park officials may mitigate some noise impacts, in the foreseeable future it is anticipated that military aircraft will continue to fly over the park. However, these study findings suggest that information may be important to help reduce the impacts of these overflights on visitor experience quality.

Educational messaging is one of several indirect management tools that may assist managers to protect, maintain, and restore the natural acoustic environment and visitor experiences in national parks. This study, along with previous soundscape messaging research (Mace et al. 2003; N. Miller et al. 1999; Manning et al. 2010; Stack 2008; Stack et al. 2011) increases managers' understanding of the strength of educational messaging as it pertains to soundscape management. Through theoretical development and elicitation testing, the strongest message was chosen and applied in this field study. This message did increase acceptability of military aircraft sounds by as much as 15%, suggesting that educational messaging may offer immediate benefits to Sequoia visitor experiences. These significant results do not necessarily suggest that the evaluated message should be implemented in Sequoia, but instead demonstrate how messaging can affect visitor perspectives and evaluations of aircraft sounds. Furthermore, we are not suggesting that NPS managers should apply more lenient standards regarding the protection of Sequoia's soundscape based on these results. We are suggesting, however, that this study demonstrated that informing visitors about sound sources that are likely to continue in Sequoia, may benefit visitor experience quality.

Given the improved relations and ensuing collaborative efforts between NPS and military staff to protect Sequoia's soundscape, the findings of this study provide these officials with additional tools to manage visitor experiences as they pertain to park soundscapes. The results suggest that this message could be implemented permanently to improve acceptability of military aircraft, or perhaps used on selective occasions when military overflights might be more prevalent. Whether NPS officials and Sequoia managers choose to employ this message or not, it is recommended that any implementation of educational messaging rely upon theoretical frameworks, such as ELM for the most effective influence. As suggested by this paper, ELM can provide greater understanding of the challenges managers may face when attempting to communicate with visitors, and can suggest strategies for stimulating central route processing. While managers cannot always reach visitors due to situational and personal variables, developing messages that are relevant, strong, and impactful (Ham 2009; Ham et al. 2009), may lead to more central route processing.

The effectiveness of soundscape-related messaging efforts should be greater when multiple methods of communication (e.g., trailhead signage, brochures, interpretive presence) are provided (Stack et al. 2011). Any message design should maintain the appearance of current NPS messaging to induce greater perceived source credibility among recipients. Messages should be implemented and evaluated through temporary placement, near areas that tend to be most problematic, to determine how they affect visitor behaviors and experiences. For example, in Sequoia, areas where visitors have reported hearing aircraft and finding those associated sounds unacceptable, may be the most appropriate locations for temporary messaging. Those messages that are found to assist with soundscape protection and improve visitor experiences could be employed more permanently. At Muir Woods National Monument, the effective results of experimental messaging through temporary signage led to designation of a permanent quiet zone within the study area (Stack et al. 2011).

LIMITATIONS AND FUTURE RESEARCH

The message used in this study was first examined during an elicitation study in a laboratory setting with undergraduate students. The elicitation sample was not diverse, and therefore may have biased message appraisal. In the field study, the evaluated message was provided only to respondents through the "primed" survey, with no additional communication diffusion (e.g., trailhead signage, brochures, interpretive ranger talk), which may have altered results. Respondents listened to the military soundclips using noise-cancelling headphones on-site at trails in Sequoia, and therefore visual cues, or external noise interruptions may have altered evaluations. Following the natural ambient soundclip, respondents were provided military soundclips in descending order, which may have produced an order effect, similar to order bias discovered through visual/photo methods (Gibson 2011). Despite trends that suggest that visitors were generally less accepting of louder than quieter military aircraft (Figure 9.1), soundclip order may have influenced perceptions. This study only tested soundclip acceptability at Sequoia, but other NPS units that experience military aircraft overflights, such as Death Valley National Park, City of Rocks National Reserve, Organ Pipe Cactus National Monument, and John Day Fossil Beds National Monument (Vicki McCusker, personal communication 11/28/11) should also be evaluated to determine the effect of communication strategies. Messaging concerning the presence of other types of aircraft (e.g., commercial or air-tour) should be tested in parks that experience those events to determine if educational information has a similar effect on visitor perceptions.

We acknowledge that these results only relate to improved social aspects pertaining to visitor experiences, and do not directly improve resource protection or preservation. However, we would hope that if messaging were implemented, it would increase visitor understanding concerning the importance of soundscape protection. While stronger messages can provide greater elaboration even when attitudes mismatch, this message may produce more fervent attitudes in individuals who hold attitudes that misalign with the concepts provided through the tested message. These individuals may in turn become agitated with the NPS for providing the message. Despite these limitations, the results of this study demonstrate the strength of a theoretically-derived message on visitor perspectives.

CONCLUSION

This study demonstrates how messaging can affect visitor perspectives concerning noise from military aircraft overflights in Sequoia. The results of this study suggest that informing visitors about the presence of military aircraft through educational messaging has the potential to increase visitor acceptability of military aircraft sounds in the park. These findings indicate that educational messaging, priming visitors to the reasons for military aircraft overflights in the park, may offer immediate benefits to Sequoia visitors' soundscape-related experiences. Thus, this study adds to the growing body of literature that has increased knowledge of soundscape management in parks, and more generally improves understanding of how educational messaging can be applied to help mitigate park issues.

10

MODELING AND MAPPING HIKERS' EXPOSURE TO TRANSPORTATION NOISE IN ROCKY MOUNTAIN NATIONAL PARK

Logan Park, Steve Lawson, Ken Kaliski,
Peter Newman, and Adam Gibson

Research conducted in a variety of national park settings suggests that the quality of visitors' experiences is tied to the naturalness of the area's soundscape. Yet human-caused noise from aircraft, roads, and other sources can make it difficult for visitors to "escape" to areas where natural sounds and quiet predominate. This chapter describes an effort to develop a modeling and mapping tool to monitor visitors' exposure to roadway noise and opportunities to experience natural sounds and quiet while hiking on trails in Rocky Mountain National Park.

INTRODUCTION

Natural and cultural sounds are integral members of the suite of resources and values that the National Park Service (NPS) is charged with preserving, restoring, and interpreting (National Park Service 2000). Results of research conducted in a variety of national park settings suggest that the quality of visitors' experiences is tied to the naturalness of the area's soundscape (Manning et al. 2007; N. Miller 2002; Tranel 2006). For example, findings from a recent study in Haleakala National Park in Hawaii suggest that the primary reason for visitors to take an overnight backcountry trip in the park is to experience the sounds of nature (Lawson et al. 2008). Human-caused sounds from aircraft, roads, maintenance activities, and other visitors, however, commonly permeate park soundscapes, making natural sounds and quiet an increasingly scarce resource (Krause 1999).

Recently, the NPS has applied an adaptive management-by-objectives framework to address soundscape management and planning (Pilcher et al. 2009). This process involves formulation of soundscape-related indicators and standards of quality. Indicators of quality are measurable, manageable proxies for desired park conditions, and standards of quality are numerical expressions of desired conditions for indicators. As an example, the NPS might specify "human-caused noise-free interval duration" as an indicator of quality related to providing visitors opportunities to experience natural sounds and quiet. A standard of quality for this indicator might specify that at least 90% of visitors will experience at least one interval of 15 minutes or more that is free of human-caused noise while visiting the park.

Soundscape-related indicators and standards of quality are now being developed at a number of national parks, but measurement of some indicators, such as highly variable soundscape metrics, is nontrivial (Ambrose and Burson 2004; Lawson and Plotkin 2006). For example, natural sound levels fluctuate because of wind, air characteristics (e.g., density, temperature), and wildlife. Furthermore, visitors' exposure to natural and human-caused sounds is difficult to observe directly or measure through visitors' self-reports in surveys. However, visitor use and noise modeling technologies are potentially useful in this situation (e.g., Lawson 2006; Lawson and Plotkin 2006; N. Miller 2004; Roof et al. 2002).

The purpose of this paper is to demonstrate the use of visitor use and noise modeling tools to provide spatially precise, integrated information about soundscape conditions within a national park setting. In particular, it presents research conducted at Rocky Mountain National Park, Colorado, to model and map visitors' exposure to transportation-related noise while visiting attractions and hiking on trails in the Bear Lake Road corridor. The results of this work are expected to provide the NPS with a monitoring tool to track soundscape-related indicators of quality in Rocky Mountain National Park that is adaptable to other national parks.

METHODS

Study Area

As noted, motor vehicles are one of the most common and widespread sound sources within national parks. Consequently, park soundscapes can be dramatically affected, both positively and negatively, by transportation planning and operations management decisions. The purpose of this project is to use noise and visitor use modeling to quantify and map the effects of shuttle bus service and private vehicle access management in the Bear Lake Road corridor on the park's soundscape. Furthermore, the project combines noise modeling outputs with visitor trip data to estimate the condition of potential soundscape-related indicators of quality.

Data Collection

For the purposes of developing the transportation noise model and generating spatially precise estimates of visitors' exposure to noise from Bear Lake Road, four primary types of data were collected in Rocky Mountain National Park during summer 2008: (1) traffic volume by vehicle classification, (2) sound level data, (3) visitor hiking routes, and (4) daily visitation by trailhead. Continuous traffic counters were installed at three locations to measure directional traffic volumes at 15-minute intervals during a two-week period selected to represent the peak period of park visitation (Figure 10.1).

Sound level data were collected at seven locations over an eight-day period during the park's peak period of visitation (Figure 10.1). The acoustic monitoring locations were selected to represent a range of soundscape environments within a typical day's hike from trailheads along Bear Lake Road. For example, monitoring sites ranged from a roadside pullout at a scenic overlook to an alpine lake 1,800 meters (5,906 ft) from the road. To collect data needed to calibrate the transportation noise model directly to traffic volumes, one of the sound level meters was collocated within approximately 55 meters (60 yds) of the traffic counter installed north of the park-and-ride lot. All eight acoustic monitors were configured to record a sound level measurement at one-second intervals, and four of the monitors were also programmed to record one-third-octave band sound levels. All the sound level meters were calibrated prior to and after sampling using a handheld calibrator.

Visitor hiking routes data were collected on 13 sampling days between 31 July and 14 August 2008 via administration of Global Positioning System (GPS) units to visitors at four trailheads along the Bear Lake Road corridor (i.e., Bear Lake, Bierstadt Lake, Glacier Gorge, and Storm Pass). The GPS units were distributed to randomly selected visitor groups at the start of their hikes and collected at the end of hikes. Daily trailhead visitation was measured with mechanical trail traffic counters, calibrated with data from direct observation (Kiser et al. 2007).

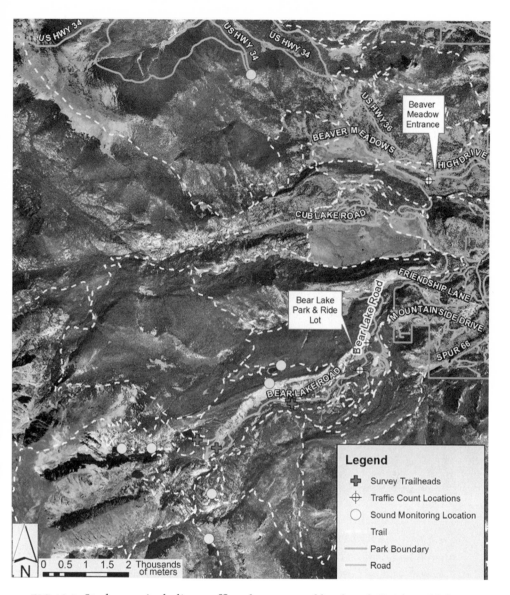

FIG 10.1: Study area, including traffic volume, sound level, and GPS-based hiking route monitoring locations.

NOISE MODELING AND MAPPING

Sound propagation modeling of the traffic noise data was conducted using Cadna/A software made by Datakustik GmbH. The geographic scope of the noise model is a 14,000-by-14,000-meter (45,934 by 45,934 ft) square, with its northeast corner just north of the park entrance and east of the eastern park boundary. The model incorporates traffic volumes for the full extent of Bear Lake Road, as recorded by the automatic traffic counters. A digital terrain model was obtained from the U.S. Geological Survey and converted into elevation contours to model the attenuation of roadway sound due to intervening terrain. Propagation algorithms found in the German RLS-90 standard are used within the software to model how vehicle sounds from the Bear Lake Road permeate the surrounding landscape (Kaliski et al. 2007). In particular, the model estimates how sound propagates from the roadway to "receiver locations" specified by the model developer, taking into account intervening terrain, absorption of sound by the ground, energy losses into the atmosphere, and losses due to geometric spreading of the sound wave emanating from the road. In this study, sound pressure level (i.e., decibel) estimates were generated for a grid of 492,000 receivers covering every 20 meters (66 ft) within the study area. The result is a grid of daytime (6:00 a.m. to 6:00 p.m.) average sound levels representing traffic sound conditions during the sampling period. The grid data were then plotted for visual display via a noise contour map to depict the study area's soundscape conditions with respect to noise from Bear Lake Road.

VISITOR USE AND NOISE EXPOSURE MODELING

The GPS tracks of visitor hikes were imported into a Geographic Information Systems (GIS) environment for error correction and analysis. The data were filtered for positional inaccuracies due to poor satellite constellations and signals interrupted by high mountain peaks. Trip data split across multiple GPS files were assembled into individual trips, and trip attributes, including hiker movement speed, initial trailhead, and intended destination, were joined to the track spatial data.

Spatial statistics tools in the GIS software were used to estimate the amount of time and distance visitors must hike from trailheads to experience alternative soundscape conditions. Estimates were also generated for the proportion of visitors who experience at least 15 minutes of natural sounds and quiet. At the time of the study, the NPS had not defined a threshold for road

TABLE 10.1. Study Area Visitation by Trailhead, Rocky Mountain National Park

Trailhead	Average daily visitation	Proportion of total visitation
Bear Lake	7,353	89.1
Bierstadt Lake	96	1.2
Glacier Gorge	638	7.7
Storm Pass	170	2.1
Total	8,257	100.0

noise beyond which natural sounds and quiet are compromised. Thus, a range of example road noise thresholds were evaluated to estimate the proportion of visitors who experience at least 15 minutes of natural sounds and quiet. The example road noise thresholds used in the analysis include = 25 dB(A) (nighttime ambient natural sound level measured in the study), = 30 dB(A) and = 35 dB(A) (daytime ambient natural sound levels), and = 65 dB(A) (the level at which noise interferes with conversational tones).

RESULTS

Results of counts conducted to measure daily visitation, by trailhead, suggest that the Bear Lake Trailhead receives the vast majority of visitor use in the study area (Table 10.1). The noise map, developed on the basis of Bear Lake Road baseline traffic conditions, in Figure 10.2, depicts higher (louder) transportation sound pressure levels in darker shades of gray and lower (softer) sound pressure levels in lighter shades of gray. Further, the noise map depicts more heavily visited trail segments with thicker brown lines, and lesser-used trails segments with thinner brown lines. This map suggests that transportation sounds from Bear Lake Road permeate the park's soundscape throughout the adjacent trail system. The noise is concentrated along the road and falls off sharply with distance. However, the extent of noise in the area requires effort on the part of visitors to reach areas of natural quiet away from Bear Lake Road. For example, model results suggest that visitors following the most direct routes to natural quiet would have to walk more than 1,000 meters (0.6 mile) from all four trailheads in the study area to reach natural quiet as defined by areas of the park with road sound levels that do not exceed 25 dB(A) (Table 10.2). Further, results in Table 10.2 suggest visitors would have to walk more than 1,000 meters (0.6 mi) from two of the four trailheads in the study area to reach areas of the park with road sound levels less than 35 dB(A).

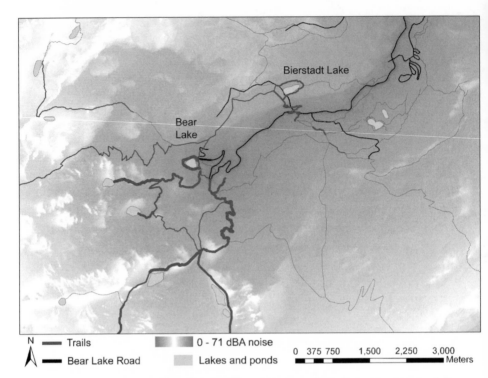

N — Trails   ▬▬ 0 - 71 dBA noise
⋀ — Bear Lake Road   ▬ Lakes and ponds

0   375 750   1,500   2,250   3,000
▬▬▬▬▬▬▬ Meters

FIG 10.2: Noise map of baseline traffic volumes on Bear Lake Road and relative intensity of hiking use on adjacent trail network.

Summaries of the GPS track data indicate that visitors' average hiking speed is 0.55 meter/second (1.2 mph). This hiking speed is somewhat lower than typical average hiking speeds for other areas (Bishop and Gimblett 2000; van Wagtendonk and Benedict 1980), because of many groups' propensity to linger or move more slowly around attraction areas such as Bear Lake and because of the relatively steep topography in the study area. This hiking rate, coupled with the hiking distance results, suggests that visitors would have to hike between 6 and 51 minutes, depending on the trailhead selected, to reach natural quiet defined by areas of the park where road sound levels are = 30 dB(A), or in some cases would never reach it (Table 10.3). As expected, the estimated travel times to reach natural quiet reported in Table 10.3 vary according to the road noise threshold used to define areas of natural quiet and sounds. Minimum distance to natural quiet varies across the trailheads in the study area by a factor of nearly 10, suggesting opportunities for management to highlight specific trails to visitors that provide greater opportunities for natural sounds and quiet.

TABLE 10.2. Hiking Distance from Trailhead Required
to Reach Closest Natural Quiet*

Trailhead	NOISE THRESHOLD/DISTANCE (M)			
	25 dB(A)	30 dB(A)	35 dB(A)	65 dB(A)
Bear Lake	1,093	206	155	0
Bierstadt Lake	1,934	1,586	1,542	23
Glacier Gorge	2,097	1,682	1,210	0
Storm Pass	1,907	1,376	973	0

*Natural quiet is defined as sound levels below noise thresholds.

TABLE 10.3. Average hiking time from trailhead required
to reach closest natural quiet*

Trailhead	NOISE THRESHOLD/TRAVEL TIME (MINUTES)			
	25 dB(A)	30 dB(A)	35 dB(A)	65 dB(A)
Bear Lake	33.1	6.2	4.7	0.0
Bierstadt Lake	58.6	48.1	46.7	0.7
Glacier Gorge	63.5	51.0	36.7	0.0
Storm Pass	57.8	41.7	29.5	0.0

*Natural quiet is defined as sound levels below noise thresholds.

The time and distance required to reach natural quiet defined by road
sound levels = 30 dB(A) may present difficulty for less mobile visitors seeking to get away from the transportation noise associated with the road. However, using the 30 dB(A) noise threshold for analysis, the results suggest that,
on average, visitors spend a majority (63.6%) of total hiking time in natural
quiet (Table 10.4). By contrast, visitors walking from Storm Pass or Bierstadt
Lake trailhead will experience elevated levels of noise for most or all of their
hike, while visitors starting from Bear Lake trailhead and hiking to more
distant lakes (e.g., Emerald Lake or Nymph Lake) will experience almost
uninterrupted escape from road sounds. The prevalence of opportunities
to experience natural quiet is also sensitive to the manner in which natural
quiet is defined. For example, "natural quiet," defined as soundscape conditions in which roadway sound levels do not exceed 65 dB(A), is experienced
by virtually all visitors in the study area.

With respect to assessing whether visitors are able to experience substantive "episodes" of natural quiet, results suggest that about half (49.6%)

TABLE 10.4. Percentage of hiking time visitors experience natural quiet*

| Trailhead | NOISE THRESHOLD/% OF HIKING TIME | | | |
	25 dB(A)	30 dB(A)	35 dB(A)	65 dB(A)
Bear Lake	54.5	68.8	77.8	100.0
Bierstadt Lake	12.1	40.1	43.7	100.0
Glacier Gorge	60.2	62.9	74.1	100.0
Storm Pass	0.6	20.1	39.5	100.0
Study area–wide	53.8	63.6	73.2	100.0

*Natural quiet is defined as sound levels below noise thresholds.

TABLE 10.5. Percentage of visitors who experience
at least 15 minutes of natural quiet*

| Trailhead | NOISE THRESHOLD/% VISITORS | | | | | % of total hikers or all trailheads |
	25 dB	30 dB	35 dB	65 dB	N	
Bear Lake	26.0	32.5	49.6	83.7	123	89.1
Bierstadt Lake	5.4	48.6	51.4	81.1	37	7.7
Glacier Gorge	45.3	55.7	59.4	85.8	106	1.2
Storm Pass	0.0	33.3	33.3	33.3	3	2.1
Total	24.1	34.1	49.6	82.6	269	100.1

*Natural quiet is defined as sound levels below noise thresholds.

of visitor groups in the study area are able to do so for at least 15 continuous minutes, using 35 dB(A) as the threshold for traffic noise (Table 10.5). When examined by trailhead, the results provide further insight into visitors' soundscape experience and how it varies across the study area. Hikers near Storm Pass do not usually experience quiet for 15 continuous minutes (33.3% of groups), but almost double that proportion do along the Glacier Gorge Trail (59.4%).

Spatial modeling results also offer insights into how soundscape experiences evolve throughout the course of specific hiking routes. For example, the noise profile depicted in Figure 10.3 is for a hiking route that begins at the Bierstadt Lake trailhead, travels to and around Bear Lake, and then heads into the backcountry. The hiker group embodied in these data experienced abrupt evolutions in their sound environment based on the hikers' route choices, encountering road noise at the trailhead (54 decibels), natural quiet

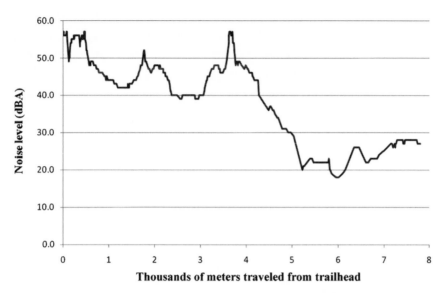

FIG 10.3: Noise level profile for hiking route from Bierstadt Lake trailhead, to and around Bear Lake, and into the park's backcountry.

on the way to Bear Lake (26 decibels), then additional road noise near Bear Lake (53 decibels).

DISCUSSION AND CONCLUSIONS

By providing insights on the noise environment, use distribution, and route decisions of visitors in Rocky Mountain National Park, results from this study demonstrate the utility of integrated visitor use and noise modeling to support indicator-based adaptive management and monitoring of park soundscapes. Furthermore, these findings suggest how visitor use and noise modeling can be used to proactively and deliberately assess the effects of transportation planning and operations on park soundscapes. Subsequent analyses with the data and models presented in this paper will be conducted to quantify and map the effects of potential modifications to the Bear Lake shuttle service and private vehicle access on soundscape conditions in the study area.

As the results of this work suggest, the modeling tools developed in this study can be used to estimate the conditions of soundscape-related indicators (e.g., percentage of visitors who experience at least 15 consecutive minutes of natural quiet) associated with baseline and alternative management scenarios. However, the NPS has not developed specific standards of quality for soundscape indicators in Rocky Mountain National Park. Formulation of empirically based standards of quality for soundscape indicators is recommended to complement the modeling tools developed in this study and to support indicator-based adaptive management of the park's soundscape.

PART 3

THE ECOLOGY
OF NATURAL DARKNESS

Part 3 of this book addresses the ecology of natural darkness. Natural darkness is a vital element of the biological and ecological integrity of national parks. However, human-caused light can affect natural darkness and other park resources as well. The five chapters in this part of the book document the wide-ranging impacts of human-caused light. As with research on the impacts of noise pollution presented in part 2, the chapters in this part of the book suggest that the impacts of light pollution can extend from individual animals to species and on to communities of species and a suite of ecological services. For example, reduced flight-to-light behavior of moths exposed to human-caused light cascades to less pollination and a reduced food source for bats and spiders.

11

THE ECOLOGICAL IMPACTS OF NIGHTTIME LIGHT POLLUTION

A Mechanistic Appraisal

Kevin Gaston, Jonathan Bennie, Thomas Davies,
and John Hopkins

Most contemporary reviews of the ecological impacts of light pollution are organized taxonomically. This synthesis presents a mechanistic framework focused on the ways in which artificial lighting alters natural light regimes, and the ways in which light influences biological systems. As artificial lighting alters natural patterns of light in space and time and across wavelengths, natural patterns of resource use and information flows may be disrupted, ultimately with consequences to the structure and function of ecosystems. This review highlights the potential influence of nighttime lighting at all levels of biological organization (from cell to ecosystem), the significant impact of even low levels of nighttime light pollution, and the existence of major gaps in research that must be filled to inform future work.

INTRODUCTION

It has been argued that the biological world is organized largely by light (Ragni and D'Alcala 2004; Foster and Roenneberg 2008; Bradshaw and Holzapfel 2010). The rotation of the Earth partitions time into a regular cycle of day and night (giving variation in light intensity of approximately 10 orders of magnitude; Table 11.1), while its orbital motion and the tilt of its axis cause seasonal variation in the length of time that is spent under conditions of light and darkness in each cycle. These major changes are overlain by more local variation caused by weather conditions, and the effect of the monthly lunar cycle on nighttime light. However, for any given latitude the light regime has been consistent for extremely long periods of geological

TABLE 11.1. Variation in levels of illuminance. Although widely used, note that lux measurement places emphasis on brightness as perceived by human vision.

	Lux
Full sunlight	103000
Partly sunny	50000
Cloudy day	1000–10000
Full moon under clear conditions	0.1–0.3
Quarter moon	0.01–0.03
Clear starry night	0.001
Overcast night sky	0.00003–0.0001
Operating table	18000
Bright office	400–600
Most homes	100–300
Main road street lighting (average street level illuminance)	15
Lighted parking lot	10
Residential side street (average street level luminance)	5
Urban sky glow	0.15

From data in British Standards Institute (2003), Rich and Longcore (2013), and Dick (2011).

time, providing a rather invariant context, and a very reliable set of potential environmental cues, against which ecological and evolutionary processes have played out.

Artificial lighting is a common characteristic of human settlement and transport networks (Boyce 2003; Schreuder 2010). The spread of electric lighting in particular has provided a major perturbation to natural light regimes, and in consequence arguably a rather novel environmental pressure, disrupting natural cycles of light and darkness (Verheijen 1958 and 1985; Outen 1998; Health Council of the Netherlands 2000; Longcore and Rich 2004; Rich and Longcore 2013a; Navara and Nelson 2007; Hoelker et al. 2010a and b; Bruce-White and Shardlow 2011; Perkin et al. 2011). Changes in light regime can be characterized as changes in the spatial distribution, the timing and the spectral composition of artificial light sources. As human communities and lighting technologies develop, artificial light increasingly encroaches on dark refuges in space, in time, and across wavelengths.

Space
Urbanization, population growth and economic development have led to rapid, and ongoing, increases in the density and distribution of artificial lighting

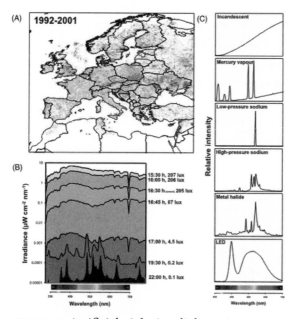

FIG 11.1: Artificial nighttime light varies in space, time and along electromagnetic spectrum. (A) Spatial variation in relative brightness trends of nighttime lights in Europe, using annual DMSP satellite data from 1992 to 2001 inclusive from NOAA National Geophysical Data Center http://www.ngdc.noaa.gov/dmsp/downloadV4composites.html. (B) Temporal change in spectral irradiance of ambient light in grassland at Tremough, UK from day (light gray) to night (black), 22.11.11; peaks at 19:30 h from indoor fluorescent lighting from nearby offices, and at 22:00 h from footpath lighting. (C) Spectral composition of main electric lighting types used since 1950, from data at http://www.ngdc.noaa.gov/dmsp/spectra.html.

over recent decades (Figure 11.1A; Riegel 1973; Holden 1992; Cinzano, Falchi, and Elvidge 2001; Cinzano 2003; Hoeker et al. 2010b). A wide variety of lighting devices contribute, including public street lighting, advertising lighting, architectural lighting, domestic lighting and vehicle lighting. The highest intensities of artificial light are experienced in the close vicinity (within meters to tens of meters) of light sources. Within illuminated urban and suburban areas, direct light from street lighting, domestic and commercial sources, and light reflected from the surrounding surfaces, can create a highly patchy light environment. Over much larger areas surrounding towns and cities, a somewhat lower intensity of diffuse background light derives from "sky glow," artificial light scattered in the lower atmosphere. Under cloudy conditions in urban areas, the sky glow effect has been shown to be of an equivalent or greater magnitude than high-elevation summer moonlight (Kyba et al. 2011a).

Time

Early municipal lighting systems often functioned only on moonless nights or prior to midnight (Jakle 2001). Throughout the 20th century, the manufacture of cheaper lighting technologies led to more persistent street lighting in developed cities, typically from dusk until dawn, 365 days a year. Lights in commercial, industrial and residential premises may be kept permanently on or switched on intermittently during the hours of darkness for reasons of security or convenience, and amenity lighting, for example floodlighting of sports pitches, is often concentrated in the hours following sunset, leading to a varying light environment throughout the night (Figure 11.1B). Economic pressures, limited energy supply and/or efforts to minimize energy consumption and carbon emissions have resulted in constraints on the timing of nighttime lighting in many regions of the world, and, led by developments in technology allowing automated timing and control, dimming or switching off of municipal lighting for periods during the night is being adopted in some developed countries (e.g., Lockwood 2011).

Spectral Composition

Different forms of artificial lighting have unique spectral signatures, each emitting light at varying intensities over a distinctive range of wavelengths (Figure 11.1C; Thorington 1985; Boyce 2003; Elvidge et al. 2010; van Langevelde et al. 2011). These spectral signatures differ from those of natural direct and diffuse sunlight, twilight and moonlight, with certain types of lighting restricted to very narrow bandwidths, while others emit over a wide range of wavelengths. Early electric street lighting relied on incandescent bulbs (Jakle 2001), emitting primarily in yellow wavelengths, while low-pressure sodium lighting, widely adopted in the 1960s and 1970s, emits a single narrow peak in the visible spectrum at 589.3 nm, giving objects a distinctive monochromatic orange hue. More recent light technologies emit over a broad range of wavelengths (high-pressure sodium lighting emits a yellow light allowing some color discrimination; high-intensity discharge lamps emit a whiter light, with significant peaks in blue and ultra-violet wavelengths, and LED-based white street lighting typically emits at all wavelengths between around 400 and 700 nm, with peaks in the blue and green; Elvidge et al. 2010). Over recent decades the spectral diversity of light sources has grown (K. Frank 1988), and the trend towards adopting lighting technologies with a broader spectrum of "white" light is likely to increase the potential for ecological impacts (including through changes in the color of sky glow; Kyba et al. 2012).

In combination, the increasing spatial, temporal and spectral distribution of nighttime light pollution provides the potential for major influences on ecological and evolutionary processes (Navara and Nelson 2007; van Langevelde et al. 2011). Substantial attention has been paid to catastrophic events, such as the mortality that can follow from the disorientation of hatchling turtles and of birds by nighttime lighting (e.g., Howell, Laskey, and Tanner 1954; Verheijen 1958, 1985; McFarlane 1963; Reed, Sincock, and Hailman 1985; Witherington and Bjorndal 1991; Peters and Verhoeven 1994; Salmon et al. 1995; Le Corre et al. 2002; Jones and Francis 2003; Black 2005; Tuxbury and Salmon 2005; Gauthreaux et al. 2006; Montevecchi 2006; W. Evans et al. 2007; Lorne and Salmon 2007; Gehring, Kerlinger, and Manville 2009; Tin et al. 2009; Rodríguez, Rodríguez, and Lucas 2012). However, a much broader set of implications has been identified (Longcore and Rich 2004; Hoelker et al. 2010b; Perkin et al. 2011). In consequence, and echoing earlier statements (e.g., Verheijen 1985), there have been several recent calls for a much improved understanding of these implications (e.g., Health Council of the Netherlands 2000; W. Sutherland et al. 2006; Royal Commission on Environmental Pollution 2009; Hoelker et al. 2010a and b; Perkin et al. 2011; Fox 2013).

Part of the challenge in providing this improved understanding lies in organizing the knowledge that already exists and in identifying the principal gaps. The literature that has been developed to date is scattered, and largely lacks synthesis within a common mechanistic framework. Previous attempts to review this material have done so by taxonomic group (Rich and Longcore 2006—with sections on mammals, birds, reptiles and amphibians, fishes, invertebrates, plants), by different processes and/or levels of biological organization (Longcore and Rich 2004—with sections on behavioral and population ecology, community ecology, ecosystem effects; Longcore and Rich 2006—with sections on physiological ecology, behavioral and population ecology, community ecology, ecosystem ecology), and by research domain (Perkin et al. 2011—with sections on dispersal, evolution, ecosystem functioning, interactions with other stressors).

Here we propose a framework that focuses foremost on the cross-factoring of the ways in which artificial lighting alters natural light regimes (spatially, temporally, and spectrally), and the ways in which light influences biological systems, particularly the distinction between light as a resource and light as an information source. Reviews of the literature to date have highlighted examples of each of the different combinations of such a cross-factoring. However, many studies do not report, for example, the spectral properties, intensity, duration and/or spatial extent of the light regime, making it hard to draw general conclusions applicable outside their geographical and taxonomic limits. For this reason perhaps, despite the global nature of increases in artificial light, the ecological impacts of light pollution are often considered to be localized and restricted to a few vulnerable species or taxonomic groups.

Considering these individual studies within our proposed framework: (*i*) helps to unify understanding of particular effects of light pollution across taxa, and to draw conclusions relevant to whole ecosystems; (*ii*) highlights the mechanisms behind the observed ecological effects of light pollution, and defines clear criteria for future ecological studies; and (*iii*) provides guidance in detecting, predicting and mitigating against current and future adverse effects of light pollution.

In the sections below we review the evidence for each of the combinations of the cross-factoring. To avoid undue redundancy, and a bias towards certain well-studied systems, we have not attempted to provide an exhaustive list of studies on the ecological effects of light pollution, but rather in each section we aim to illustrate the key issues and identify progress and opportunities for further work.

LIGHT AS A RESOURCE

Both light and darkness can act as a resource for organisms (Kronfeld-Schor and Dayan 2003; Gerrish et al. 2009). Through photosynthesis, energy is captured by autotrophs in the form of light and cycled through ecosystems; furthermore, many physiological processes and behavioral activities require either light or dark conditions to operate. The balance between hours of light and of darkness constrains the time available for these processes and so changes in the availability of both light and darkness as a resource can have positive or negative effects on an organism, dependent on whether it is the presence or absence of light that poses the greater constraint.

Photosynthesis

In green plants, light is absorbed for photosynthesis by chlorophylls and carotenoids at wavelengths between 400 and 700 nm. While this range encompasses much of the visible emissions by artificial lights, in most cases the levels of photosynthetically active radiation (PAR) associated with nighttime light pollution are extremely low relative to sunlit conditions (typically less than 0.5 μmol m^{-2} s^{-1} compared with between 100 and 2000 μmol m^{-2} s^{-1} for sunlit conditions) and the effect of light pollution on net carbon fixation is likely to be negligible in most cases. Although Raven and Cockell (2006) calculate that the combined PAR flux from sky glow in an urban area and moonlight from a full moon could theoretically exceed the lower limit for photosynthesis, in most cases only direct illumination in the close vicinity of light sources, for example the leaves of trees within a few centimeters of streetlights, is likely to be sufficient to maintain net carbon fixation during nighttime and at lower light levels offset nocturnal respiratory losses. The consequences of this highly localized effect on individual plants and on ecosystems are largely unexplored.

One environment in which light pollution is known to have marked effects on ecosystems through photosynthesis is in artificially lit cave systems. The introduction of lighting into caves used as visitor attractions promotes highly localized growth of "lampenflora" communities completely dependent on artificial light as a source of energy. These communities may include autotrophs such as photosynthetic algae, mosses and ferns growing in the vicinity of light fixtures, as well as fungi and other heterotrophs utilizing the input of organic matter (K. Johnson 1979). These communities may displace or disrupt the trophic ecology of energy-limited cave ecosystems. Algal growth on the walls can also seriously damage and obscure objects of geological and archaeological interest within caves (Lefèvre 1974), and is an issue of some concern.

Partitioning of Activity between Day and Night

Partitioning of time has been thought to be a major way in which the ecological separation of species is promoted (Kronfeld-Schor and Dayan 2003). Temporal niche partitioning between diurnal, crepuscular and nocturnal species occurs as they avoid competition by specializing in a particular section along the light gradient (Gutman and Dayan 2005). Indeed, while ecological and evolutionary studies have focused foremost on diurnal species, a substantial proportion of species is adapted to be active during low-light conditions (Lewis and Taylor 1964; Hoelker et al. 2010a). Natural variation in nighttime lighting, particularly in moonlight due to the phase of the moon and cloud-cover conditions, has been shown to affect the timing of activity in a range of species (e.g., Imber 1975; Morrison 1978; Gliwicz 1986; Kolb 1992; Tarling, Buchholz, and Matthews 1999; G. Baker and Dekker 2000; Fernandez-Duque 2003; Kappeler and Erkert 2003; Beier 2006; Woods and Brigham 2008; Gerrish et al. 2009; Penteriani et al. 2010 and 2011; Smit et al. 2011). Spatial gradients in the amount and seasonal distribution of biologically useful semi-darkness (including moonlight and twilight) have been proposed as drivers of patterns of behavior (Mills 2008). Visually orienting predators have a reduced ability to detect prey in dark conditions, and may increase their activity or achieve higher rates of predation success under lighter conditions; prey species may reduce activity in lighter conditions in response to a perceived increased risk of predation. Some shorebird species use visual foraging by day but tactile foraging during hours of darkness—nighttime light may allow them to use visual foraging throughout the night (Rojas et al. 1999).

Moonlight-driven cycles in predator-prey activity have been observed in such taxonomically diverse species as zooplankton and fish (Gliwicz 1986), predaceous arthropods (Tigar and Osborne 1999), blue petrels *Halobaena caerulea* and brown skuas *Catharacta skua* (Mougeot and Bretagnolle 2000), owls and rodents (Clarke 1983), and lions *Panthera leo* and humans (Packer et al. 2011). Prey species may respond to the increased risk of predation at night by decreasing their activity (e.g., Kotler 1984; Daly et al. 1992; Vásquez 1994; Skutelsky 1996; Kramer and Birney 2001) or changing their microhabitat to utilize dark spaces such as the shelter of bushes (e.g., Price, Waser, and Bass 1984; Kolb 1992; Topping, Millar, and Goddard 1999), and may compensate by greater activity at dawn and/or dusk; Daly et al. (1992) have shown how such "crepuscular compensation" in response to high nocturnal predation rates can lead to increasing rates of predation by diurnal predators as prey activity encroaches into daylight hours. Diurnal and crepuscular predators may become facultative nocturnal predators under suitable light conditions (e.g., Milson 1984; Combreau and Launay 1996; Perry and Fisher 2006). Conversely, nocturnal predators that rely on non-visual clues to hunt, such as snakes, may decrease activity during lighter nights in order to avoid detection by prey and their own predators (Bouskila 1995; Clarke, Chopko, and Mackessy 1996). Behavioral changes are likely to induce changes in energetic costs; Smit et al. (2011) have shown that freckled nightjars *Caprimulgus tristigma* respond to dark nights by entering torpor, while moonlit nights allow foraging as food availability is sufficient to overcome the energetic costs of thermoregulation.

Despite the large number of studies that demonstrate the effect of moonlight in altering the behavior of species, there have been relatively few that have formally examined the effect of artificial light in altering behavior or restructuring temporal niche partitioning. Reports have long existed that some diurnal species exploit the "night-light niche" and become facultatively nocturnal in urban environments, for example jumping spiders (Wolff 1982; K. Frank 2009), reptiles (Garber 1978; Perry and Fisher 2006), and birds (Martin 1990; Negro et al. 2000; Santos et al. 2010). In rodents, Bird, Branch, and Miller (2004) have shown that foraging behavior in beach mice *Peromyscus polionotus* is restricted by artificial lighting, while Rotics, Dayan, and Kronfeld-Schor (2011) have shown that while the nocturnal spiny mouse species *Acomys cahirinus* restricted activity under artificial light, its diurnal congener *Acomys cahirinus* did not expand its activity to compete during the hours of artificial illumination.

There are few known examples of artificial light as a resource directly mediating behavior; although some species have been found to increase foraging activities and antipredator vigilance under such conditions (e.g., Biebouw and Blumstein 2003), the vision of some nocturnal predators has been shown to be impaired by artificial lighting and their foraging success reduced (e.g., Buchanan 1993). Reports of the effects of light in providing resources by attracting concentrations of prey are more frequent (e.g., Heiling 1999; Buchanan 2006). Increased foraging around streetlights has been widely reported for some species of bats (e.g., Rydell 1991, 1992, 2006; Blake et al. 1994; Polak et al. 2011), particularly around lamps which emit at low wavelengths, attract large numbers of insects, and which may interfere with prey defenses (Svensson and Rydell 1998); Rydell (2006) regards the habit of feeding around lights by bats as having become the norm for many species. Other bat species avoid lights (Kuijper et al. 2008; Stone, Jones, and Harris 2009), possibly to minimize the risk of avian predation (Speakman 1991; Rydell, Entwistle, and Racey 1996). Similarly, nocturnal orb-web spiders *Larinioides sclopetarius* preferentially build webs in areas which are well lit at night, where higher densities of insect prey are available; a behavior that appears to be genetically predetermined rather than learned (Heiling 1999). This suggests the possibility of evolutionary responses to utilize novel niches created by artificial lighting.

The relative lengths of night and day can influence foraging opportunities, predation and/or competition costs and the trade-offs among these (e.g.,

Clarke 1983; Falkenberg and Clarke 1998; Berger and Gotthard 2008). In turn this can influence the abundances of organisms (e.g., Carrascal, Santos, and Tellería 2012). Presumably nighttime lighting that served effectively to change perceived night and day lengths could amplify these effects.

Dark Repair and Recovery

It has been suggested that continuous periods of darkness are critical for certain processes controlling repair and recovery of physiological function in many species, and hence that darkness can be seen as a resource for physiological activity. Seeking an explanation for an observed increase in ozone injury in plants at high latitudes, Vollsnes et al. (2009) have shown that dim nocturnal light, simulating the northern Arctic summer, inhibits recovery from leaf damage caused by atmospheric ozone in subterranean clover *Trifolium subterraneum*. Futsaether et al. (2009) found a similar result in red clover *Trifolium pretense* but not in white clover *Trifolium repens*. In *Arabidopsis thaliana*, Queval et al. (2007) have shown links between day length and the rate of oxidative cell death. Since the patterns of anthropogenic light pollution and ozone pollution are spatially correlated on a global scale (see e.g. Cinzano et al. 2001; Ashmore 2005), the extent to which low-intensity nighttime light could affect repair and recovery from ozone damage requires further investigation.

Gerrish et al. (2009) argued that hours of darkness provide organisms with time for repair to DNA damage to cells caused by solar UV-B radiation (285–315 nm). However, light in the blue to UV-A portion of the spectrum is necessary for DNA repair through photoreactivation *via* the photolyase enzyme (with maximum absorption at 380 and 440 nm), while "dark repair" through the excision repair pathways is independent of light (B. Sutherland 1981; Britt 1996; Sinha and Hader 2002). The role of darkness here is presumably limited to the lack of damage due to solar UV-B radiation during the night. Since artificial lighting typically emits negligible amounts of UV-B radiation it is unlikely that light pollution either increases DNA damage or inhibits the processes of repair in this instance; indeed, light sources emitting in the blue and UV-A may have an effect in promoting DNA repair through photoreactivation.

LIGHT AS AN INFORMATION SOURCE

The direction, duration and spectral characteristics of natural light are widely used by organisms as sources of information about their location, the time of day and year, and the characteristics of their natural environment (Neff et al. 2000; Ragni and D'Alcala 2004). Indeed, considerable energetic costs are often borne in order to maintain the necessary sensory systems (Niven and Laughlin 2008). Artificial lighting can disrupt this flow of information and provide misleading cues. The wavelengths of light are critical to its efficacy as an information source due to the varying spectral sensitivity of organisms' receptors. In vascular plants, for example, the most well-studied photorecep-tors are phytochromes, which exist in two photo-interconvertible forms—a biologically inactive red-light-absorbing form (Pr) which upon absorption of red light is converted to a biologically active form (Pfr). Pfr is converted back to Pr on absorbing far-red photons, so under steady light of a given red/far-red ratio the active form of phytochrome reaches equilibrium (Lin 2000; Neff et al. 2000; H. Smith 2000). The phytochrome system plays a key role in detecting shade and measuring day length, and has been shown to influence vegetative growth and architecture, the timing of germination, flowering, bud burst and dormancy and senescence, and the allocation of resources to roots, stems and leaves (H. Smith 2000). In addition, blue and ultra-violet light receptors called cryptochromes influence light responses in many species of algae, higher plants, and animals (Cashmore et al. 1999), and have been shown to play a role in regulating circadian clocks in mam-mals (Thresher et al. 1998). In animals with vision, complex information on the spectral composition of light may be derived from several photorecep-tors with varying spectral sensitivities (Kelber, Vorobyev, and Osario 2003), and in mammals retinal ganglion cells that are independent of the visual system may be involved in entraining circadian clocks (Berson, Dunn, and Takao 2002). In many cases organisms have been shown to be sensitive to extremely low levels of light at night, well within levels of anthropogenic light pollution (Kelber and Roth 2006; Bachleitner et al. 2007; J. Evans et al. 2007; D. Frank, Evans, and Gorman 2010).

Circadian Clocks and Photoperiodism

Three natural periodic cycles in the light regime are detected by organ-isms—the daily cycle of day and night, seasonal changes in day length, and the monthly lunar cycle. The daily and seasonal cycles in particular provide cues that can be used to anticipate regular changes in the environment such as temperature or humidity that also follow a daily or annual cycle. The lunar cycle has importance for activity and reproduction in some species, which may be responding directly to the availability of light as a resource (see section "Light as a Resource" above) alternatively they may utilize the lunar light cycle to anticipate environmental changes connected with nighttime light or tidal conditions (M. Taylor et al. 1979), or purely as a regular cue to synchronise reproductive activity (e.g., Tanner 1996; G. Baker and Dekker 2000; Takemura et al. 2006).

Light may influence circadian patterns of behavior in two ways, entrainment and masking, which may be difficult to distinguish in natural systems. Virtually all plants and animals possess a circadian clock, an endogenous system that regulates aspects of their activity and physiology on a cycle that approximates 24 h, but which in the absence of external cues may drift out of phase with day and night (Sweeney 1963). In order for the clock accurately to track the diurnal cycle, it is regulated by "zeitgebers," environmental cues that entrain or reset the clock. The light environment is critical in providing such cues in many species. Entrainment occurs when regular patterns of light and darkness regulate the phase and frequency of the endogenous clock (Menaker 1968). Artificial light after dusk or prior to dawn can cause phase shifts in the circadian rhythm, delaying or advancing the cycle. Low levels of light at night may disrupt melatonin production in fish, birds and mammals, with a wide range of downstream physiological consequences (Navara and Nelson 2007; for example see Cos et al. 2006; J. Evans et al. 2007; Reiter et al. 2007; Bedrosian et al. 2011a and b). Since light pollution typically occurs both before dawn and after dusk, it is difficult to predict the effect of any shift in the circadian clock. In laboratory experiments, entrainment has been shown to occur at both persistent levels of low light and with short pulses of relatively bright light (Brainard et al. 1983; Haim et al. 2005; Zubidat, Ben-Shlomo, and Haim 2007; Shuboni and Yan 2010). The duration and intensity of light required to disrupt circadian rhythms under field conditions is unknown, but these studies suggest potential for impacts on species affected by widespread low-level light such as urban sky glow or less often considered transient lighting sources such as vehicle lights (Lyytimaki, Tapio, and Assmuth 2012).

Exposure to light at night has been shown to disrupt the circadian cycle of hormone production in humans, particularly melatonin, which has been linked to an increase in cancer risk in shift-workers (Stevens 1987, 2009; Megdal et al. 2005; Reiter et al. 2011). Melatonin production is regulated by the circadian clock, which in mammals is entrained by retinal ganglion cells with a peak sensitivity in blue light at around 484 nm (Berson et al. 2002). Melatonin production is similarly reduced in rats under nighttime light levels of 0.2 lux (Dauchy et al. 1997), and in hamsters at levels above 1 lux (Brainard et al. 1982), and has been shown to suppress immune responses and increase the rate of tumor growth (Dauchy et al. 1997; Bedrosian et al. 2011b). Similar melatonin-mediated effects of nighttime light on immune function are seen in laboratory studies of birds (C. B. Moore and Siopes 2000). The requirement for continuous periods of darkness to entrain the circadian clock and regulate hormone activity may be widespread among animals, yet the ecological effects of potential disruption of the circadian clock are unknown.

By contrast, masking occurs when a light stimulus overrides the endogenous clock; for example artificial light at night may increase activity in diurnal or crepuscular species (positive masking) or suppress it in others (negative masking; see e.g., Santos et al. 2010; Rotics et al. 2011). The ecological effects of direct entrainment of circadian clocks by artificial light may be difficult to distinguish from opportunistic changes in light-resource use or direct effects of light on behavior through masking. For example, light pollution has been shown to advance the initiation of dawn singing considerably in some temperate bird species in urban areas (M. Miller 2006), with implications for breeding success (Kempenaers et al. 2010). The extent to which this effect of light on behavior is mediated by circadian rhythms, or whether light triggers this behavior independently of an endogenous clock through masking is unknown.

In temperate and polar ecosystems, organisms frequently use day length as a cue to initiate such seasonal phenological events as germination, bud formation and burst, reproduction, senescence, eclosion, diapause, molt, embryonic development, and migration (e.g., Gwinner 1977; Densmore 1997; Dawson et al. 2001; Niva and Takeda 2003; Heide 2006; Cooper et al. 2011). By contrast, species whose ranges are restricted to lower latitudes are likely to be less dependent on day length to regulate annual cycles of activity (although in dry seasonal climates near the equator even very small differences in seasonal day length can be utilized by plants to trigger phenological events; see Rivera et al. 2002). Over evolutionary time species have adapted to wide variation in the range of day length that they encounter—in the Permian period deciduous forests existed in Antarctica at latitudes of 80–85°S, experiencing total darkness for months in the winter and 24 h daylight during summer, a light environment without analogue in modern forests and unlikely to be within the survivable range of extant tree phenotypes (E. Taylor, Taylor, and Cuneo 1992). Photoperiod, and therefore presumably changes in what is perceived as photoperiod as a result of artificial lighting, has consequences for a variety of physiological traits. It has long been observed that certain species of deciduous tree maintain their leaves for longer in autumn in the vicinity of streetlights (Matzke 1936), potentially leaving them exposed to higher rates of frost damage in late autumn and winter. Experiments in horticultural systems have shown a wide range of responses to artificial nighttime lighting, depending both on the species and the spectral composition of the light source, including delay and promotion of flowering, and enhanced vegetative growth (Cathey and Campbell 1975; Kristiansen 1988). Animal species, including lizards (*Sceloporus occidentalis*; Lashbrook and Livezey 1970) and rodents (*Microtus socialis*; Zubidat et al. 2007) control their thermoregulatory activity in response to seasonal changes in photoperiod. Plant physiologists draw a distinction between "long-day" responses, in which a long dark period suppresses an effect, and "short day" responses, in which a long dark period promotes an effect. In animals, both day length and the relative change in day length may act as proximal triggers (Vepsäläinen 1974). Species with a wide latitudinal range show local adaptation in their photoperiodic response (Bradshaw 1976), and photoperiodic control allows species to coordinate key events in their life cycle with suitable weather conditions. Photoperiodic response has been shown to evolve rapidly in an invasive species expanding into different latitudes, reflecting changing relationships between the seasonal climate and the information given by day-length cues (Urbanski et al. 2012). Disruption of this control may lead to organisms becoming out of step with their climate, with the timing of other organisms (such as pollinators or food sources), or unable to adapt to climatic change (Bradshaw, Zani, and Holzapfel 2004; Bradshaw and Holzapfel 2010).

The biological rhythms of organisms are known to be linked across different levels of food webs, with, for example, plant-herbivore-parasitoid rhythms being synchronized both as a consequence of bottom-up and top-down processes (S. Zhang et al. 2010). This raises the likelihood that disruptions to the rhythms of individual species by nighttime lighting can ramify widely.

Visual Perception

A wide range of adaptations exists throughout the animal kingdom to make use of reflected light at different levels and wavelengths, allowing the recognition of important features of the environment (Land and Nilsson 2002; Warrant 2004; Warrant and Dacke 2011); discoveries about the breadth of the abilities of organisms in this regard continue to be made (e.g., Kelber, Balkenius, and Warrant 2002; Grémillet et al. 2005; Allen et al. 2010; Baird et al. 2011; Hogg et al. 2011). A substantial proportion of animal species are adapted to see at light levels well below those at which human vision is effective, in which they can often see color and navigate well (Warrant 2004;

Warrant and Dacke 2010, 2011). The interaction between the intensity and spectral composition of artificial light and the adaptation of an organism's eyes will affect whether visual perception is enhanced, disrupted or unaffected by light pollution, and hence the potential downstream behavioral and ecological effects.

The intensity of light at which animals are able to identify objects varies considerably among species. Many are able successfully to navigate visually and locate resources at light levels at which human vision is impossible (e.g., Dice 1945; Larsen and Pedersen 1982). A considerable proportion of nocturnal activity occurs during periods of "biologically useful semi-darkness" (Mills 2008), making use of the relatively low light intensities during twilight and moonlight; however, nocturnal species may also modify or reduce activity during such periods to avoid competition or predation (Clarke et al. 1996). Light intensities recorded from artificial sources, from both direct illumination a considerable distance from a source and diffuse sky glow, are well within the range shown to be effective in enhancing animal vision and triggering behavioral changes. Less well known is the extent to which artificial nighttime light may disrupt vision systems adapted to dark conditions.

The light-sensitive photoreceptor pigments of animal eyes vary in the wavelengths of light to which they are most absorbent. Color is perceived as a representation in a limited number of dimensions of the multi-dimensional spectral reflectance of an illuminated surface, and the information content of color perception varies as a function of the number and spectral sensitivity of different types of photoreceptor pigments. The human eye contains three photoreceptors (trichromatic) that are used in photopic (daytime) vision and maximally absorb light at wavelengths of 558 (red), 531 (green) or 419 nm (blue) (Dartnall, Bowmaker, and Mollon 1983). Reptiles and birds commonly possess four photoreceptor pigment types, increasing the information content of color perception across much of the spectrum [including ultraviolet (uv) light] compared to the majority of mammals which possess two photoreceptor pigment types (Osorio and Vorobyev 2008). The mantis shrimp *Odonatodactlyus* represents an extreme case of color sensitivity, with 12 photoreceptor pigment types (Marshall and Oberwinkler 1999). Large numbers of types potentially allow organisms better to discriminate between objects of contrasting spectral reflectance in their environment, and the relative distribution of photoreceptor sensitivities determines the portions of the electromagnetic spectrum in which color vision is most sensitive.

Changing the spectral properties of artificial lights is therefore likely to alter the environment which individual organisms are able to see in different ways. Broader spectrum light sources such as light-emitting diodes (LEDs) are often likely to provide improved color discrimination. This may allow animals better to navigate, forage for resources, locate and catch their prey, and identify or display for mating (such as in the plumage feathers of birds; Hart and Hunt 2007). The trichromatic and tetrachromatic visual systems of many hymenopteran and lepidopteran insects allow them to recognize and compare between the nectar sources provided by flowering plants (Chittka and Menzel 1992). The color of a flowering plant as perceived by an insect, and the ease with which the insect can recognize different flowers, are likely to be improved under broad-spectrum compared to narrow-spectrum lighting conditions. Changing the spectral composition of artificial light could therefore affect the competitive fitness of animals in a variety of ways. Given the current shift in lighting technology towards broader spectrum light sources, future research into the impact of different artificial light sources on the recognition of important environmental signals by animal groups is clearly necessary.

Spatial Orientation and Light Environment

Many organisms use lightscapes as cues for directional movement (Tuxbury and Salmon 2005; Ugolini et al. 2005; Warrant and Dacke 2011). The restructuring of these lightscapes by light pollution can thus result in these movements being disrupted. Examples of such disruption have been documented for moths and other insects (e.g., K. Frank 1988), frogs (B. Baker and Richardson 2006), reptiles (e.g., Salmon et al. 1995), birds (e.g., Gauthreaux et al. 2006; Rodríguez et al. 2012), and mammals (Beier 1995; Rydell 2006).

The widespread attraction of moth species to nighttime lights has long been exploited in the design of traps for their capture. The reasons for such disruption of their natural movement patterns remain to be fully determined, although interference with the use of moonlight for navigation is likely important (Warrant and Dacke 2011). Many insects, including members of the Hymenoptera, Lepidoptera and Coleoptera, can navigate using the pattern of polarized celestial light in the sky (e.g., Dacke et al. 2003). The use of UV light as opposed to other wavelengths to detect polarized light patterns has been postulated to be advantageous because the degree of polarized light scattered downwards from clouds and forest canopies is higher in the UV (Barta and Horvath 2004). The natural signal is diminished by urban sky glow (Kyba et al. 2011b), and through this effect variation in sky glow may potentially explain geographic differences in the response of moth-trap catches to phases of the moon (Nowinszky and Puskás 2010). Whether flight-to-light behavior is driven by the disruption of natural polarized light patterns alone seems unlikely as this behavior occurs even with artificial lights which emit no uv component (van Langevelde et al. 2011). However, the use of polarized uv light detection for navigation by insects may explain why flight-to-light behavior is disproportionately associated with emissions at shorter wavelengths (van Langevelde et al. 2011). Polarized light patterns reflected back from the ground can also be used to locate water bodies due to the polarizing nature of their surfaces. Indeed, a number of cases exist where insects have been attracted to sources of polarized light reflected back from anthropogenic structures such as wet asphalt roads, leading to increasing concern over the deleterious effects of these and other light polarizing anthropogenic structures (Horvath et al. 2009). It seems likely that such effects may be exacerbated by the introduction of artificial lighting, although ecological case studies have not to our knowledge been documented.

Beetles of the family Lampyridae are notable for their use of bioluminescence in mate location. It is possible that artificial light is playing a significant role in the decline of these taxa, due to disruption of mate location (Lloyd 2006).

Migrating birds utilize at least two mechanisms for navigation that may be disrupted by artificial lighting. Magnetoreception is considered to be the principal mode of orientation. The detection systems for magnetoreception include the magnetic-field-dependent orientation of paired radical molecules in the photopigment that forms during photon absorption, and the presence of magnetite within the beak (R. Wiltschko et al. 2010). Migration direction has been demonstrated to be determined using the blue and green photoreceptors in European robins *Erithacus rubecula* (R. Wiltschko et al. 2007), while red light disrupts migration direction in silvereyes *Zosterops l. lateralis* (W. Wiltschko et al. 1993). This has led to calls for the spectral composition of artificial lighting to be managed to mitigate against disorientation of birds (Poot et al. 2008), however the level of disorientation caused by particular wavelengths of light appears to vary according to intensity, and is not restricted to red lights alone (R. Wiltschko et al. 2010).

In addition to possessing a magnetic compass for orientation, birds are also thought to calibrate this compass using celestial light during twilight or at night (Cochran, Mouritsen, and Wikelski 2004). In some species the mechanism of calibration has been demonstrated to be the detection of polarized light patterns during sunrise and sunset (e.g., Muheim, Phillips and Akesson 2006). However, as is the case with insects, whether artificial lighting can affect these patterns, and the consequences this may have for navigation, are currently unknown.

In addition to the above examples of movement towards light, many motile organisms exhibit light-avoidance behaviors (e.g., M. Moore et al. 2000; Buchanan 2006; Boscarino et al. 2009). It seems extremely likely that for many such taxa the avoidance of artificial illumination will result in reduction in the space and other resources available to them (e.g., Kuijper et al. 2008). One of the most ecologically significant consequences of negative phototropic behavior is the widespread diel migration of zooplankton in aquatic systems (e.g., M. Moore et al. 2000) which would appear to be sensitive to levels of light oscillation well below those produced by artificial illumination (Berge et al. 2009).

CONCLUSIONS

As human communities and lighting technologies develop, artificial light increasingly encroaches on dark refuges in space, in time, and across wavelengths. At a given latitude, natural light regimes have been relatively consistent through recent evolutionary time, and the global rapid growth in artificial light represents a potentially significant perturbation to the natural cycles of light and darkness. Natural light is utilized by organisms both as a resource and as a source of information about their environment, and artificial light has the potential to disrupt the utilization of resources and flow of information in ecosystems.

A broad set of case studies of ecological implications of light pollution has been documented. Across a wide range of species, there is evidence that artificial light affects processes including primary productivity, partitioning of the temporal niche, repair and recovery of physiological function, measurement of time through interference with the detection of circadian, lunar and seasonal cycles, detection of resources and natural enemies and navigation. However, the effects on population- or ecosystem-level processes, such as mortality, fecundity, community productivity, species composition and trophic interactions are poorly known. Furthermore, the studies identifying these processes to date are scattered within literature from a wide range of disciplines, are strongly weighted towards higher vertebrates and ecosystems and largely lack synthesis within a common mechanistic framework.

We propose a framework that focuses foremost on the interactions between the ways in which artificial lighting alters natural light regimes (spatially, temporally, and spectrally), and the mechanisms by which light influences biological systems, particularly the distinction between light as a resource and light as an information source. Such a framework focuses attention on the need to identify general principles that apply across species and ecosystems, and integrates understanding of physiological mechanisms with their ecological consequences.

Reviewing the evidence for each of the combinations of this cross-factoring particularly highlights: (*i*) the potential influence of nighttime lighting at all levels of biological organization (from cell to ecosystem); (*ii*) the significant impact that even low levels of nighttime light pollution can have; and (*iii*) the existence of major research gaps in understanding of the ecological impacts of light pollution.

Future research on the ecological impacts of light pollution needs to address several key issues: (*i*) to what extent does the disruption of natural light regimes by artificial light influence population and ecosystem processes, such as mortality and fecundity rates, species composition and trophic structure; (*ii*) what are the thresholds of light intensity and duration at different wavelengths above which artificial lighting has significant ecological impacts; and (*iii*) how large do "dark refuges," where the intensity and/or duration of artificial light falls below such thresholds, need to be to maintain natural ecosystem processes?

12

REDUCED FLIGHT-TO-LIGHT BEHAVIOR OF MOTH POPULATIONS EXPOSED TO LONG-TERM URBAN LIGHT POLLUTION

Florian Altermatt and Dieter Ebert

This chapter examines the evolutionary consequences of light pollution on moth populations. Moths from urban areas that have had high, globally relevant levels of light pollution for several decades show a significantly reduced flight-to-light behavior compared with populations of the same species from pristine dark-sky habitats. As nocturnal insects are of great significance as pollinators and the primary food source of many vertebrates, an evolutionary change of the flight-to-light behavior could cascade across species interaction networks.

BACKGROUND

Anthropogenic light pollution is recognized as a global environmental change negatively affecting humans (Falchi et al. 2011), animals (Hoelker et al. 2010b; Longcore and Rich 2004) and microbes (Hoelker et al. 2010b). The negative effects are well-known and even captured in the Shakespearian saying "Thus hath the candle singed the moth," referring to someone tempted by something that eventually will lead to his downfall. As a global phenomenon (Bennie et al. 2015; Cinzano et al. 2001), light pollution is a by-product of urbanization (Cinzano et al. 2001) and interferes with the natural day-night cycle, resulting in changed phenologies of organisms in the polluted habitat (Gaston et al. 2013). An even more direct effect of light pollution is the widely observed attraction of organisms ranging from migrating birds to insects to the light sources, which can result in increased mortality through direct burning or increased exposure to predators (Jones and Francis 2003; A. Warren 1990). Furthermore, attraction to light may lure organisms out of their native habitat and interfere with normal feeding behavior, mating and reproduction (Longcore and Rich 2004). The negative effects of light pollution on community composition (T. Davies et al. 2012; Hoelker et al. 2015), population dynamics and organisms' Darwinian fitness (Gaston et al. 2014) have been globally documented, and mortality at artificial light sources can be 40- to 100-fold higher than in dark-sky populations (Jones and Francis 2003; Longcore and Rich 2004).

Surprisingly, however, little is known about potential evolutionary consequences. Existing data show that there is a high variability among species and sexes to be attracted by light (Altermatt et al. 2009; R. Baker and Sedovy 1978), and anecdotal evidence from naturalists even indicates within-species variation (K. Frank 1988). Given the detrimental effect of the flight-to-light behavior in polluted areas it has been postulated that natural selection should favor individuals with a lower propensity of being attracted by light (Altermatt et al. 2009; Gaston et al. 2013).

Organisms' phenotypic and genetic responses to anthropogenic environmental changes are well known, such as the classic example of industrial melanism of the peppered moth. However, possible selective responses to light pollution have not yet been documented, even though artificial night lighting has strongly increased over the last decades (Longcore and Rich 2004), has a pronounced geographical variation (Cinzano et al. 2001), and is having increasingly negative effects on many species (Longcore and Rich 2004), creating the necessary conditions for directional selection and adaptation to a novel anthropogenic stressor. Here, we report an experimental quantification of a reduced flight-to-light behavior of small ermine moths from populations occurring in areas with high light pollution compared with populations with low pollution.

METHODS

Study Organism and Study Sites

We used the small ermine moth (*Yponomeuta cagnagella*), which is widely distributed across Europe. Its larvae feed strictly on European spindle (*Euonymus europaeus*) (Menken et al. 1992). Larvae live in loose silk-webs, forming gregarious full-sib groups on their host plants. In order to test for an effect of multi-generational exposure to artificial light pollution on flight-to-light behavior, we collected larvae of the small ermine moths in areas with low-to-medium levels of light pollution (less than 3×10^{-9} W sr^{-1} cm^{-2}; subsequently called dark-sky populations) and in areas which have been exposed to strong light pollution for several decades (current light pollution 20 to more than 40×10^{-9} W sr^1 cm^{-2}; subsequently called light-polluted populations, using data from the Earth Observation Group (2017) (Cinzano et al. 2001). The latter populations were always sampled in the immediate vicinity of street-lights in urban parks or along streets (maximum distance to light source 20–50 m).

Laboratory Conditions

We collected second-instar larvae, such that organisms were independently sampled with respect to the subsequent experimental test method, in 10 independent populations from locations in northwestern Switzerland and in eastern France in spring 2007. These populations were not connected by suitable habitat and represented independent units. At each location, larvae from multiple full-sib family clutches were collected. Some of these populations had been used in a previous analysis on sex-specific flight-to-light behavior (Altermatt et al. 2009). Larvae were reared in the laboratory under standardized, common garden conditions in 500 ml plastic boxes at 80% relative humidity and a daily light-dark cycle of L16:D8 h, including 1 h dusk and dawn phases. The temperature was 23° C during the light phase, and 19° C during the dark phase. Food was provided ad libitum.

Flight-to-Light Experiment

We experimentally tested the flight-to-light behavior of small ermine moths in an indoor flight cage, measuring 5.7 × 2.5 m at the bottom, 5.7 × 1.8 m at the top and of 3 m height. In this cage, moths can be attracted from every position, as the effective range of the light traps used is about 3–5 m (R. Baker and Sadovy 1978). Temperature and humidity in the flight cage were held constant at 20–22° C and 60–80% relative humidity, respectively.

A standard Heath trap equipped with a fluorescent tube (Philips, Eindhoven, the Netherlands, TL 6 W/05) was placed at the long-end side of the flight cage. Previous to the experiment, adult moths were individually marked and kept at L16:D8 cycle such that they adapted physiologically to the day-to-night change. For the flight-to-light experiment, we used individuals 2–3 days post-eclosion before they were sexually mature. We thereby pooled individuals from different families and populations of origin. Moths were released at the opposite end to the lamp, and the tests started at the onset of dusk. The species flies generally at night and is not only crepuscular. We recorded the number and identity of all moths captured by the light trap and all moths not attracted after 8 h.

Analysis

We used a generalized linear mixed effect model to test for a difference in flight-to-light behavior of moths originating from light-polluted areas versus dark-sky areas. As a response variable, we used the odds ratio of individuals attracted versus individuals not attracted, and a model with a binomial error distribution. Light pollution, sex and their interaction were included as fixed effects in the model, whereas populations and family nested within populations were included as random effects. Analyses were done in R v. 3.0.1 (Core Development Team 2013).

RESULTS

We tested the flight-to-light behavior of 1,048 adult moths. Individuals from the populations experiencing high light pollution had a significantly lower propensity to fly to light (Z-value = -2.1, $p = 0.036$; Figure 12.1) than individuals from the populations from dark-sky areas. Overall, the mean reduction in flight-to-light behavior was 30% (36% reduction for males and 28% reduction for females). Furthermore, females were overall significantly less attracted by light (Z-value = 4.3, $p < 0.001$, see also Altermatt et al. 2009). There was no significant interaction between population of origin and sex on the flight-to-light behavior, and the interaction term was removed from the model. The sex ratio within populations did not deviate from 1:1 (mean + or - s.e. of females relative to males was 1.04 + or - 0.11, $p > 0.8$).

DISCUSSION

We experimentally showed that small ermine moths from populations exposed for many generations to high levels of light pollution common to urban and suburban areas have a significantly reduced propensity to fly to light compared with moths from dark-sky populations. As moths and other insects attracted to artificial light are experiencing high levels of mortality and reduced fitness (Hoelker et al. 2015; Jones and Francis 2003; A. Warren 1990), we interpret this as evidence for an adaptive response following Darwinian fitness selection caused by human-induced change in environmental conditions. We cannot completely exclude that some of

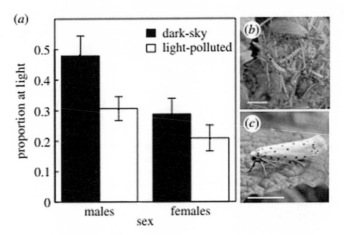

FIG 12.1: Flight-to-light behavior of small ermine moths from dark-sky and light-polluted populations. (a) Proportion (mean ± s.e. across all populations) of small ermine moths attracted by artificial light under controlled experimental conditions. Overall, moths from populations which have been experiencing extensive light pollution are significantly less attracted to the light than moths from populations without light pollution (dark sky; $p = 0.036$). (b) Larvae of the small ermine moth (*Yponomeuta cagnagella*) on its host plant European spindle (*Euonymus europaeus*). (c) Adult small ermine moth. Scale bars, 5mm.

the observed effect was due to developmental differences based on the site of origin affecting adult behavior or other selective agents associated with urbanization. However, as we collected the larvae at a very early stage (second instar) and raised them in a common garden, we think this is less likely than a common selective difference with respect to light pollution linked to urbanization.

The implications of our finding are important and manifold. First, a reduced flight-to-light behavior may release natural populations from some of the negative consequences of artificial light pollution (Hoelker et al. 2015). Specifically, there may be less direct mortality (Jones and Francis 2003; K. Frank 1988) and lower predation risk (T. Davies et al. 2012; A. Warren 1990). Second, the lower propensity for flight-to-light in city populations suggests that the original function of this behavior is in fact not essential for survival in urban landscapes. This might be because the adaptive value of this behavior is generally weaker than the benefits of reducing this behavior in light-polluted environments. Alternatively, in illuminated urban environments, the need to exhibit flight-to-light behavior might be reduced. Third, a difference in the flight-to-light propensity between rural and urban populations may bias biodiversity surveys based on commonly used light traps (R. Baker and Sadovy 1978), in particular if populations of different species respond differently to changes in light condition.

Consistent with many other species, males are the more mobile sex in small ermine moths, and are thus also more often attracted to light (Altermatt et al. 2009). Of note, males from light polluted habitats exhibited an almost identical flight-to-light behavior to females from dark-sky habitats (Figure 12.1a). The reduced flight-to-light propensity, however, may come with a cost. Explanations for the reduced flight behavior of moths originating from light-polluted populations include that these moths have a behaviorally or physiologically reduced perception or reaction to the light (K. Frank 1988). For example, size of the eyes, light receptors in the eyes or information processing of perceived light might be altered. Alternatively, the reduced flight-to-light response may be caused by a decrease in overall mobility. Such a change could have far-reaching consequences, as a reduced mobility is known to negatively interact with habitat fragmentation and dispersal, to reduce flower visiting rates and pollination or take the moths out of the sky, and thus reduce prey availability for the nocturnal insectivores, such as bats or spiders.

Given the recent recognition of the pollution effects acting on individuals to ecosystems, there have been calls to reduce light pollution and streetlight intensity (Longcore and Rich 2004). While this is commendable, our demonstration of a presumably adaptive change in flight-to-light behavior in populations having been exposed to long-term light pollution suggests that the effects of light pollution on natural populations may not necessarily vanish immediately when the lights are turned off, and that light pollution has already resulted in systematic changes in organisms' behavior. This advocates for preventing light pollution, event at low levels, from spreading to areas that are so far unaffected, in order both to reduce the direct negative effects of light pollution as well as to prevent long-term evolutionary changes.

13

ARTIFICIAL NIGHT LIGHTING AFFECTS DAWN SONG, EXTRA-PAIR SIRING SUCCESS, AND LAY DATE IN SONGBIRDS

Bart Kempenaers, Pernilla Borgstroem, Peter Loes,
Emmi Schlicht, and Mihai Valcu

This chapter examines the effects of artificial night lighting on dawn song in five common forest-breeding songbirds. In four species, males near streetlights started singing significantly earlier at dawn than males elsewhere in the forest, and this effect was stronger in naturally earlier-singing species. The authors compared the reproductive behavior of blue tits breeding in edge territories with and without streetlights to that of blue tits breeding in central territories over a seven-year period. Under the influence of streetlights, females started egg laying on average 1.5 days earlier. These and additional findings suggest that light pollution has substantial effects on the timing of reproductive behavior and on individual mating patterns. Over the long term, there may be evolutionary consequences by changing the information embedded in previously reliable quality-indicator traits.

INTRODUCTION

Urbanization and related anthropogenic activities have a strong impact on ecosystems (Foley et al. 2005; Fischer and Lindenmayer 2007), particularly through habitat destruction and chemical, noise, and light pollution (Grimm et al. 2008). Studies in avian urban ecology have shown effects of urbanization on population dynamics and community composition (Marzluff et al. 2001; Devictor et al. 2007), and behavioral ecologists recently emphasized the impact of anthropogenic noise on avian communication (e.g., Slabbekoorn and Peet 2003) and species interactions (Francis et al. 2009b). In contrast to the general awareness of the consequences of noise pollution for terrestrial organisms (A. Chan et al. 2010a; Barber et al. 2010), the effects of artificial night lighting on natural populations have received much less attention (Navara and Nelson 2007), although it is becoming an emerging area of research (Rich and Longcore 2013). In this study, we examined the influence of light pollution on the reproductive behavior of a population of blue tits over seven breeding seasons. The study site has edges next to roads with and without streetlights. Comparing reproductive behavior of birds in edge territories with and without streetlights to that of birds in central territories allowed us to differentiate between edge effects per se and effects of artificial night lighting. We also made daily recordings of the dawn song of five common songbird species over a 19-day period in spring, comparing locations near and away from streetlights.

EFFECTS ON TIMING OF DAWN SONG

Males from four of the five species started singing significantly earlier in locations close to streetlights than in locations away from streetlights (Figure 13.1). A study on American robins (*Turdus migratorius*) also found that males began singing earlier in areas with high levels of artificial night lighting (M. Miller 2006), but another study suggested that this was due to a response to daytime noise, not night lighting (Fuller et al. 2007). It is perhaps not surprising that noise can have a stronger effect on singing than light (Fuller et al. 2007), because noise makes communication difficult, if not impossible. A response to daytime noise is an unlikely explanation for our results, however, because noise levels are very low throughout the study area. Furthermore, an effect of noise should be most pronounced for species in which males start singing close to the start of human activities, which is after dawn during the recording period. In fact, our results show the opposite pattern: the advance in the timing of dawn singing was stronger in species in which males generally started dawn song earlier relative to sunrise. This suggests that species that start singing long before dawn (e.g., the robin *Erithacus rubecula*) are more sensitive to light pollution.

In several previously studied species, including the willow tit *Poecile montanus* and the blackbird *Turdus merula*, the dawn song started earlier during peak female fertility (Sexton et al. 2007; Welling et al. 1995). For the majority of the singing individuals, we do not have information on their females' fertility, but the differences in start of dawn song between the locations with and without artificial night lighting were present over most of the recording period (Figure 13.1).

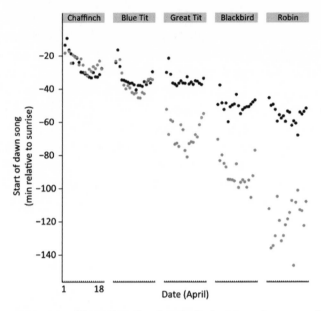

FIG 13.1: Effect of artificial night lighting on the start of the dawn chorus in five songbird species. Shown is the average start of dawn song relative to the time of sunrise on each day from March 31 until April 18 in territories affected by artificial light (light dots; $n = 6$) and in territories without artificial light (dark dots; $n = 6$).

EFFECTS ON TIMING OF REPRODUCTION

We compared the lay dates of female blue tits that bred on different territory types. Females from edge territories with streetlights started egg laying on average 1.5 days earlier than females that bred either in a neighboring territory or in a central forest territory. This was not simply an edge effect, because females in territories along the non-lighted edge did not start laying earlier. The effect was similarly strong when lay date was plotted against the distance from the nearest streetlight to the nest, whereby the effect decreased exponentially, as expected from the decreasing influence of light (generalized linear mixed model [GLMM], effect of log-transformed distance on lay date: 0.77 plus or minus 6 0.29, n = 321, z = 2.62, p = 0.009). Although female age has a strong effect on lay date, the effect of artificial night lighting was independent of female age, and territories near streetlights were not more likely to be occupied by adult females. Our findings are consistent with a study on captive blue tits that showed that females advanced their laying date when exposed to artificially extended photoperiods (Lambrechts et al. 1997). Note that although clutch size generally increased with earlier laying in our population (Foerster et al. 2003), females that bred on territories near streetlights did not lay more eggs compared to females breeding elsewhere, suggesting that they were not of higher quality or in better condition.

EFFECTS ON EXTRA-PAIR PATERNITY

Artificial night lighting had strong effects on patterns of extra-pair paternity. Males that occupied edge territories with streetlights were twice as successful in obtaining extra-pair mates (females with whom they sired at least one extra-pair offspring) than their close neighbors or males occupying central forest territories. Again, this was not an edge effect per se, because males in non-lighted edge territories were in fact less successful than other males in the population, probably because of a lower local breeding density at the edge of the study area. Similarly, male paternity gain decreased exponentially with the distance of the male's territory center to the nearest streetlight (GLMM, effect of log-transformed distance on number of extra-pair mates: -0.50 plus or minus 0.12, n = 321, z = -4.06, p < 0.0001). As shown previously, male age also had a strong effect on paternity gain: older males were more likely to sire extra-pair offspring, sired more extra-pair offspring (Delhey et al. 2006), and had more extra-pair partners. Our results indicate that artificial night lighting affected males of both age classes, but it had a stronger effect on the success of yearling males. Yearling males rarely sired extra-pair offspring when occupying territories that were not influenced by light. However, under the influence of artificial night lighting, they became almost as successful in obtaining extra-pair mates as adult males occupying nonlighted territories (Figure 13.2).

An alternative explanation for these results is that edge territories near streetlights were occupied by males of higher quality. However, our analyses do not support this. First, males in streetlight territories did not differ in age, size, or condition from males breeding elsewhere. Second, males in streetlight territories were equally likely to lose paternity in their brood as males breeding either in a neighboring territory or in a central forest territory. In general, males occupying edge territories lost less paternity than neighboring males, probably because they had fewer neighbors (local density effect). Third, females paired to males in streetlight territories did not lay larger clutches (though they started laying earlier) or produce more fledglings.

We propose that the observed increase in extra-pair success is caused by

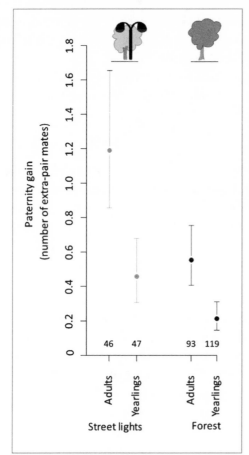

FIG 13.2: Effect of artificial night lighting on paternity gain. Paternity gain is the number of extra-pair females with whom an individual sired at least one offspring. Shown are data for yearling and adult males occupying edge territories with and without influence of artificial night lighting.

the effect of light on dawn song, which in turn influences female extra-pair behavior. A previous study showed that adult male blue tits started singing earlier at dawn than yearling males and that early-singing males had higher extra-pair success (Poesel et al. 2006). Our results suggest that night lighting induces males to sing earlier at dawn. Early dawn song may be a signal of male quality to which females are attracted for extra-pair copulations (Cockburn et al. 2009; Suter et al. 2009). This is supported by studies on dawn song in eastern kingbirds (*Tyrannus tyrannus*): early-singing males were larger and had longer flight feathers (Murphy et al. 2008), and early-singing, large males sired most extra-pair young (Dolan et al. 2007). Effects of light on singing behavior could be expected to extend to neighboring males through behavioral interactions. If so, this clearly did not translate into an effect on extra-pair behavior: males in territories next to night-lighted territories were not more successful at obtaining extra-pair mates than other males in the population.

DIRECT OR INDIRECT EFFECTS OF LIGHT

Our results suggest that under the influence of artificial night lighting, females laid eggs earlier during the season and males started dawn song earlier in the morning. Both of these effects may be a direct consequence of the exposure to night light. It is also possible that the effect on male behavior was influenced by the effect on female reproductive status, or vice versa. Night lighting may have caused earlier development of female gonads, and hence earlier laying (Lambrechts et al. 1997). For individual males, the start of dawn song is influenced by the reproductive status of the female, such that the earliest singing is observed on the days closest to the start of laying (Poesel et al. 2006). Males influenced by night light may thus sing earlier in the morning because of a shift in the timing of dawn song with date, such that the earliest dawn song occurs earlier during the season for these males. However, Figure 13.1 suggests that males in lighted territories started singing earlier over most of the recording period, and the results remain unchanged when variation in female reproductive state is partly removed by including for each day only the earliest singing male—presumably closest to the start of laying—among the six microphone locations (data not shown). Nevertheless, in future studies on the timing of male dawn song, it will be important to take the reproductive status of the female into account. An alternative scenario is that night lighting caused males to sing earlier at dawn, which in turn induced their females to lay earlier during the season. These alternative hypotheses could perhaps be tested experimentally by providing artificial night lighting to one sex but not to the other (e.g., night lighting inside the nest box with the roosting female). Finally, we note that males in lighted territories may have started performing dawn song earlier in the season, which may have influenced both their females' lay dates and their extra-pair success. Data on individual variation in the seasonal timing of dawn singing are lacking.

CONCLUSION

Our findings indicate that light pollution may have important long-term consequences by affecting the timing of reproductive behavior, individual mating patterns, and the information embedded in previously reliable quality-indicator traits. Artificial night lighting caused female blue tits to start egg laying earlier, which may lead to a mismatch between the time of peak food demand from the offspring in the nest and the peak in food availability (Lambrechts et al. 1997). Thus, artificial night lighting potentially leads to maladaptive timing of reproduction. Artificial night lighting also caused males to sing earlier and led to an increase in their extra-pair success. Earlier observations in blue tits and other passerines showed that females leave their territory early in the morning to perform extra-pair copulations (Cockburn et al. 2009; Kempenaers et al. 1992). Our analyses support the hypothesis that females target the earliest-singing males for extra-pair copulations, suggesting that the timing of dawn song is a quality indicator (Dolan et al. 2007). Artificial night lighting would then disrupt the link between quality and dawn song, making yearling males more attractive than they would otherwise be. Thus, light pollution potentially leads to maladaptive mate choice decisions of females with respect to extra-pair behavior, thereby altering selection pressures on mating behavior. Whether earlier singing, earlier laying, and altered extra-pair mating patterns come at a cost to the individuals involved remains to be shown, but using artificial light to experimentally manipulate these reproductive behaviors opens avenues for future research.

EXPERIMENTAL PROCEDURES

General Field Procedures

We studied a population of blue tits breeding in nest boxes in Kolbeterberg, Vienna (48°13' N, 16°14' E) between 1998 and 2004. We checked all nest boxes at least weekly during nest building, daily just before and during laying and close to hatching, and again at least weekly during the nestling stage. We caught the feeding parents in the box when the young were 8–10 days old, banded them with a metal band and three-color bands, and took measurements and a blood sample. We banded 14- to 15-dayold nestlings with a metal band and took a blood sample. Unhatched eggs and dead nestlings were collected to obtain a DNA sample. For the purpose of this study, we only included first breeding attempts in which the male identity was known. In total, we monitored 508 breeding attempts, caught 693 unique individuals (351 males and 342 females), and assigned paternity to 5,165 offspring.

Position and Intensity of Streetlights

Part of our study site borders streets in a quiet suburban residential area with streetlights with high-pressure sodium lamps that are turned on during the entire night. We measured the light intensity at night at ground level using a digital luminance meter (PCE-172; resolution: 0.01–100 lux, accuracy: plus or minus 5%), starting just below the streetlight and at regular distances into the forest. The amount of artificial light declined exponentially and was not detectable in the forest at distances above 50 m.

Territory Mapping and Assignment

To analyze the effect of artificial night lighting, we contrasted territories influenced by streetlights (defined as falling within 50 m of a streetlight) against neighboring territories under a natural light regime. Territories were estimated as Dirichlet tiles (e.g., Sibson 1980). Dirichlet tile construction and both polygon-polygon and point-polygon analyses were performed using the R 2.9.0 software system (R Core Development Team 2009) with the packages sp, spdep, maptools, and rgdal.

PARENTAGE ANALYSIS

We used five to eight polymorphic microsatellite markers to determine parentage of offspring, following standard procedures described in detail elsewhere (Delhey et al. 2006; Foerster et al. 2006).

STATISTICAL ANALYSIS

We used GLMMS with individual and nest box as random factors. We performed all analyses with the R 2.9.0 software system (R Core Development Team 2009), package lme4 (Bates et al. 2011), and multcomp (Hothorn et al. 2008) Valcu Valcu. To investigate the effects of artificial night lighting on paternity, we used two types of GLMM: (1) for paternity loss (dependent variable is the proportion of extra-pair young within the brood), we constructed models with a binomial error structure and a logit-link function, and (2) for paternity gain (dependent variable is the number of extra-pair females with whom a male sired at least one offspring), we constructed models with a Poisson error structure and a log-link function. For the analysis of breeding onset (lay date) and other phenotypic traits, we used GLMMS with a Gaussian error structure.

14

ARTIFICIAL LIGHT PUTS ECOSYSTEM SERVICES OF FRUGIVOROUS BATS AT RISK

Daniel Lewanzik and Christian Voigt

In this chapter, an experiment with captive bats suggests that food was less often explored and consumed in dimly illuminated areas than in total darkness, suggesting that artificial light alters the foraging behavior of fruit-eating bats. Additional observations in free-ranging bats found that they were less likely to harvest the fruiting heads of plants when the plants were illuminated by a street lamp rather than being in natural darkness. These findings have implications for biodiversity and natural succession in ecosystems affected by artificial light at night, via a reduction in bat activity and thus a reduction in nocturnal seed distribution in lit areas.

INTRODUCTION

Ecological light pollution, the alteration of the natural light and dark cycle by artificial light at night (Longcore and Rich 2004), has received increasing attention since it became evident that artificial light at night may be detrimental for many animals and ecosystem processes (reviewed in Rich and Longcore 2013) but continues to spread at unprecedented rates (Hoelker et al. 2010b). Obligatorily nocturnal animals such as bats are particularly prone to night lighting, since they may be exposed to artificial light during their entire activity period. Yet, light intensities as low as moonlight can potentially reduce the foraging behavior of bats (e.g., Morrison 1978; Fleming 1988).

So far, only a few experimental studies have addressed the effects of light pollution on bats and all of those dealt with insectivorous bats mainly in the temperate zone. These studies have shown that some species abandon traditional commuting routes when illuminated by either high-pressure sodium (orange) or light-emitting diode (LED; white) streetlights, which potentially deterred bats from reaching their preferred foraging habitat (Stone, Jones, and Harris 2009, 2012). *Eptesicus bottae* flew faster and ceased hunting insects when exposed to artificial light (Polak, et al. 2011) and obstacle avoidance capabilities of free-ranging *Myotis lucifugus* were altered by experimental illumination (Orbach and Fenton 2010). Only a very few insectivorous species were shown to make use of insect accumulations at artificial lights (e.g., Rydell 1991) though, in these instances, their foraging effort could be reduced significantly.

In the tropics, feeding habits of bats are much more diverse than in the temperate zone. Many tropical bats consume nectar and fruits, thus offering pollination and seed dispersal services to several hundreds of plant species (Ghanem and Voigt 2012). Next to birds, frugivorous bats constitute the most numerous seed-dispersing agent in the Neotropics where they are particularly important for the dispersal of seeds during the early stages of succession (Medellin and Gaona 1999; Muscarella and Fleming 2007). Due to this important role for ecosystem functioning, bats may represent a keystone taxon in the tropics (Willig et al. 2007).

In contrast to insectivorous bats, fruit-eating species do not benefit from foraging at lights and therefore should preferentially stay in dark areas to avoid being visibly exposed to predators (e.g., Fleming 1988). Accordingly, indirect evidence suggests that, for example, nectar- and fruit-eating lesser long-nosed bats *Leptonycteris curasoae* avoid lit areas (Lowery, Blackman, and Abbate 2009). Yet, since artificial light conditions were not experimentally altered in that study, it was not possible to determine whether this effect is due to artificial light at night or to some confounding factor of urbanization, such as altered vegetation cover and/or increased noise levels. Thus far, experimental evidence for light avoidance behavior of frugivorous bats is lacking, even though this feeding guild plays an essential role in the succession and maintenance of plant diversity especially in fragmented landscapes of the Neotropics (Muscarella and Fleming 2007). When human populations encroach in natural habitats, areas that were previously dark at night might become artificially illuminated, which may repel frugivorous bats. If these effective dispersal agents refrain from foraging in illuminated areas, artificial light at night may not only disrupt the habitats of light-sensitive species but also jeopardize the ecosystem services fruit-eating bats provide. This problem may become increasingly urgent in tropical countries with a prospering economy and an exponential growth of their human populations (Central Intelligence Agency 2011; United Nations Population Fund 2011).

Both growing economy and increased urbanization are known to correlate strongly with the degree of light pollution by street lamps (e.g., Elvidge et al. 2001).

We asked whether artificial light at night diminishes the harvesting activity of frugivorous bats at food plants and thus reduces the likelihood of seeds to be dispersed by bats. We focused on the effects of the widespread high-pressure sodium vapor light because high-intensity discharge lamps such as sodium lamps accounted for more than 80% of the global outdoor lighting market in 2010 (Baumgartner et al. 2011). Though the penetration rate of LED lights might increase, for example in Europe and North America during the forthcoming decades due to government initiatives, we believe that sodium lights will remain predominant in many developing countries of the tropics because they are cost efficient. Sodium lights have both low initial and low operating costs (Rea, Bullough, and Akashi 2009), and LEDS have not yet reached a competitive cost position (Baumgartner et al. 2011). Further, LED streetlights have been shown to repel several insectivorous bat species to a similar degree as high-pressure sodium lights (Stone, Jones, and Harris 2012). To test the effect of artificial light on the harvesting activity of bats, we conducted a binary choice experiment during which we simultaneously offered fruits to Sowell's short-tailed bats *Carollia sowelli* in a dark and in a dimly illuminated compartment of a flight cage. We used Sowell's short- tailed bats because they are the primary disperser of pepper seeds (genus *Piper*), a key plant group during early succession in the Neotropics (Muscarella and Fleming 2007). We expected *C. sowelli* to evade artificial light and consequently to use the dimly illuminated compartment less often and to harvest fewer fruits from it than from the dark compartment. To ascertain the relevance of our experiment for free-living populations, we also video-recorded the feeding activity of bats at individual ripe *Piper* infructescences under dark and illuminated conditions in the wild in order to test whether the light treatment reduced the removal rate of ripe infructescences.

MATERIALS AND METHODS

Experiments were conducted at La Selva Biological Station (Heredia Province, Costa Rica, 10°26'N, 83°59'W) in November–December 2011 and in March 2012. Monitoring of wild *Piper* plants was also conducted in November–December 2012. For both experiments, we used a custom-made street lamp to illuminate either one choice compartment or free-living *Piper* plants (see below). The lamp consisted of a high-pressure sodium light bulb which was covered by a translucent beaker glass and operated by an electronic control gear. The lamp was mounted at a height of 3–5 m on a pole and powered via a wall socket. The necessary voltage of 220 V was produced by a series of transformers. High-intensity discharge lamps such as high-pressure sodium lamps accounted for more than 80% of the global outdoor lighting market (Baumgartner et al. 2011) and are commonly used as street lamps across the world (enLighten. n.d.). The particular light bulb used was manufactured for the use in street lamps.

Choice Experiment

We captured bats in a Costa Rican lowland rain forest reserve (La Selva) by setting up 6-m and 9-m mist nets (height: 2–5 m, mesh: 16 9 16 mm; Ecotone, Gdynia, Poland) from dusk until at latest 2300 h. Bat species were identified according to Timm and LaVal (1998), and all other than adult *C. sowelli* were released immediately after capture. We transferred *C. sowelli* into a shared keeping cage (6.1 × 3.4 × 2.5 m) that was situated at a distance of about 50 m from the closest clearing and surrounded by mature forest such that bats in the keeping cage were not exposed to any artificial light but to the natural light/dark cycles. Captive bats were supplied with banana, papaya and water *ad libitum* and kept together in captivity for a maximum of 5 days before being transferred to the choice experiment.

For the choice experiment, we released individuals singly in a flight cage (Figure 14.1) that was situated at a linear distance of about 250 m from the keeping cage. The experimental flight cage consisted of three compartments, the release area (5 × 3 × 2 m) and two choice compartments of equal size (2 × 1.5 × 2 m) which were separated from the release area by a retractable mesh curtain. One choice compartment was dimly illuminated by our custom-made street lamp. Since the lamp could not be dimmed sufficiently to have only low light intensities inside the choice compartment when fixing the lamp inside the choice compartment, we set up the streetlight outside of the flight cage at a distance of about 3 m from the rear end of the choice area (Figure 14.1). Except for the front (the "entrance"), the other choice compartment was shielded from the light by black plastic foil. To produce the same echo-acoustic environment for both choice compartments, we covered the illuminated compartment with transparent plastic foil. Between experiments, we randomly switched between illuminating the right and the left choice compartment. We set up two infrared-sensitive cameras and three infrared lights to record the behavior of bats (Figure 14.1).

Depending on fruit availability, we equipped the choice compartments

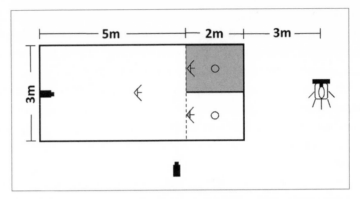

FIG 14.1: Scheme of the flight cage set-up. Bats could enter two choice compartments in which fruits were offered on a platform (circle). We randomly chose one of the two choice compartments to be shielded from the experimental light. We used two infrared-sensitive cameras (one in the back of the flight cage and one in line with removable mesh dotted line) to observe bat behavior. Infrared lights were installed on the ground of the release area pointing towards the choice compartments (open triangle) and one each on the ceiling of each choice compartment directed downwards (filled triangles).

with ripe infructescences or fruits that local *C. sowelli* are known to forage on, namely *Piper sanctifelices*, *Solanum rugosum* or *Ficus colubrinae*. For a given dual choice experiment, we always used same numbers of fruits of the same plant species in both compartments, in most trials this was four *Piper* infructescences. *Piper* infructescences were put with their basal part in a small plastic bowl filled with silica gel and placed centrally in the choice compartments on a platform around 80 cm in height such that bats could harvest them in flight. Branches of *Solanum* and *Ficus* with an equal number of fruits (5–15) were fixed at the ceiling of the choice compartments when we did not find enough ripe *Piper* infructescences. During some trials, we also offered banana on the central platform because we either lacked other ripe fruits or bats were not motivated to forage on fruits other than banana. Light intensity at the *Piper* infructescence was below the threshold of the luxmeter in the dark compartment and 4.5 plus or minus 0.4 lux (mean plus or minus SD) in the illuminated compartment, measured horizontally towards the lamp. This light intensity (4–5 lux) corresponds to a distance of approximately 8 m from the lamp if the light was not dimmed, assuming an isotropic light source and optimal conditions.

Experimental trials were conducted between 1830 and 0200 h. The entrance to the choice compartments was closed when we released a bat in the release area, yet the fruit scent could pass through the dividing mesh. Bats were habituated to the flight cage until they either clearly switched from flying in circles to flying back and forth in front of the choice compartments or until they stopped flying and continuously clang to the mesh for at least 30 s. We then lifted the curtain that separated the choice compartments from the release area and recorded the bat's behavior for at least 15 min with the video cameras. After experiments, all bats were released at the site of capture.

Based on the video recordings, we counted the number of explorative flights, that is, the number of entries in each choice compartment, within 15 min after opening the choice area. To account for differences in total numbers of flights between individuals, we used a weighted regression (generalized linear model with family = binomial and link = logit) on the number of explorative flights in either choice compartment. For the regression, we incorporated the independent variables "gender" and "side-of-light," indicating which of the two choice compartments was illuminated, as well as the interaction between "gender" and "side-of-light." The weighing was achieved in R using a two-vector object combining the number of flights in both left and right choice compartment as the dependent variable for the GLM fit.

Further, we determined from the video recordings whether bats harvested fruits/infructescences in either the dark or the lit compartment. Usually, bats harvested only one infructescence and became torpid afterwards for the remaining of the recording period. In a few trials, however, bats fed on more than one fruit. For those individuals, we only included the compartment of the first feeding activity in the analysis. To evaluate whether bats harvested fruits less often under illuminated than under dark conditions, we conducted a generalized linear model for a binary response variable (family = binomial, link = logit) also incorporating "gender," "side-of-light" and the interaction between the two factors as predictor variables.

Harvest of Wild Piper Infructescences

To verify the relevance of the flight cage experiment for free-ranging populations of bats, we also conducted a field-based light experiment. We regularly checked 14 P. sancti-felices plants for ripe infructescences. Thirteen of these plants grew at the edge between secondary forest/abandoned agroforestry and the clearing (c. 2 ha) of the biological station (at a maximal distance of 25 m from the forest edge). One additional plant was monitored at the edge between a smaller clearing (c. 150 m²) and secondary forest. Plants were chosen according to the site's accessibility to electric power to run the high-pressure sodium light. However, due to numerous wall sockets at the buildings in the clearing, most *Piper* plants at the forest edge were within the range of our extension cable (c. 25 m), but we focused only on those that were more than 25 m apart. There is a potential lack of spatial independence in these samples due to the proximity of the *Piper* plants to each other or the identity of the foraging bats. Ideally, we would have worked on replicate study plots that were at several kilometers apart or even in different countries, but unfortunately, this approach was not feasible. Our choice of monitored plants aimed at minimizing spatial dependence given the constraints for setting up experimental lights, yet we cannot rule out the possibility that our data may suffer to some extent from a lack of independence. However, due to the high abundance of *C. sowelli* at our study site (Rex et al. 2008) and the overall distance of monitored *Piper* plants, we suggest that harvest events were almost independent.

At smaller plants, we were able to mark every ripe infructescence when monitoring the respective plant since *Piper* plants produce only a few ripe infructescences each night over extended periods of time. At large plants with many ripe infructescences, we randomly chose a subset of the ripe ones. Every *Piper* plant was used at least twice, once under naturally dark conditions and once when it was illuminated by the experimental street-light. At most plants, however, we increased the number of infructescences monitored by marking ripe infructescences on more than one dark and one illuminated night (n = 63 marked infructescences for dark and light condition, respectively). On average, we marked 5 plus or minus 4 and 5 plus or minus 3 (mean plus or minus SD) infructescences per plant during dark and illuminated conditions, respectively (min to max = 1.14 and 1.10), shortly before sunset by knotting a short piece (c. 5 cm) of thin orange thread to the branch at a distance of about 5 cm from the respective ripe infructescence. Due to the orange color of the sodium vapor light, the thread was only distinguishable from the plant by its color during daylight but not during dark or artificially lit conditions. Three hours following sunset, we counted the number of marked infructescences that were harvested.

The light was placed at a mean (plus or minus SD) distance of 2.5 plus or minus 0.7 m from the observed infructescences. It was switched on before sunset and ran until midnight. The mean light intensity (plus or minus SD) was 57.0 plus or minus 19.1 lux at the monitored *Piper* infructescences under illuminated conditions which is comparable to light intensities measured underneath or in proximity to high-pressure sodium streetlights (e.g., Stone,

Jones, and Harris 2009: 52 lux). During the dark treatment, light intensity was below the threshold of the luxmeter (0.01 lux). Light intensities were measured horizontally at a height of 18 m towards the lamp using the lux-meter LX-1108. The nature of the first treatment (either dark or light) was assigned randomly to experimental plants. After each illuminated monitoring, we waited at least three nights before using the same plant again under dark conditions to avoid any sequential effects on the outcome of the experiments.

We used a logistic regression framework to analyze the influence of light on the probability of fruits to be harvested by bats. The dependent variable y of the model was a binary variable, indicating whether a given fruit had been harvested (y = 1) or not (y = 0). We considered the light treatment as a binary variable (defined by: 0 = dark, 1 = light) modeled as a fixed effect, and we modeled the plant identity as a random effect to account for the lack of independence of fruits marked at the same bush. As such, the model corresponds to a generalized linear mixed effect model (GLMM) that we fitted using the function glmer from the packages LME4 v. 0.999999–2 (Bates et al. 2011). We tested the effect of the light treatment by comparing the observed likelihood ratio test statistic measured for this covariate to its distribution under the null hypothesis obtained by parametric bootstrap (referred as PBtest in the results). This was done using the function PBmodcomp from the package PBKRTEST v.0.3–5 (Halekoh and Højsgaard 2013) that we used through the wrapper package AFEX v. 0.5–71 (Singmann 2013).

At a subset of 12 plants, we also video-recorded a randomly chosen ripe infructescence under both dark and illuminated conditions from sunset until midnight. From those recordings, we determined the time (minutes after sunset) at which the respective infructescence was harvested. We then tested for significant differences between the two treatments using the paired-samples t-test in PASW statistics 18.0 (SPSS Inc., Chicago, IL, USA).

If not mentioned otherwise, all analyses were conducted in R (R Core Development Team 2012). We used an alpha value of 5%.

RESULTS

In our dual choice experiment, we conducted 56 experimental trials using 39 male and 17 female *C. sowelli*. The number of explorative flights in either choice compartment was affected by light treatment ($z = 8.87$, $p < 0.001$) but not by gender ($z = -0.84$, $p = 0.402$) nor by the interaction between gender and light treatment ($z = 0.94$, $p = 0.349$). Bats performed less explorative flights in the dimly illuminated than in the dark compartment

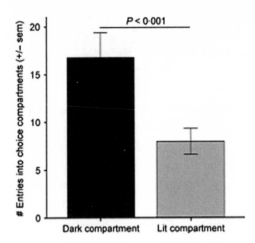

FIG 14.2: Observed mean number of entries per bat from 56 *Carollia sowelli* bats in either the dark or the dimly illuminated ("lit") choice compartment.

(Figure 14.2). On average, bats entered the dimly illuminated compartment four times (median; min/max = 0 and 41, respectively) and the dark compartment eight times (median; 0–88). The light treatment also affected in which compartment bats harvested food (z = 2.29, p = 0.022), but neither gender (z = -0.48, p = 0.35) nor the interaction between gender and light treatment (z = 1.16, p = 0.247) had an effect on this decision. Bats harvested food almost twice as often in the dark than in the dimly illuminated compartment (n dark = 36, n light = 20). In the free-ranging population, our camera recordings (n = 40) revealed that after sunset no other vertebrates besides bats harvested infructescences of *P. sancti-felices* at our study site.

We found that the light treatment exerted a significant influence on the probability of a fruit being harvested (PBtest: likelihood ratio test statistics = 19.2, 666 simulations reaching convergence, p < 0.009, Figure 14.3a). In the naturally dark environment, 100% (n = 63) of fruits were harvested within 3 h after sunset, while the model predicts that only 89.5% of fruits were harvested on each plant under illumination. This estimate deviates slightly from the 77.8% (49 of 63) of infructescences that were harvested across all plants during the experiment because the removal rates differed between plants (variance of the random effect expressed in the logit scale = 4.77) and the data collection was not balanced with respect to plants, while model estimates are.

If harvested at all, infructescences under illumination were harvested about 2 h later than infructescences from the same plants but in a dark surrounding (mean plus or minus SD = 84 plus or minus 42 min and 196 plus

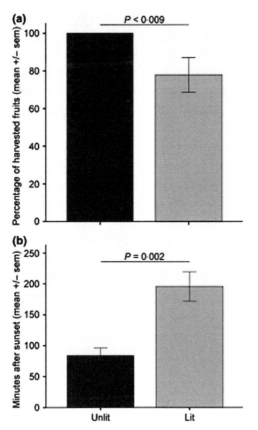

FIG 14.3: (a) Observed percentage of harvested *Piper sancti-felices* infructescences among all marked ones ($n = 14$ plants) and (b) for infructescences that were harvested, the minutes after sunset when infructescences were harvested by free-ranging bats from plants in either a naturally dark surrounding ("unlit") or from the same plants under illumination of a street lamp ($n = 12$ infructescences each from a different *P. sancti-felices* plant for dark and illuminated conditions, respectively).

or minus 82 min after sunset, respectively; paired-samples $t = -4.1$, $n = 12$, $p = 0.002$; Figure 14.3b).

DISCUSSION

Our study provides first evidence that frugivorous bats are repelled by artificial light at night, indicating that light pollution interferes with valuable ecosystem services provided by nocturnal seed dispersers. In particular, experiments with captive *C. sowelli* highlighted that bats performed more explorative flights and harvested fruits more often in a dark than in an illuminated environment. Given the low light intensities used in the experiment, we infer that *C. sowelli* was repelled by intensities even lower than those measured underneath streetlights. We therefore suggest that the rapid spread of light pollution might severely affect the spatial foraging behavior of frugivorous bats. Nocturnal seed dispersers may visit fruiting plants or entire feeding areas less often when these are illuminated by artificial light. Particularly frugivorous bats such as *C. sowelli* depend on many fruiting plants because each plant individual produces only a few ripe infructescences per night. Consequently, bats of the genus *Carollia* search ripe infructescences at numerous plants each night and switch frequently between distant feeding areas when foraging (Fleming 1988).

Our findings with captive bats were consistent with those obtained from free-ranging bats. Wild bats harvested fewer *Piper* infructescences from illuminated *Piper* plants and, when foraging did occur, they removed infructescences from illuminated plants about 2 h later than from plants in complete darkness. This delay in foraging activity may drastically reduce the likelihood of seed dispersal for a plant, particularly when additional adverse conditions reduce the activity of bats later at night, for example during tropical rainfalls (Voigt et al. 2011). Further, if a *Piper* infructescence is not harvested during the first night after ripening, it may not be removed and may fall to the ground (Thies and Kalko 2004). Irrespective of whether an illuminated infructescence is harvested later at night or whether it is completely neglected and not removed at all, in both circumstances, the avoidance behavior of frugivorous bats towards artificial light at night reduces the probability of successful seed dispersal. This has major implications for ecosystem functioning when tropical habitats are increasingly exposed to artificial light. Bat-dispersed successional plants in particular, such as *Piperaceae* and *Solanaceae*, might suffer from a reduced visitation rate in an illuminated environment. Due to their preference for disturbed areas, pioneer plants are more likely exposed to artificial light, for example, when streetlights are established along roads or when lights at buildings illuminate the surroundings at night.

Anthropogenic disturbance per se may not necessarily reduce bat abundance and the associated ecosystem services, because some bat species are relatively resistant to fragmentation. Many frugivorous bat species fly up to 2–5 km across open areas in the Neotropics (Bernard and Fenton 2003) and some species which are specialized on pioneer plants might even be more abundant in disturbed habitats (Willig et al. 2007). These bats are important for the rapid succession in clearings because they produce a copious seed rain even in deforested areas such as abandoned pastures (Medellin and Gaona 1999). In the Neotropics, the majority of cleared lowland forest becomes pasture but more than 50% of the clearings in the Amazon are abandoned within 10 years because of the poor fertility of tropical soils (Hecht 1993).

Here, bat-mediated seed intake could promote reforestation and reduce the many negative outcomes associated with abandonment such as pronounced land erosion which may cause landslides, runoff, water loss, leaching and siltation of streams and rivers. However, the ability of a species to resist anthropogenic disturbance depends on the nature and the level of disturbance. Although frugivorous bats might easily traverse open areas between forest fragments in naturally dark nights, our results suggest that they are less likely to use habitats which are "polluted" by artificial light at night. It appears that artificial light constitutes a severe anthropogenic disruptive factor which affects even species that are tolerant of fragmentation or other anthropogenic changes to ecosystems. Accordingly, succession with pioneer plants may slow down in areas with artificial light and habitat loss may be aggravated for light-sensitive species. This may result in cascading effects that could prove expensive for landowners and communities.

Artificial light from villages and street lamps may serve as a "light barrier" that inhibits light-sensitive bats from conducting long-distance seed dispersal and pollination services between remaining forest fragments and therefore increases the degree of isolation. The light-barrier argument goes beyond what can be directly inferred from our experiment, but it seems plausible given the fact that streetlights are usually brighter than the 4–5 lux used in our experiment. Also, bats of the genus Carollia usually fly at low heights above-ground (Rex et al. 2011) and may therefore be unwilling to cross illuminated streets above the glare of lamps. Some support for a light-barrier effect comes from a study which showed that the few frugivorous bat species which do occur in urban areas can rarely be captured along roads (Oprea et al. 2009). Further, even some insectivorous bats that could potentially benefit from feeding on insects attracted to streetlights avoid roads more than other urban land cover classes when commuting (G. Davies, Hale, and Sadler 2012) or do not commute in the catchment area of streetlights at all (Stone, Jones, and Harris 2009, 2012). If commuting of frugivorous bats is affected in a similar way by light barriers, then artificial light at night might not only lead to genetic isolation of illuminated plants and to a loss of suitable habitats for light-sensitive species but could also hinder seed exchange and genetic connectivity between whole forest fragments (Jordano et al. 2011). Then, maintenance of biodiversity and finally ecosystem functioning could be at risk in areas composed of forest remnants embedded in a matrix without sufficiently dark corridors. Possibly, such a scenario may be realized in many tropical countries, as both deforestation and light pollution proceed at high rates across the tropical climate domain (Hoelker et al. 2010a; Food and

Agriculture Organization of the United Nations and European Commission Joint Research Centre 2012).

On a global scale, bats are known to disperse seeds not only of *Piper* but also of hundreds of other tropical tree and shrub species that support biodiversity (Thomas 1991). In addition, many agriculturally produced fruits such as mango and shea as well as many economically relevant timber species are pollinated or dispersed by bats (Ghanem and Voigt 2012). The production of shea trees (a bat-dispersed species) was estimated to exceed 25 million metric tons each year (Lovett 2005), highlighting the relevance of bats as seed dispersers for species used by humans. Artificial light at night may severely affect these economies when pollinating and seed-dispersing services of bats are reduced.

Problems associated with artificial light may become even more aggravated on a larger geographical scale, considering that light pollution is increasing rapidly at an annual rate of about 6% worldwide (Hoelker et al. 2010b). Since the degree of light pollution parallels population growth and economic development (e.g., Elvidge et al. 2001), it can further be expected that artificial light at night increases at exceptionally high rates in many tropical countries. For example, the outdoor lighting market in Latin America is estimated to nearly double between 2010 and 2020 (Baumgartner et al. 2011). Due to the exponential growth rate of human populations in many tropical countries (United Nations Population Fund 2011), people will encroach further into formerly pristine habitats than ever before. Since this encroachment is probably accompanied by an intensified use of artificial light, it might have deleterious consequences for nocturnal seed dispersal and habitat connectivity.

CONCLUSIONS

We conclude that the detrimental effects of light pollution are likely to increase and may have a great impact on biodiversity, particularly in the tropics where artificial light follows human encroachment in natural habitats at unprecedented rates. Policy makers should pay attention to the ecological impacts of artificial light, and policy should ensure that artificial light is not excessively used. To mitigate the negative effects, artificial light should be restricted to (i) where it is needed, (ii) when it is needed and to (iii) an illumination level that achieves its purpose but does not exceed it. Particularly in the tropics, where nocturnal seed dispersers are crucial for ecosystem functioning, maintaining unlit habitats large enough to guarantee viable populations of light-sensitive species should be a high priority, since even very low light intensities were sufficient to reduce the foraging activity of fruit-eating bats. To achieve this, it is essential to raise awareness of the ecological impacts of artificial light by informing people and policy makers about the deleterious effects light pollution can have on a wide range of taxa (reviewed in Rich and Longcore 2006).

15

LIGHT POLLUTION IS ASSOCIATED WITH EARLIER TREE BUDBURST ACROSS THE UNITED KINGDOM

Richard ffrench-Constant, Robin Somers-Yeates,
Jonathan Bennie, Theodoros Economou, David Hodgson,
Adrian Spalding, and Peter McGregor

The potential of light pollution to change plant phenology and its corresponding effects has not been fully examined to date. The study reported in this chapter examined the effect of artificial lighting on the timing of budburst in trees. A thirteen-year data set on spatially referenced budburst data across the United Kingdom of four deciduous tree species was examined in relation to satellite imagery of nighttime lighting and average spring temperature. Budburst occurs up to 7.5 days earlier in brighter areas, with the relationship being more pronounced for later-budding species. Excluding large urban areas from the analysis showed an even more pronounced advance of budburst, confirming that the urban "heat-island" effect is not the sole cause of earlier urban budburst. As light pollution is a growing global phenomenon, the findings of this study are likely to be applicable to a wide range of species interactions around the world.

INTRODUCTION

Most organisms have evolved for millions of years under predictable cycles of light and dark resulting from the earth's rotation and orbit. Ambient light plays an important role in natural systems, acting as an abiotic cue organizing both daily and seasonal patterns in activity (Gaston et al. 2013). At higher latitudes, changes in day length are therefore an accurate indicator of the progression of the season, and specifically the onset of more favorable spring conditions (Brasler and Körner 2012). However, the extent to which these fundamental light-driven processes are being influenced by light pollution is unclear. We wanted to assess whether UK-wide data on nighttime lighting could be correlated with advances in tree budburst.

Vascular plants use phytochrome photoreceptors, sensitive to the red:far red ratio of light, to effectively determine the day length, and this ability assists them in timing key phenological events such as budburst, flowering and bud set, so that they coincide with favorable environmental conditions (e.g., H. Smith 2000). An experiment reducing the red:far red ratio of light at twilight advanced budburst in silver birch (*Betula pendula*) by approximately 4 days (Linkosalo and Lechowicz 2006). For many organisms, the accurate timing of such events has important fitness effects. Further, in multi-trophic systems, the period of optimal conditions is often governed in part by species phenology at the underlying trophic level (Visser and Holleman 2001). This is well exemplified by the interaction between the Pendunculate oak tree (*Quercus robur*) host plant and its winter moth caterpillar (*Operophtera brumata*) herbivore, which has been extensively examined in the context of the likely impacts of anthropogenic climate change on phenology (Buse et al. 1999; Buse and Good 1996). Oak trees are thought to use both temperature and photoperiod as abiotic cues, to unfurl their buds at a time that will maximize the length of the growing season, while at the same time reducing the risk of frost damage (Bennie et al. 2010; Brasler and Körner 2014). In turn, the winter moth herbivore is under pressure to match its egg hatch with the timing of budburst. Thus, if the eggs hatch too early, the larvae may face starvation, and if they hatch too late, they will be forced to eat less digestible, and better protected tannin-rich leaves (Buse and Good 1996; Feeny 1970; Visser and Holleman 2001). A combination of photoperiod and temperature forcing is considered to be important for determining budburst phenology in most temperate trees, with the temperature forcing requirement for budburst decreasing to a minimal value when accumulated winter chilling and/or increases in photoperiod have been detected (Vitasse and Basler 2013). Opportunistic, early-successional species, and tree species that come into leaf earlier in the spring, tend to be more sensitive to temperature alone with little influence of photoperiod. Late-successional species, which also tend to break bud later in the spring, tend to have a more marked response to photoperiod (Brasler and Körner 2014). Observations suggest that the leaf phenology of several urban tree species is altered in the direct vicinity of street lighting, both in terms of earlier budburst and later leaf fall (Bennie et al. 2014).

Over the last 150 years, the natural nighttime environment has been drastically altered by the proliferation of man-made artificial lighting. In 2001, it was estimated that almost a fifth of the earth's land surface was polluted by light (Cinzano et al. 2001), and subsequently the amount of artificial light has been increasing at approximately 6% annually. The increasingly large amount of artificial nighttime lighting and the known importance of light to natural systems have led to widespread concern over the potential ecological impacts of light pollution (Hoelker et al. 2010a; Longcore and Rich 2004). Specifically, concern has been expressed about the potential of light pollution to disrupt trophic interactions through artificially altering the day length as perceived by living organisms (T. Davies et al. 2012; Longcore and Rich 2004). Spring phenology including tree budburst is advanced in urban areas, and it is generally considered that the main cause is the urban "heat island" (UHI) effect of enhanced temperature regimes (e.g., Jochner et al. 2012; X. Zhang 2004). However, experiments that artificially altered photoperiod have shown that budburst of a number of species of late-successional trees was delayed when the photoperiod was shortened (Brasler and Körner 2012 and 2014). The nighttime light environment of urban and suburban areas is extremely heterogeneous, with light intensities varying across several orders of magnitude over horizontal and vertical distances of a few meters (Bennie et al. 2016). We therefore examined the hypothesis that increasing photoperiod, via artificial lighting, will hasten the earliest recorded date of budburst, and that this effect will be greater in late budding than early budding species. To test this hypothesis, we analyzed spatial-temporal data on budburst and satellite imagery of nighttime lighting to investigate whether light pollution is correlated with budburst date. Strikingly, we find that earlier budburst is associated with nighttime lighting, that the effect is larger in late- than early budding species, and that magnitude of this advance is likely to be too great to be explained by residual urban temperature effects alone.

MATERIAL AND METHODS

Budburst Data

Spatially referenced budburst data were collected from 1999 to 2011 by "citizen scientists" and submitted to the UK phenology network (www.natures calendar.org.uk). We used data from four available deciduous species: European sycamore (*Acer pseudoplatanus*, known as sycamore maple in North America), European beech (*Fagus sylvatica*), Pedunculate oak (*Q. robur*) and European ash (*Fraxinus excelsior*). Recorders were asked to note "budburst" as the date when the color of the new green leaves is just visible between the scales of the swollen or elongated bud; they were advised, if they were having difficulty in deciding when to record, to wait until the event was occurring in three plants of the same species within close proximity to each other, to record the trendsetters rather than the extraordinary. Each record from a single observer is georeferenced and is treated as a separate point observation, therefore even if several observers record at the same or nearby points (as is likely to be the case in more densely populated areas), then there should be no systematic bias towards earlier records, assuming that the distribution of recording effort and accuracy made by individual observers is independent of recorder density. The potential recorder error within the data collection protocol was deemed unlikely to be problematic in the present analyses, as there is no reason to expect that the distribution of recording effort and accuracy of individual observers is affected by observer density or the amount of artificial nighttime lighting.

Light Pollution Data and Calibration

The global dataset of annual nighttime satellite images for 1999–2011 from the Defense Meteorological Satellite Program's Operational Linescan System (DMSP OLS) was used to quantify the amount of artificial light at the locations of the spatially referenced budburst dates. These data are produced and made publicly available by the NOAA National Geophysical Data Centre (K. Baugh et al. 2010) and have previously been used to map the extent of light pollution (Cinzano et al. 2001; Imhoff et al. 1997; Sutton 2003). These satellite images depict a global, cloud-free composite of stable nighttime light at approximately 1 km resolution, resampled from data at a resolution of approximately 2.7 km. Each pixel is represented by a value of 0–63; a value of zero represents areas of relative darkness, whereas brightly lit urban areas usually saturate at a value of 63. (Given the coarse resolution of these data, a spatially referenced budburst date within a bright pixel, for example, will not necessarily be located in a bright area; it is just assumed to be more likely to be.) These data will here be referred to as either DMSP data or DMSP values.

Accurate inter-annual comparisons of the DMSP data are difficult because the data have been collected by multiple satellites with a lack of onboard intercalibration between the satellite sensors and the gain control of their optical sensors is changed continually to generate consistent imagery of clouds. This means that a specific pixel value in a given year may not represent the same actual level of brightness as a pixel of the same nominal value in another year. In addition, there are inaccuracies with the geolocation of the DMSP data which result in apparent differences in the location of pixels between years; up to 3 pixels (approx. 3 km) between some years. In order to compare images between years, the geolocation errors must therefore be rectified and the images intercalibrated. In this study, correction of geolocation errors and intercalibration of images followed the methods described in a previous study (Bennie et al. 2014). Intercalibrated DMSP data were resampled, using bilinear interpolation, to a 5 km grid to match the resolution of the temperature data.

Gridded Temperature Data

Owing to the increased amount of artificial light in urban areas, and the fact that urban areas are known to be warmer than surrounding rural areas because of the UHI effect (Arnfield 2003), we anticipated that temperature would positively covary with the amount of artificial light. To control for this potential covariance, 5 × 5 km gridded mean monthly air temperature data were incorporated into the analysis. These gridded air temperature data cover the majority of the UK and were created using weather station data, through an interpolation process that takes into account topographical, coastal and urban features (Met Office n.d.; Perry and Hollis 2005). For the present analysis, a new 5 × 5 km gridded dataset of average spring air temperatures was created for 1999–2011 from the monthly gridded temperatures. This was performed by averaging the temperatures for February to April, as the timing of first leaf date is known to strongly correlate with temperatures within this period (Sparks and Carey 1995).

Spatial Matching and Statistical Analysis

Budburst data were spatially matched with both resampled DMSP light pollution values and mean air temperature values within years, giving 11,968 data points for Acer, 10,061 points for Fagus, 8,908 data points for Quercus and 10,899 for Fraxinus. The light and temperature data for each budburst sampling point were extracted using bilinear interpolation from the 5 km resolution grids. The budburst data were transformed from British National Grid to WGS1984 before extracting the DMSP values (Figure 15.1). A generalized additive mixed model with a scaled t-distribution was used to analyze the relationship between the amount of light pollution and the date of budburst. Budburst date, quantified as the number of days from 1 January in the corresponding year, was incorporated into the model as the response variable and was assumed to follow a t-distribution (a symmetric distribution like the Gaussian but with heavier tails). The mean of the response μ was

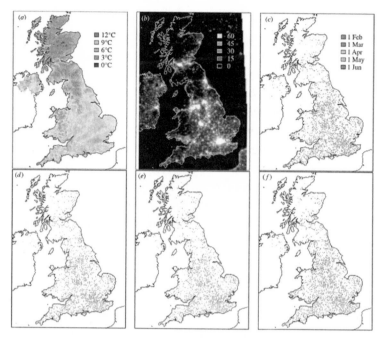

FIG 15.1: (a) Average spring temperatures in 2011, (b) DMSP night-time lights in 2011, (c–f) locations of budburst data for all years, for (in order of budburst) sycamore (c), beech (d), oak (e) and ash (f).

modeled in terms of additive influences from the various predictors; specifically, smooth non-parametric functions of the DMSP value, mean spring air temperature and their interaction. Calendar year was incorporated into the model as a random effect to account for inter-annual variation of budburst date. To allow for latitudinal variation in day length and other spatial trends in the data, parametric linear and quadratic terms of Easting, Northing and their interaction were also incorporated additively in the mean of the response. An interaction between DMSP value and temperature was included to analyze whether the relationship between budburst date and DMSP value varied at different temperatures.

The analysis was repeated excluding data points found within large urban areas to remove any residual effect of the UHI not captured by the temperature dataset, along with other potential effects of urbanization on budburst date. The Ordnance Survey Meridian 2 dataset was used to define the boundaries of urban areas, and the analysis was carried out on the data points that fell outside of settlements with a population of greater than or equal to 125,000. As a further check to reduce the possibility of inflated degrees of freedom owing to non-independence of observations in close proximity, the analyses were repeated using only observations that were at least 5 km distant from another observation in the same year. Data from Northern Ireland were also excluded from these further analyses as the Ordnance Survey Meridian 2 dataset does not cover this area.

Predictions from these models, within the limits of the data used for model calibration, were carried out to aid inference. The gam function from the R package mgcv (v. 1.8-4) was used for fitting the generalized additive mixed (GAM) models (S. Wood 2006 and 2011), and approximate tests of significance were carried out on the model terms by using the function anova.gam; this function carries out Wald tests of significance on the smooth and parametric terms within a single fitted GAM object. All statistical analyses were carried out using R (v. 64 3.0.3) (R Core Development Team 2014) (Table 15.1).

As a test that any observed relationship between DMSP and budburst date was robust to different forms of analysis, for each species we carried out a partial regression on the residuals from a GAM of DMSP (response) and spring temperature (predictor) and the residuals from a GAM of budburst date (response) and spring temperature (predictor).

TABLE 15.1. Terms and properties of generalized additive mixed models fitted to data excluding data points found within large urban areas (population 12,500)

Response variable	Explanatory terms				Model statistics
Acer psuedoplatanus budburst date	Smooth terms	EDF	χ^2	p-value (approx.)	R^2 (adj.)
	DMSP value	2.48	17.73	<0.001	0.121
	Spring temperature	2.81	100.43	<0.001	Deviance explained
	DMSP value, spring temperature (interaction)	0.00160	224.73	<0.001	12.1%
	year (random factor)	9.98	257.39	<0.001	REML
	parametric terms	DF	χ^2	p-value	28,572
	Northing	1	13.525	<0.001	No. of obs.
	Easting	1	3.623	0.0570	7,024
	Northing: Easting (interaction)	1	3.750	0.0528	
	Northing²	1	26.231	<0.001	
	Easting²	1	1.679	0.195	

(continued)

Response variable	Explanatory terms				Model statistics
Fagus sylvatica	Smooth terms	EDF	χ^2	p-value (approx.)	R^2 (adj.)
budburst date	DMSP value	2.574	3.03	0.36	0.191
	Spring temperature	2.864	276.9	<0.001	Deviance explained
	DMSP value, spring temperature (interaction)	0.003	433.5	<0.001	19.1%
	year (random factor)	9.295	325.5	<0.001	REML
	parametric terms	DF	χ^2	*p*-value	23,708
	Northing	1	0.216	0.642	No. of obs.
	Easting	1	95.857	<0.001	6,053
	Northing: Easting (interaction)	1	19.916	<0.001	
	Northing^2	1	2.868	0.0904	
	Easting^2	1	18.337		
Quercus robur	Smooth terms	EDF	χ^2	P-value (approx.)	R^2 (adj.)
budburst date	DMSP value	2.62	8.63	0.032	0.360
	Spring temperature	2.91	580.3	<0.001	Deviance explained
	DMSP value, spring temperature (interaction)	0.000543	86.27	<0.001	20.5%
	year (random factor)	9.891	394.2	<0.001	REML
	parametric terms	DF	χ^2	*p*-value	19,686
	Northing	1	7.68	0.00557	No. of obs.
	Easting	1	12.8		5,296
	Northing: Easting (interaction)	1	39.4		
	Northing^2	1	6.18	0.0129	
	Easting^2	1	0.020	0.887	

RESULTS

Our analysis showed no significant effect of DMSP value on the species with earliest budburst, *A. pseudoplatanus* but significant effects of the DMSP value on budburst date in three of the four species; listed here in order of budburst, *Fa. sylvatica* ($\chi2^2 = 1190.5$, $p \leq 0.001$, n = 10,061, *Q. robur* ($\chi^2 = 7093.8$, $p \leq 0.001$, n = 8,908) and *Fr. excelsior* ($\chi^2 = 953.2$, $p \leq 0.001$, n = 10,899). In all three cases, the relationship was negative (areas with brighter lights

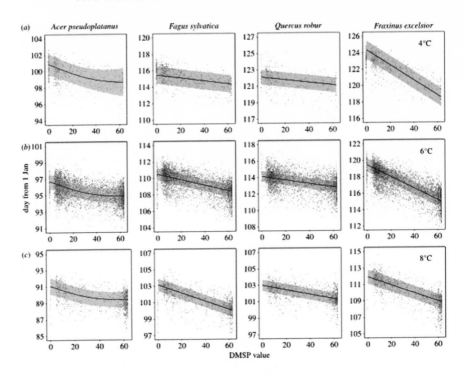

FIG 15.2: Plotted model predictions of the relationship between DMSP night-time lights and budburst date at different spring temperatures; 48°C (a), 68°C (b) and 88°C (c), and from left to right (in order of budburst), Acer, Fagus, Quercus and Fraxinus. Predictions are made for budburst at the mean latitude of data points included in the model. The black line represents the predicted mean and the shaded gray area the predicted 95% CIs. Points represent residuals of individual data points where the spring temperature lies within 0.58°C of the prediction temperature in each panel.

typically experienced earlier budburst) and there was a significant interaction between temperature and DMSP (Figure 15.2). The largest magnitude of effect was for *Fr. excelsior* at lower temperatures, where the difference in fitted model predictions between the darkest rural and most brightly lit urban sites was 7 days (Figure 15.2). When large urban areas are excluded from the analysis, the predictions show a qualitatively similar, but nonlinear relationship between budburst date and the DMSP value, and the effect of DMSP value was here also significant for *A. pseudoplatanus* (Figure 15.3; χ^2 = 17.73, p ≤ 0.001, n = 6,053). For *Fa. sylvatica*, the effect of DMSP was non-significant, but there was a significant interaction with spring temperature (χ^2 = 433.5, p ≤ 0.001, n = 6,053), with earlier budburst only associated with

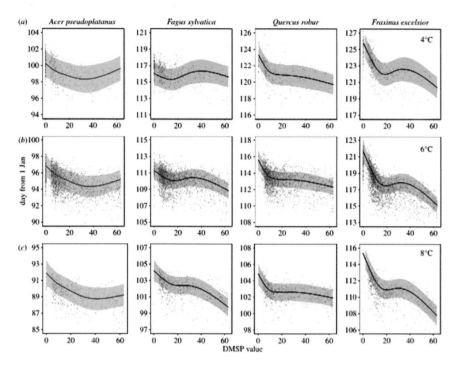

FIG 15.3: (a–c) As for Figure 15.2, but with urban areas (populations exceeding 125,000) removed.

artificial light at higher temperatures. Both *Q. robur* and *Fr. excelsior* had significant relationships between budburst date and DMSP (*Q. robur* χ^2 = 8.63, p = 0.032, n = 5,296 and *Fr. excelsior* χ^2 = 37.4, p ≤ 0.001, n = 6,762), and also significant interaction terms with temperature (Figure 15.3). Excluding observations from urban areas from the analysis, *Fr. excelsior* buds in areas with average spring temperatures of 4°C are likely to burst approximately 5 days earlier in the brightest areas compared with the darkest areas, and buds that experience average spring temperatures of 8°C are predicted to burst approximately 7.5 days earlier in the brightest areas. As a conservative test for spatial non-independence, further excluding all observations in the close vicinity (within 5 km) of any other observations recorded in the same year had no qualitative effect on the significance levels of the results reported here.

DISCUSSION

The results highlight, for the first time, to our knowledge, and at a national scale, a relationship between the amount of artificial nighttime light and the date of budburst in deciduous trees. This relationship is unlikely to be caused by the UHI effect, as it is robust to the exclusion of large urban areas where temperatures are known to be elevated. Similarly, this effect is unlikely to be related to an increase in temperature alone; the maximum magnitude of effect size predicted between the brightest and darkest sites (7.5 days) is roughly equivalent to that predicted due to 2°C. Specifically, it has already been shown that urban areas are both brighter (Imhoff et al. 1997; Sutton 2003) and warmer (UHI effect; Arnfield 2003) but this is, to our knowledge, the first study explicitly investigating the relationship between the amount of nighttime light and budburst while controlling for the temperature increases within urban areas. In summary, similar predictions were obtained from a model fitted to budburst data points found outside of large urban areas suggesting that it is nighttime lighting causing the advance in budburst as opposed to other factors which can vary owing to urbanization, such as temperature, humidity, water availability and chemical pollution levels (Honour et al. 2009; Jochner et al. 2013; Kozlov et al. 2007). In addition, for trees experiencing average spring temperatures of 4°C, the model predicts that budburst will be advanced by up to 7.5 days in the brightest areas compared to the darkest areas. The exposure of plants to artificial light at night is highly heterogeneous at a fine scale. Skyglow, diffuse light scattered in the atmosphere from city lights, can illuminate areas of many square kilometers to levels exceeding moonlight, but effects of artificial light on phenology have to date only been recorded as a consequence of direct illumination in the vicinity of light sources, which can be several orders of magnitude brighter (Bennie et al. 2016). As the spatial data for this study was aggregated to 5 km resolution, and the DMSP data have no direct calibration, the DMSP value for each pixel cannot be easily related to an illuminance or irradiance that any individual tree is exposed to at night. Moreover, even in dark pixels, an individual tree adjacent to a streetlight may be exposed to bright light, while a tree in a large unlit urban park might be relatively dark despite being located in a bright pixel. However, the DMSP pixel brightness is probably a good indication of the density of outdoor light sources, and hence the probability of any tree within that pixel experiencing a relatively high level of direct illumination; observers recording the first bud break in three trees in close proximity will therefore be considerably more likely to be recording trees exposed to artificial light in "bright" than "dark" pixels.

Our finding that phenology of woodland tree species may be affected by light pollution, suggests that smaller plants growing below the height of streetlights are even more likely to be affected. Such results highlight the need to carry out experimental investigation into the impact of artificial nighttime lighting on phenology and species interactions. It also suggests that looking at other aspects of phenology, such as leaf senescence, would be highly worthwhile. Importantly, further studies should also try and take into account differences in light quality such as the specific wavelengths of light generated by different lighting types.

PART 4

THE EXPERIENCE
OF NATURAL DARKNESS

Part 4 of this book addresses the impacts of light pollution on the quality of the visitor experience in parks and protected areas. The opportunity to experience natural darkness can be a vital part of visiting many national parks, and many visitors report that experiencing natural darkness, along with interpretive programming about the night sky, is an important reason they visit national parks. As with the degree of natural quiet, the degree of natural darkness can be strongly related to the natural environment and the experience of visiting national parks. For example, reductions in night sky quality caused by light pollution can reduce visitors' opportunity to see and appreciate iconic nocturnal species that depend on dark conditions.

16

PRESERVING PRISTINE NIGHT SKIES IN NATIONAL PARKS AND THE WILDERNESS ETHIC

Dan Duriscoe

This chapter presents a cultural manifesto for the protection of dark night skies, or what the author calls "the unfettered view of the universe on a dark, clear, moonless night." Concern over the loss of wilderness is deeply rooted in the American conservation movement, extending back to the establishment of the first national parks in the latter half of the nineteenth century. The concept of wilderness must now be extended to include natural darkness. Recent decades have seen dramatic increases in anthropogenic light and associated light pollution, which have markedly diminished humans' ability to see and appreciate dark night skies. The author argues that dark night skies should be added to a growing list of park resources, as demanded by contemporary interpretation of both the Organic Act of 1916 that created the National Park Service and the Wilderness Act of 1964.

INTRODUCTION

In the American West throughout the latter part of the 19th century, the conquest of the frontier by American industrial culture led to a profound sense of loss among conservation-minded individuals. They mourned the passing of a way of life and of an unspoiled grand landscape that fostered individual freedoms, simple rewards for hard work, and an intimacy with the land that was required for mere survival. The last 20 or 30 years have seen a similar or analogous rapid disappearance of a resource that was once taken for granted: the unfettered view of the universe on a dark, clear, moonless night. Today, we are on the verge of losing the pristine night sky entirely in the 48 contiguous states. However, unlike losing a species to extinction, topsoil to erosion, or yet-to-be-explored virgin lands to development, the night sky is 100% recoverable.

The alarm signaling the potential irretrievable loss of wildlands was sounded in large part by the preservationists of the infant wilderness movement, such as Henry David Thoreau and John Muir. This alarm awakened the public and politicians, and the idea of wilderness preservation eventually became one of the hallmarks of American values. Professional and amateur astronomers were the first to cry out for preservation of night skies in the 1970s, and an effort to save what's left of visual astronomy from earth is now well underway. Loosely collected under the umbrella organization of the International Dark-Sky Association, those wishing to preserve night skies have become more eclectic, and now include lighting engineers, landscape architects, urban planners, federal land managers, and the general public. In the Far West, some of the darkest and clearest of night skies in the world are found in national parks such as Yellowstone, Glacier, Bryce Canyon, Canyonlands, and Death Valley. Many of these parks are now experiencing or are threatened by light pollution. I propose that the preservation of the right to view the universe from America's wilderness national parks is a duty assigned to the National Park Service under the Wilderness Act of 1964 and the Organic Act of 1916. Furthermore, the present-day effort to save night skies is not merely analogous to, but an integral part of, the wilderness ethic.

The idea that wild and beautiful lands with all their appealing attributes should be preserved *for their own sake* has sometimes been described as "the esthetic of the sublime" (Rodman 1983, 86). Since the Romantic movement, people of European background have equated the feeling of awe such places bring about with sacredness, or a place that is beyond and far greater than humanity. Perhaps no landscape has promoted such feelings more than that which includes the night sky, which has been described as "that most glorious and compelling and inspiring of nature's faces" (Schaaf 1988, 205). While the environmental ethic has evolved to focus on a more holistic approach to ecosystem health rather than exclusively on the preservation of sacred, sublime places, the preservationist tradition is still very much a part of American culture. It is an idea that is particularly appropriate when applied to a place that is primarily beyond earth's boundaries, as no management practices humans have yet devised are affecting the health of extraterrestrial resources. The *ability to observe* that heavenly landscape is what is in danger.

Firsthand observations are exceedingly important to the development of values, philosophies, and matters of a spiritual nature. It may be argued that technological improvements in the observation of the universe beyond earth have rendered firsthand observations unnecessary or even inferior. The Hubble Space Telescope, placed above light pollution and atmospheric distortion of ground-based observatories, has provided views of space of unprecedented detail. One may merely log on to Web sites such as "Astronomy Picture of the Day" and enjoy an unparalleled view of the universe from the vantage point of a personal computer. While there is no reason to believe that such images have been artificially created or manufactured, such an experience is still essentially virtual reality. It is equivalent to watching a church service on television rather than attending in person, or watching a nature film on wild animal behavior rather than observing them firsthand. Part of the intent of the Wilderness Act of 1964 was to provide all Americans access to "primitive and unconfined" recreation and opportunities for the spiritual enlightenment and personal development such experiences provide. The view of a dark night sky can certainly be interpreted as an integral part of that experience, and remote wilderness parks are among the few places left where it can be seen.

It is illustrative to review events surrounding the creation of two of the most famous and cherished national parks, Yellowstone and Yosemite. Although Yellowstone is the nation's oldest national park, created in 1872, Yosemite Valley was set aside as a park or preserve first by a federal grant of land to the state of California in 1864. Roderick Nash, in his classic work *Wilderness and the American Mind*, describes both these great milestones in America's cultural development. He attributes the creation and early maintenance of the Yosemite park largely to Frederick Law Olmsted, the leading landscape architect of his time. Olmsted's words in an advisory report to the California Legislature in 1865, as quoted and commented on by Nash, make a strong case for the esthetic of the sublime:

> It [the report] opened with a commendation of the preservation idea which precluded "natural scenes of an impressive character" from becoming "private property." Olmsted next launched a philosophical defense of scenic beauty: it had a favorable influence on "the health and vigor of men" and especially on their "intellect." . . . Capping his argument, Olmsted declared: "the enjoyment of scenery employs the mind without fatigue and yet exercises it; tranquilizes it and yet enlivens it; and thus through the influence of the mind over the body, gives the effect of refreshing rest and reinvigoration of the whole system" (Nash 1982, 106).

While Yosemite's unique and striking beauty was virtually indisputable and embraced by nearly all who visited the valley, the new state preserve was relatively small in area and was intended primarily to preserve the scenery in and around the valley only. Yellowstone National Park was initially thought of as a "museum of wonderful and natural curiosities," but it was realized later by members of Congress that this vast area of the Montana Territory was indeed a wilderness preserve. It was in this spirit that the language of the Wilderness Act of 1964 was crafted. The value of a wilderness area lies not only in the scenery and geologic oddities, but in its "primeval character," essentially "untrammeled by man." A challenge to the notion of the need for a large, wild preserve occurred in 1886 when a railroad was proposed to cross park land to support mining ventures in the area. Nash recounts the debates in the *Congressional Record* and comments on this major victory for the preservation ethic:

> Representative [Lewis E.] Payson [of Illinois] . . . understood the park's function as the protection of curiosities. "I can not under-stand the sentiment," he admitted, "which favors the retention of a few buffaloes to the development of mining interests amounting to millions of dollars."
>
> But to Representative William McAdoo of New Jersey, Yellow-stone performed a larger function. Answering Payson, he pointed out that the park also preserved wilderness which the railroad would destroy even if it did not harm the hot springs. He added that the park had been created for people who might care to seek "in the great West the inspiring sights and mysteries of nature that elevate mankind and bring it in closer communion with omniscience" and that it "should be preserved on this, if for no other ground." McAdoo continued with a vindication of the principle of wilderness preserva-tion: "the glory of this territory is its sublime solitude. Civilization is so universal that man can only see nature in her majesty and primal glory, as it were, in these as yet virgin regions."
>
> A vote followed in which the railroad's application for a right-of-way was turned down 107 to 65. Never before had wilderness values withstood such a direct confrontation with civilization (Nash 1982, 114–15).

These accounts demonstrate two important perspectives on wilderness preservation that appear to have a long history in our culture. In the Yosemite example, Olmsted's view that certain landscapes have such intrinsic beauty that they can never become "private property" can inarguably be applied to the view of the night sky. If an artificial light is erected and maintained that compromises or interferes with the view of the night sky from a wilderness preserve, that light is in violation of one of the basic premises of the wilderness ethic: namely, obvious evidence of human technology becomes visible on the landscape. Such a situation is known as "light trespass," and may be regarded as just as serious a violation of the wilderness character as the trespass of domestic livestock or off-highway vehicles onto wilderness lands. If Olmsted's philosophy is extended to night sky "scenery," skies visible from wilderness preserves should receive the same protection as the terrestrial landscape within the preserve itself. A similar concept was put forth in the Clean Air Act Amendments of 1977, whereby vistas that were considered integral to the meaning and purpose of federal Class I (wilderness) areas were given special protection from visibility impairment, even if the vista overlooked lands outside the preserve.

The Yellowstone example is remarkable in that it shows the firm conviction by members of Congress in the idea of wilderness preservation over 100 years ago, and in the realization that preservation of such areas requires a large buffer of protection from development. The proposed railroad would almost certainly have been routed to avoid the "natural curiosities" of the geysers, waterfalls, and canyons of Yellowstone, but Congress recognized that it would violate the character of "these as yet virgin regions" by voting to keep its path entirely outside of the park. Regions now possessing "virgin" night skies that also fall within designated wilderness can be seen to be experiencing a similar threat to that facing Yellowstone in the 1880s. To protect pristine sky quality, however, the buffer must extend beyond the park's boundaries because of the potential for the glow of lights from cities, towns, or industrial areas to pollute the skies within a hundred-mile or more radius of the source.

Exactly what is lost when the view of the night sky is less than pristine? Astronomers, professional and amateur, have addressed this question from a technical or scientific standpoint for at least three decades. An increase in sky brightness leads to a decrease in contrast between faint or diffuse astronomical objects and the sky background, in many cases rendering them invisible. Many of the more sublime features of the night sky are subtle or diffuse in nature, such as the zodiacal light and the Milky Way. A glow near the horizon from distant cities or towns, while not significantly affecting the sky quality near the zenith, may cause a "light dome" to become silhouetted against the horizon and foreground objects in the direction of the city, washing out stars and other astronomical objects. This leads to a significant degradation of the wild appearance of the landscape for the nighttime wilderness visitor. An unbroken carpet of stars extending nearly to the horizon in all directions with no evidence of artificial light is one of the more impressive features of a wilderness landscape, especially in the high-mountain or desert regions where the air is commonly very transparent.

The human eye-brain combination is well adapted to allow us to function in all but the densest of forests or deepest of canyons at night, even when the moon is not in the sky. According to Schaaf (1988, 167), "the faintest stars the human eye can glimpse without optical aid are about 100,000,000,000,000 (100 trillion) times dimmer than the brightest light it can perceive without suffering damage." A landscape illuminated by the full or nearly full moon is relatively easy to negotiate for the properly dark-adapted human eye. In open country, even with nothing more than starlight at the most remote of wilderness locations, it is possible for humans to get around without the use of an artificial light. It became second nature to our ancestors to use what are now known as "astronomer's tricks" (such as averted vision) to enhance night vision. Wilderness travelers of today almost entirely restrict their wanderings to the daytime hours. Occasionally, however, there arises a need or desire to travel at night, planned or unplanned, and often without the aid of a flashlight. It is during these episodes that much is learned about the eye's true capabilities, and the natural light sources in the night sky are fully appreciated. There are those who actually *prefer* to travel at night, especially in hot desert regions or exposed subalpine or alpine environments, to avoid some of the harmful effects of full sunlight on the body. For them, the wilderness preservation ethic applies twenty-four hours a day.

It is no mere coincidence that amateur astronomers commonly pursue their hobby in national parks. What better environment to appreciate the beauty of the night sky than from protected and pristine earthly landscapes? The San Francisco Sidewalk Astronomers, led by John Dobson in the 1970s and 1980s, promoted the glories of the night sky with star parties at Glacier Point in Yosemite National Park and in Death Valley National Park, gatherings which continue to the present day. Most national parks have interpretive programs on the night sky for the public. It is a simple step to include the preservation of night sky visibility in programs concerned with wilderness values or air quality–related values.

The most important value of firsthand observations of the night sky wilderness may well be the possibility that such experiences will lead to an expansion of an individual's perception. Aldo Leopold, in his famous book *A Sand County Almanac* (1949, 204), put forth the idea of a land ethic, which "simply enlarges the boundaries of the community to include soils, waters, plants, and animals, or collectively the land." This concept is introduced only after Leopold relates many observations of animal behavior in wild places, so that it is obvious to the reader that such an ethic is an inevitable consequence of intimate knowledge of a place. The more one knows about a place, its features and inhabitants, the more one is likely to advocate preservation of those features, because they have led to one's personal and spiritual development. Rodman elaborates on this remarkable character of Leopold's work that has universal appeal: "When perception is sufficiently changed, respectful types of conduct seem 'natural,' and one does not have to belabor them in the language of rights and duties. Here, finally, we reach the point of 'paradigm change.' What brings it about is not exhortation, threat, or logic, but a rebirth of the sense of wonder that in ancient times gave rise to philosophers but is now more often found among field naturalists" (Rodman 1983, 90).

Leopold writes of trying to put oneself in the position of various wild animals—the skunk, the mouse, the muskrat—and trying to imagine what their perception of their environment is. Those who would observe the universe under pristine skies from the platform of a wilderness preserve are ideally situated to place themselves *out there*. This type of perception expansion inevitably leads to the sense of wonder that fosters a better understanding of the place of humans, not only on this planet but in the universe as a whole. While the objects, events, and processes that go on outside earth may seem to some only to have practical value to theoretical physicists and fortune tellers, there can be little argument as to the cultural and spiritual significance of the face of the night sky. If it can be left "unimpaired for future generations," the opportunity for the development of such intimate knowledge of the universe may in fact lead to respectful types of conduct, one of which might just be the judicious and conservative use of artificial lighting.

17

A SYSTEM-WIDE ASSESSMENT OF NIGHT RESOURCES AND NIGHT RECREATION IN THE U.S. NATIONAL PARKS

Brandi Smith and Jeffrey Hallo

Concern about natural darkness is conventionally focused on night skies. However, this chapter encourages readers to broaden this concern to include an array of recreation activities that are conducted at night. Examples include camping, interpretive programs, and observing nocturnal wildlife. The authors encourage the adoption of the terms "night resources" and "night recreation" as a way to address the broad spectrum of recreation activities that are dependent on natural darkness. The study reported in this chapter is supported by an examination of Web sites for national and state parks and a survey of national park managers. Study findings suggest that nighttime recreation activities are diverse and that many visitors participate in them.

INTRODUCTION

The nighttime environment has historically included darkness in outdoor settings, brightened only to the degree that celestial objects and human-sourced light allowed. Human-caused lighting has increased in intensity and use over the last several decades, producing what is known as light pollution, or nuisance lighting. It is estimated that much of the world's skies are now deemed light-polluted, and the severity and extent of light pollution are expected to increase substantially (Cinzano et al. 2001). A key trait of nuisance lighting is that it shines where it is not wanted (Brons et al. 2008), creating light trespass, or is deemed problematic in some other way. The U.S. National Park Service (NPS) has documented light pollution up to 200 miles (322 km) from its source in the form of sky glow: the orange or milky-gray glow characteristic of many metropolitan areas at night. Remote locations that have few or no nuisance light sources of their own can be affected by distant light sources via sky glow.

The NPS has a small team of scientists dedicated to addressing what it calls "natural lightscapes," or increasingly, "natural darkness." The Night Sky Team (NST) uses science and technology to better understand the impact of anthropogenic light on the view of the celestial sky and to develop management recommendations for protection of these nighttime resources. Since its inception in 1999, the NST has expanded its scope to address the cultural, historical, ecological, and experiential (i.e., recreational) value of the night in the national parks. NPS management policies paralleled this change and in 2001 incorporated discussion of ecological and cultural values of natural lightscapes (natural resources and values found in the absence of human-caused light). Yet the bulk of nighttime stewardship remains focused on the celestial view and stargazing. This narrow bias may be a result of the decades of outreach by professional and amateur astronomers or the appearance of other park-related efforts and organizations. For example, the Starlight Initiative, the International Union for Conservation of Nature's (IUCN) World Commission on Protected Areas, and the International Dark-Sky Association remain focused on the view of the sky, whether on scientific, aesthetic, or cultural grounds.

A consequence of this institutional narrow focus is that a park manager may dismiss or minimize the value of the nighttime environment if the desire for stargazing is low, and overlook the wide range of other recreational activities that are linked to a naturally dark nighttime environment. Additionally, the fraction of the public that enjoys stargazing per se is likely smaller than the fraction that enjoys other nighttime recreational activities. Nighttime recreation may include other activities such as nocturnal species observation, historical or cultural learning, night fishing, camping, and night hiking. Night resources include nocturnal flora and fauna, the relative quiet of the night, and a natural dark environment. No accepted definitions of night recreation or night resources exist. This is problematic because an incompletely or incorrectly defined activity or resource cannot be properly managed, protected, or fully appreciated.

Empirical examinations of night resources—other than the night sky—and night recreation are just beginning to occur from a social science perspective. A need exists to better understand the diversity of activities, experience opportunities, and use levels of night recreation in parks and protected areas. This paper presents (1) a census of night recreation activities offered in U.S. national and state parks; (2) proposed, expanded, and formalized definitions of *night recreation* and *night resources*; and (3) an assessment of opportunities for, access to, and visitor participation in night recreation activities in the U.S. national park system.

METHODS

Census of Night Recreation and Expanded Definitions

Because of the narrow, incomplete, and informal definitions of *night recreation* and *night resources*, a census of night activities offered in national ($n = 392$) and state park ($n = 3,500$) units with Web sites was conducted as a preliminary step. We visited and searched each park unit Web site by exploring the site systematically (i.e., home page, visitor activities information, and activity calendars) and using specific search terms (i.e., night, dark, star, moon, and nocturnal). We assumed that Web site content and calendar listings of activities and educational programs were current and accurate. The census included both national and state parks to enhance the breadth of investigation of potential forms of night recreation and night resources. For each Web site visited, we recorded night-dependent or night-related recreation activities. This list then served as the basis for more complete definitions and examinations of night recreation and night resources.

Night Opportunities and Activities in the National Parks

Based on the census, we created a paper-based questionnaire to assess the opportunities for, access to, and visitor participation in night recreation activities in the national park system. Also, the questionnaire allowed the activities identified through the Web site census to be examined for validity and completeness. We sent questionnaires to superintendents (or equivalent) of the national park units. We included only those parks solely managed by the NPS. This yielded a final study population of 390 national park units.

We distributed questionnaires using a modified Dillman (2007) approach. This approach involved an initial mailing with the questionnaire and a cover letter, followed by a postcard reminder to nonrespondents, a second mailing of the questionnaire and a modified cover letter to nonrespondents, and a final contact by telephone. The cover letter and questionnaire contained a definition of *night recreation* and *night resources* (presented later in this paper) and a request that the survey be forwarded to the park employee who the superintendent felt would best be able to answer the questions.

Parks were asked to complete the survey even if they did not consider themselves a "night park" to ensure a complete assessment of night activities in the national parks.

In the questionnaire, respondents were asked to indicate whether their park is ever open during dark hours. This question was intended to assess the number of parks that potentially could offer night activities or that may use night resources for visitor enjoyment. Additionally, the number of parks whose information facilities, such as visitor centers, are ever open during dark hours was captured. We then asked respondents whether the listed night resource activities occurred in their park and whether visitors could engage in the activity on their own or as part of a park program. Respondents were also able to indicate whether an activity is prohibited in their park. Finally, respondents were asked to note the number of both campers/lodgers and other nighttime visitors in their park.

RESULTS

Census of Night Recreation and Defining Night Resources

The census of night activities yielded 15 night-dependent or night-related recreation activities or categories of activities (Table 17.1). This broad range of night activities is evidence that night resources and night activities go beyond the night sky and stargazing and supports the need for more comprehensive definitions of the terms *night resources* and *night recreation*. We note that no definition of these terms or concepts is given or implied in the National Park Services' *Management Policies 2006*, and the term *lightscape* used in this document is a limited and vague concept described as "natural resources and values that exist in the absence of human-caused light" (2006, 57). Based on the variety of night resource activities found in our census of Web sites, we propose that the terms *night resources* and *night recreation* be more comprehensively defined as follows:

> Night resources: anything that either enhances the visitor experience after sunset (including safety measures, recreational opportunities, and interpretive programs), or that is most active or prominent at night, including animals, plants, and features of the night sky.
>
> Night recreation: any recreational activity occurring after sunset, including camping.

TABLE 17.1. Night-dependent or light-sensitive night recreation activities recorded in a census of state and national park Web sites

Activity	A participating park	Activity example
Campfires	Patapsco Valley State Park, Md.	Campfire programs with park-sponsored entertainment (i.e., cooking campfire food for audience, storytelling)
Camping	Whitewater State Park, Minn.	Overnight "I Can Camp" program that teaches participants how to set up tents, build campfires, and cook outdoors
Interpretive programs at night park	New Bedford Whaling National Historical Park, Mass.	"AHA! Night" (Art, History, Architecture), held throughout the district in collaboration with the community
Night bike riding	Riverside State Park, Wash.	Nighttime mountain bike riding allowed within park boundaries (self-facilitated)
Night boating, canoeing, kayaking, or rafting	Lake Catherine State Park, Ark.	Full-moon kayak tours
Night concerts or plays	Cape Disappointment State Park, Wash.	"Waikiki Beach concert series" throughout summer months
Night fishing	Bill Burton Fishing Pier State Park, Md.	Fishing from piers specially lit for night fishing
Night hiking or walking	Rocky Mountain National Park, Colo.	"Walk into Twilight" (2 hours, ranger-led), observing sights and sounds of night in the park
Night hunting	Big South Fork National Recreation Area, Tenn.	Self-facilitated hunting of specified game
Night photography	Glacier National Park, Mont.	"Astrophotography of Glacier's Night Sky" (ranger-led)
Night snow skiing or snowshoeing	Voyageurs National Park, Minn.	"Night Light Snowshoe Hike") (ranger-led
Special night events or festivals	Antietam National Battlefield, Md.	"Civil War Soldier Campfire Program"
Stargazing, star parties, or viewing the Northern Lights	Blackwater Falls State Park, W.Va.	"Astronomy weekend" featuring speakers, workshops, and stargazing parties
Viewing natural, cultural, or historical resources at night	Hawai'i Volcanoes National Park, HI	Identifies "Night Glow" viewing areas for visitors based on current lava flow locations
Wildlife viewing at night (excluding spotlighting)	Congaree National Park, S.C.	"Owl Prowls" (ranger-led)

Night Opportunities and Activities in the National Parks

A total of 313 national park system units returned completed questionnaires, yielding a response rate of 80.3%. Of those, 80.2% (251 units) indicated that their park is open at least sometimes after sunset. Just over 54% of respondents indicated that information facilities in their park, such as visitor centers, are ever open to visitors at night.

Respondents were asked to indicate which of the 15 previously identified night activities visitors could participate in, under what conditions, and whether visitors engage in these activities (Table 17.2). Results show that each night resource activity listed occurs and is pursued by visitors in at least one park. Also, each activity was prohibited in at least one park. Night interpretive programming is the most widely offered ($n = 210$) and pursued ($n = 181$) night activity. (This difference in the number of parks in which programs are offered versus participated in may be partially due to measurement error—many respondents indicated that an activity is participated in at their park, but did not indicate whether or not visitors could do this activity on their own, as part of a park-facilitated program, or both.) Second to this, night hiking or walking was permitted as a self-facilitated activity in 190 parks, with 179 parks indicating that visitors engage in this activity.

Respondents were asked to indicate their best estimate of visitors (both lodgers/campers and other nighttime visitors) who use their park at night on an annual basis (Table 17.3). A majority of respondents did not supply a number, choosing either "Not Applicable" or "Don't Know," or did not respond to the item at all. Of those who did supply a numeric response, 56 (17.9%) estimated that fewer than 500 people camp or lodge in their park in an average year. Likewise, 43 (13.7%) indicated that fewer than 500 nighttime visitors (noncampers/lodgers) use their parks in a given year. Other response ranges were indicated with less frequency, but some parks indicated that hundreds of thousands or millions of visitors either stay in their parks overnight or visit during nighttime hours annually.

DISCUSSION

The majority of national park units responding to the survey reported that they are open during night hours at some point. This figure includes parks that only occasionally grant visitors access during night hours, such as for historical reenactments or holiday programs. However, just over half of responding national park units indicated that information facilities, such as visitor centers, are open during night hours. In these places, nighttime vis-

TABLE 17.2. Frequency of night recreation activity availability and reported visitor participation in the national park system

Activity	Specifically prohibited	Permitted to do this activity on their own	Permitted to do this activity as part of program	Visitors engage in this activity in my park
Campfires	162 (51.4%)	90 (28.6%)	51 (16.2%)	123 (39.0%)
Camping	165 (52.4%)	97 (30.8%)	37 (11.7%)	132 (42.2%)
Interpretive programs at night	25 (7.9%)	64 (20.3%)	210 (66.7%)	181 (57.5%)
Night bike riding	127 (40.3%)	149 (47.3%)	8 (2.5%)	88 (27.9%)
Night boating, canoeing, kayaking, or rafting	151 (47.9%)	104 (33.0%)	12 (3.8%)	76 (24.1%)
Night concerts or plays	64 (20.3%)	41 (13.0%)	156 (49.5%)	91 (28.9%)
Night fishing	153 (48.6%)	105 (33.3%)	3 (1.0%)	98 (31.1%)
Night hiking or walking	65 (20.6%)	190 (60.3%)	77 (24.4%)	179 (56.8%)
Night hunting	280 (88.9%)	17 (5.4%)	2 (0.6%)	32 (10.9%)
Night photography	56 (17.8%)	193 (61.3%)	51 (16.2%)	153 (48.6%)
Night snow skiing or snowshoeing	108 (34.3%)	112 (35.6%)	16 (5.1%)	74 (23.5%)
Special night events or festivals	25 (7.9%)	53 (16.8%)	209 (66.3%)	143 (45.4%)
Stargazing, star parties, or viewing the Northern Lights	37 (11.7%)	162 (51.4%)	109 (34.6%)	168 (53.3%)
Viewing natural, cultural, or historical resources at night	36 (11.4%)	179 (56.8%)	125 (39.7%)	166 (52.7%)
Wildlife viewing at night (excluding spotlighting)	69 (21.9%)	177 (56.2%)	61 (19.4%)	137 (43.5%)
Other	1 (0.3%)	5 (1.6%)	1 (0.3%)	4 (1.3%)

Note: Frequencies represent the number and percentage of park units responding affirmatively.

itors may not have access to information about park resources and may not have the opportunity to interact with park personnel to learn about activities or resources not featured in printed information sources. Therefore, it is likely that nighttime visitors are not given information that would allow them to experience night resources, including simply being made aware of those resources. With the exception of scheduled campfire or evening programs, nighttime visitor use is often allowed but not supported by open facilities, available staff, or readily available information. Parks seldom create

TABLE 17.3. Number of night visitors (annually) reported
by units of the national park system

Quantity	CAMPERS AND LODGERS		NONCAMPERS/LODGERS	
	Frequency	Percentage	Frequency	Percentage
Less than 500	56	17.9%	43	13.7%
500 to 999	4	1.3%	9	2.9%
1,000 to 4,999	9	2.9%	11	3.5%
5,000 to 9,999	11	3.5%	4	1.3%
10,000 to 19,999	9	2.9%	5	1.6%
20,000 to 49,999	8	2.6%	3	1.0%
50,000 to 99,999	9	2.9%	1	0.3%
100,000 to 199,999	5	1.6%	1	0.3%
200,000 to 499,999	9	2.9%	9	2.9%
500,000 to 999,999	1	0.3%	1	0.3%
More than 1,000,000	2	0.6%	1	0.3%
Not Applicable	118	37.7%	156	49.8%
Don't Know	46	14.7%	43	13.7%
No Response	25	8.0%	21	6.7%

areas intended for stargazing, actively encourage nighttime use of trails, or accommodate nighttime cultural events.

Most respondents did not know the number of nighttime visitors to their park unit. This may reflect a difficulty in counting visitors, but may also suggest that nighttime use of parks and demand on night resources are not well monitored. When provided, estimates of use suggest that night recreation in park units is often low, but some parks reported nighttime visitor use levels that are substantial. This variation is likely due to factors such as the night resources that a park contains, the uniqueness of these resources, how they are promoted or used, and the type and number of visitors to a park. Some parks seem more night-focused than others. For example, Golden Gate National Recreation Area has thousands of visitors who come to participate in night concerts and other performances. Other park units may not offer or recognize particular night recreation activities because they have no indication that it would appeal to visitors and have not identified any other reason to offer certain experiences.

Night recreation activities may require facilitation by park personnel and may therefore add to the demand for park personnel in time and cost. Parks may find assistance from outside volunteers or organizations that are aligned with a given activity. For example, a park that does not have personnel to facilitate a night hike may find volunteers in a nature-based organization who are able to lead such an activity. Likewise, astronomy groups may be a rich source of assistance for night sky programs.

Respondents were able to indicate whether any of the 15 listed night activities were prohibited in their park unit. Night hunting, camping, and campfires were most often prohibited, reflecting the philosophy and policy of many national parks, a lack of campground facilities, and wildfire threats, respectively. Other night recreation activities may be prohibited because of the inherent dangers of a given park or activity. For example, hiking and walking in parks during daytime hours are permitted in most parks, yet 20% of respondents reported that night hiking and night walks are explicitly prohibited in their park. This may be partially because of increased perceptions of risk (e.g., tripping, hostile wildlife, disorientation) associated with hiking at night. Several parks indicated that night access is limited in an effort to protect their night resources, such as sea turtles that nest at night.

Respondents were also able to indicate whether a given night recreation activity could occur as part of a park program or whether a visitor could engage in the night activity without supervision. Results suggest that a majority of night activities most often occur individually (i.e., "on your own"), rather than with a ranger or as part of a formal program. However, a substantial percentage of activities did occur with a ranger or as part of a park program. This makes sense because many night activities (e.g., nocturnal species observation, astronomy, night concerts/events) require technical expertise, specialized equipment, or knowledge that makes participating in these activities as an individual less feasible. In such cases the park interpretive ranger or performer might be considered a park unit's night resource.

We also note that findings from this study are an incomplete picture of night recreation and night resources because they represent only managers' observations and management policies. Visitors must be polled about their perceptions of night recreation and night resources. It is likely that NPS managers do not have a completely accurate perception of which night recreation activities or related night resources are of value to visitors. Research demonstrates that park managers and visitors often have distinct and divergent attitudes, values, and beliefs (Manning 2011).

In many cases night may not be perceived as a distinct condition but rather as a gradual transition from or to daytime lighting. This may include crepuscular periods immediately before, during, or after sunrise or sunset. Likewise, some resources or recreation activities may not be distinctly night-focused, but are influenced heavily by natural light conditions. For example, the bat flight at Carlsbad Caverns National Park and the sunrise at Haleakala National Park are both substantially night- and light-related, but may not occur wholly while the sun is below the horizon. Also, nature photographers often seek out and take advantage of special lighting conditions associated with the "golden hour" that occurs immediately before sunset or after sunrise. Resources and recreation activities such as these may be considered crepuscular resources or recreation activities. Likewise, caving and visiting pre-electricity-era historical structures could be considered light-dependent resources or recreation activities.

Perhaps the most substantial outcomes of the research presented here are the proposed definitions of both *night recreation* and *night resources*. Survey results show that these definitions are more inclusive and accurate than those informal and implied definitions that now limit consideration of night in parks to the night sky or night sky viewing. These proposed broadened definitions may enhance recognition of night resources, their use and enjoyment by visitors, and their management.

18

INDICATORS AND STANDARDS OF QUALITY FOR VIEWING THE NIGHT SKY IN THE NATIONAL PARKS

Robert Manning, Ellen Rovelstad, Chadwick Moore,
Jeffrey Hallo, and Brandi Smith

In the introduction to this book, we introduced a management-by-objectives framework for managing park resources and experiences. This is an adaptive process that can be applied to natural, cultural, and recreational resources. A key component of this management approach is the formulation of indicators and standards of quality for the resource under consideration. This chapter describes a two-phase program of research at Acadia National Park designed to help guide formulation of indicators and standards of quality for night sky viewing in the park.

NIGHT SKIES AS A "NEW" PARK RESOURCE

The emerging importance of natural darkness and night skies is a function of the intersection of a growing consciousness about their values and a crisis over their rapid disappearance. For millennia, people have gazed upon the cosmos in their enduring efforts to understand both the physical and metaphysical worlds, and this suggests that night skies are an important cultural resource (Bogard 2013). Human culture is conventionally organized around the rhythms of the sun, moon, and stars; observations of the night sky are embodied in the religions and mythologies of cultures around the world; and the celestial world has been the inspiration for art, literature, and other forms of cultural expression (Rogers and Sovick 2001a; Collison and Poe 2013). Modern science has extended the importance of night skies by demonstrating the relevance of darkness in the biological world; many of the world's species rely on the absence of artificial light for breeding and feeding patterns and other behaviors (Lima 1998; Witherington and Martin 2000; Le

Corre et al. 2002; Alvarez del Castillo et al. 2003; Longcore and Rich 2004; Pauley 2004; G. Perry and Fisher 2006; Rich and Longcore 2013; Wise and Buchanan 2006; López and Suárez 2007; Navara and Nelson 2007; Chepesiuk 2009; Luginbuhl et al. 2009). Light pollution can even affect humans through sleep disturbance and other health effects (Nicholas 2001; Clark 2008; Chepesiuk 2009).

Unfortunately, the night sky is disappearing from view primarily because of "light pollution" that reduces the brightness of the stars and prevents the human eye from fully adapting to natural darkness. Outdoor lighting that is excessive, inefficient, and ineffective can produce light pollution that degrades the quality of natural darkness and the night sky by creating "sky glow." Cinzano et al. (2001) estimated that more than 99% of the U.S. population (excluding Alaska and Hawaii) lives in areas that are light polluted and that two-thirds of Americans can no longer see the Milky Way from their homes. Light pollution is caused by increasing development, but may be more related to lighting that is oriented upward or sideways rather than down at the intended target. Light from urban areas can reduce the brightness of the night sky over 200 miles (322 km) away (http://www.nature.nps .gov/sound_night/; B. Smith and Hallo 2013).

National parks, especially those far from urban areas, are some of the last refuges of dark night skies, and the importance of night skies is increasingly reflected in National Park Service (NPS) policy and management. Current NPS management policies include a requirement for managing "lightscapes," or natural darkness and night skies (National Park Service 2006), and a relatively new NPS administrative unit, the Natural Sounds and Night Skies Division, was created to help carry out this responsibility. Night sky interpretive programs are now conducted in an increasing number of units of the national park system, as manifested in night sky festivals and "star parties" at Yosemite, Acadia, and Death Valley National Parks; creation of a night sky ranger position at Bryce Canyon National Park; and development of an observatory at Chaco Culture National Historical Park. The NPS established its Night Sky Team, a small group of scientists, in 1999 and this has led to rigorous measures of night sky quality and associated monitoring in the national park system. Night sky quality is included as a "vital sign" by many of the 32 NPS Inventory and Monitoring Networks that cover the national park system. The recent influential NPS report, "A Call to Action," includes a recommendation that the NPS "lead the way in protecting natural darkness as a precious resource and create a model for dark sky protection" (National Park Service 2014). A recent survey of managers across the national park system found that night skies (and "night resources" more broadly, including the opportunity to observe nocturnal species) are frequently used by visitors and that managers are interested in identifying and managing night resources more actively (B. Smith and Hallo 2011).

INDICATORS AND STANDARDS OF QUALITY
FOR NIGHT SKY VIEWING

Contemporary approaches to park and outdoor recreation management rely on a management-by-objectives approach (Manning 2007; Whittaker et al. 2011; Interagency Visitor Use Management Council 2016). This management approach relies on formulation of indicators and standards of quality that serve as empirical measures of management objectives (such as protection of natural darkness). Indicators of quality are generally defined as measurable, manageable variables that are proxies for management objectives, while standards of quality (sometimes called "reference points" [Manning 2013] or "thresholds" Interagency Visitor Use Management Council 2016]) define the minimum acceptable condition of indicator variables (Manning 2011; Whittaker et al. 2011; Interagency Visitor Use Management Council 2016). For example, a conventional indicator of quality for a wilderness experience is the number of groups encountered per day along trails, and a standard of quality is the maximum acceptable number of groups encountered, such as five. Once indicators and standards of quality have been formulated, indicator variables are monitored and management actions implemented to help ensure that standards of quality are maintained. This is an adaptive process that has been incorporated into NPS visitor use planning and management (Interagency Visitor Use Management Council 2016).

Formulation of indicators and standards of quality that address recreational use of parks can include engagement of park visitors. A growing body of research illustrates how this can be done through visitor surveys and associated theoretical and empirical approaches (Manning 2011). Several recent studies have concluded that there is a need for this type of research applied to night sky viewing or stargazing. For example, reflecting on their recent survey of park managers about nighttime recreation, B. Smith and Hallo conclude that "visitors must be polled about their perspectives of night recreation and night resources" (2013, 58). In their evaluation of night sky interpretation at Bryce Canyon National Park and Cedar Breaks National Monument, Mace and McDaniel conclude that "additional research could lead to development of standards and indicators of quality for night skies in parks and protected areas, a perspective that has been very successful in the field of park and outdoor recreation management" (2013, 55).

THE STUDY: VISITOR SURVEYS

We conducted this study to help guide formulation of indicators and standards of quality for night sky viewing in the national parks. The program of research included two visitor surveys conducted at Acadia National Park (Acadia). Acadia is located primarily on Mount Desert Island, Maine. Many visitors stay overnight in one of the park's two campgrounds, Blackwoods and Seawall. Because of its location away from large metropolitan areas, Acadia prides itself on being a premier location to view the night sky in the eastern United States. The importance of the night sky at Acadia is manifested in the park's annual Night Sky Festival, a four-day event featuring special presentations and star parties in which amateur astronomers set up their telescopes and allow visitors to see many celestial bodies. Acadia's regularly scheduled ranger programming also features night walks and astronomy evening programs.

The first survey addressed the importance of night sky viewing and associated indicators of quality. The survey instrument included two batteries of questions. The first addressed the importance of night sky viewing to park visitors by posing a series of statements (shown in Table 18.1) and asking respondents to report the extent to which they agreed or disagreed with each statement using a five-point response scale that ranged from -2 ("strongly disagree") to 2 ("strongly agree"). The second battery of questions presented a series of items (shown in Table 18.2) that visitors might see after dark in the park. The list included celestial bodies and human-caused sources of light. We asked respondents to report which items they did or did not see and indicate the extent to which seeing or not seeing these items added to or detracted from the quality of their experience in the park. A nine-point response scale that ranged from -4 ("substantially detracted") to 4 ("substantially added") was used. This latter battery of questions is adapted from a "listening exercise" that has been used to assess natural and human-caused sound in national parks and its effects on the quality of the visitor experience (Pilcher et al. 2009; Manning et al. 2010).

We administered the survey to park visitors at the two campgrounds in Acadia. We sampled campground visitors because they were the most likely to be in the park at night (there are no other accommodations in the park). We intercepted groups of campers as they entered the campgrounds and gave

TABLE 18.1. The importance of viewing the night sky
to Acadia National Park visitors

| | FREQUENCY OF RATING (%) | | | | | | | |
| | Strongly disagree | | Unsure | Strongly agree | | | Standard | |
Statement	−2	−1	0	1	2	Mean	deviation	n
Viewing the night sky (stargazing) is important to me.	0.0	1.0	9.4	35.4	54.2	1.4	0.7	192
The National Park Service should work to protect the ability of visitors to see the night sky.	0.0	1.1	8.9	36.8	53.2	1.4	0.7	190
The National Park Service should conduct more programs to encourage visitors to view the night sky.	0.5	2.1	27.1	37.5	32.8	1.0	0.9	192
Acadia has a good reputation as a place to view the night sky.	1.6	2.6	44.4	25.9	25.4	0.7	0.9	189
One of the reasons chose to visit Acadia is to view the night sky.	4.2	12.1	39.5	26.3	17.9	0.4	1.1	190
I would visit Acadia less often if it became more difficult to see the night sky.	7.9	17.9	40.0	22.6	11.6	0.1	1.1	190

a questionnaire to the group for a self-identified group leader to complete. We asked respondents to complete the questionnaire before they went to sleep that night or early the following morning, and then return the completed questionnaire to a drop box as they left the campground the next morning. We administered the survey for 13 days in August 2012. We contacted 277 groups and 273 agreed to participate; 194 completed questionnaires were returned representing a 70% response rate.

The second survey addressed standards of quality for night sky viewing. We prepared a series of eight visual simulations of the night sky at Acadia as

TABLE 18.2. Questionnaire list of items seen and not
seen at Acadia

The Milky Way
Constellations
Stars or planets
Meteors/shooting stars
The moon
Satellites
Aircraft
Lights from distant cities
Lights from nearby towns
Campfires
Automobile lights
Flashlights
Lanterns
Streetlights
Portable work lights
Park building lights
Emergency vehicle lights

---

shown in Figure 18.1. These simulations portrayed equally spaced degrees of light pollution. We asked respondents to rate the acceptability of each of the simulations using a seven-point response scale that ranged from -3 ("very unacceptable") to 3 ("very acceptable"). We asked an additional suite of questions based on the series of visual simulations, as follows:

- Which image shows the night sky you would prefer to see in the park?

- Which image represents the maximum amount of human-caused light the National Park Service should allow in and around this park?

- Which image is so unacceptable that you would no longer come to this park to stargaze or view the night sky?

- Which image is so unacceptable that you would not stargaze or view the night sky when visiting this park?

- Which image looks most like the night sky you typically saw in this park during this trip?

Image 1

Image 2

Image 3

Image 4

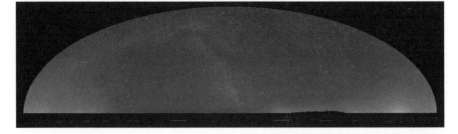

FIG 18.1: Visual simulations of night sky quality at Acadia National Park. These are panoramas of "light domes" as seen from Cadillac Mountain in Acadia. Image 1 is a natural night sky based on observations taken in the park in 2008. Each of the following images shows a three times increase in artificial light. Image 8 shows a severely light-polluted sky.

These simulations were prepared by the NPS Natural Sounds and Night Skies Division.

Image 5

Image 6

Image 7

Image 8

- Which image looks most like the night sky you think is "natural" in this park?

- Which image looks most like the night sky you typically see from your home?

We administered the survey to park visitors at seven attraction sites in Acadia. Visitors were sampled if they had spent at least one night on Mount Desert Island in the vicinity of the park. We intercepted visitors as they entered the attraction sites and gave a questionnaire to the group for a self-identified group leader to complete. We instructed respondents to complete the questionnaire at that time and return it to the survey attendant stationed there. The survey attendant answered any questions respondents had about the questionnaire. We administered the survey for nine days in August and September 2013. We contacted 274 groups and 137 visitors agreed to participate and completed questionnaires representing a 50% response rate.

Surveying visitors about the night sky can be challenging. One of the survey objectives was to ensure that survey participants had spent at least one night in or just outside the park to help make certain they had had an opportunity to view the park's night sky. A pilot test recruited visitors at the park's evening campfire programs, but few visitors were willing to participate at this late hour. The two other sampling approaches described earlier were more successful in reaching the target population while attaining an acceptable response rate. Another challenging issue is determining the night sky conditions that respondents experienced, since these conditions can be highly varied and transitory. In this study, we asked respondents to report the study photograph that was most like the conditions they typically experienced in the park.

Visual research methods are an effective approach to measuring standards of quality for parks and related areas (Manning and Freimund 2004; Manning 2007). For example, visually based studies can be especially useful for studying standards of quality for indicator variables that are inherently difficult or awkward to describe in conventional narrative/numerical terms, such as trail erosion. A visual approach has been used to study a wide variety of indicators of quality, including crowding, conflict, resource impacts, and management practices (Manning 2011). Several studies have addressed multiple dimensions of the validity of visual research methods, and findings are generally supportive (Manning 2007). However, findings are mixed on the issue of the order in which study photographs should be presented to respondents and the range of potential standards of quality presented (Manning 2011; Gibson et al. 2014). This study addresses these issues by presenting study photographs on posters, allowing respondents to see all photos at the same time (rather than one at a time), and presenting a complete range of night sky conditions from pristine to severely light polluted (see Figure 18.1).

STUDY FINDINGS

Importance of Night Skies

Findings from the battery of questions addressing the importance of night sky viewing are shown in Table 18.1 and indicate that the vast majority of visitors feel that (1) night sky viewing is important, (2) the NPS should protect opportunities for visitors to see the night sky, and (3) the NPS should conduct more programs to encourage visitors to view the night sky. Most visitors also reported that Acadia has a good reputation for night sky viewing and that this is one of the reasons they chose to visit Acadia. However, feelings were mixed as to whether respondents would visit Acadia less if it became more difficult to see the night sky (40% reported that they were unsure about this).

Indicators of Quality for Night Sky Viewing

Findings from the battery of questions addressing indicators of quality for night sky viewing are presented in the form of an importance-performance framework as shown in Figure 18.2. Importance-performance analysis is a way to evaluate visitor desires and associated experiences and has been used to identify indicators of quality in a range of park and outdoor recreation settings and for several recreation activities (Guadagnolo 1985; Mengak 1986; Hollenhorst and Stull-Gardner 1992; Hollenhorst et al. 1992; K. Hunt et al. 2003; Pilcher et al. 2009). For example, importance-performance analysis was used to identify indicators of quality for natural quiet in national parks (Pilcher et al. 2009). Similarly, a study of visitor experiences in wilderness used importance-performance analysis to reveal indicators of quality for resource and experiential conditions on trails and in campgrounds (Hollenhorst and Gardner 1994).

Figure 18.2 graphs the percentage of visitors who did or did not see the items listed in Table 18.2 X-axis) and how seeing or not seeing these items affected the quality of visitors' experiences (Y-axis). Generally, the figure shows that most visitors did not see many of the celestial objects included

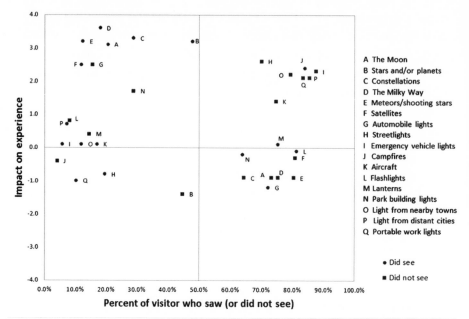

FIG 18.2: Importance-performance framework. The graph shows (1) the percentage of respondents who saw (circles) and did not see (squares) items A through Q (listed in the key on the right) on the x-axis, and (2) the positive or negative impact of this on the y-axis. For example, about 50% of respondents reported seeing stars and/or planets (item B circle) and also reported that this had a very positive impact on the quality of their experience (registered a scale value of 3.2).

in the questionnaire, but that when they did, it substantially added to the quality of their experience. Likewise, most visitors did not see many of the sources of human-caused light and this also substantially added to the quality of their experience. Campfires are an exception to these generalizations: most visitors saw campfires and this added to the quality of their experience. Overall, the findings suggest that the brightness of celestial bodies and, therefore, light pollution is an important indicator of quality at Acadia.

Standards of Quality for Night Sky Viewing

Findings from the questions addressing standards of quality for night sky viewing as manifested in the brightness of celestial objects (or alternatively, the amount of light pollution) are shown in Figure 18.3 and Table 18.3. The graph in Figure 18.3 is derived from the average (mean) acceptability ratings for each of the eight visual simulations. This type of graph has been used to

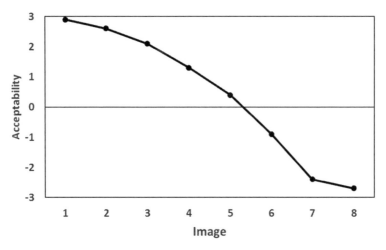

FIG 18.3: Social norm curve for night sky viewing at Acadia National Park. This curve graphs the average acceptability scores for each of the images shown in Figure 18.1.

help formulate standards of quality for resource and experiential conditions in a number of national parks (Manning et al. 1996; Shelby et al. 1996; Freimund et al. 2002; Hsu et al. 2007). It is clear from the graph that increasing amounts of light pollution are increasingly unacceptable. Average acceptability ratings fall out of the acceptable range and into the unacceptable range at around image 5 in the series presented in Figure 18.1, and this represents a potential standard of quality (defined earlier as the minimum acceptable condition of an indicator of quality). However, the data in Table 18.3 suggest a range of other potential standards of quality. For example, Acadia managers have identified night skies as an especially important resource and this suggests that a higher standard of quality—closer to what visitors feel is the maximum amount of human-caused light the NPS should allow (between images three and four in Figure 18.1)—may be appropriate.

CONCLUSION

Night skies are increasingly recognized as an important resource—biologically, culturally, and experientially—in the national parks, and this is reflected in recent NPS policy and management. This study documents this importance to national park visitors. The importance of night skies will require more explicit management in the national parks, including formu-

TABLE 18.3. Alternative standards of quality of night sky viewing

Study question	Image number	Standard deviation
The point at which the social norm curve crosses the neutral point of the acceptability scale (from fig. 4)	5.2	—
Which image shows the night sky you would prefer to see?	1.1	0.4
Which image shows the maximum amount of human-caused light the National Park Service should allow?	3.7	1.8
Which image is so unacceptable that you would no longer come to this park to view the night sky?	6.1	1.4
Which image is so unacceptable that you would no longer view the night sky when visiting this park?	6.1	1.4
Which image looks most like the night sky you typically saw in this park during this trip?	2.3	1.2
Which image looks most like the night sky you think is "natural" at this park?	2.0	1.2
Which image looks most like the night sky you typically see from your home?	4.9	2.1

lating indicators and standards of quality for viewing the night sky. The program of research described in this paper suggests how park visitors and other stakeholders can be engaged in this process. Findings from this study suggest that the amount of light pollution is a good indicator of quality for management of night skies, and that standards of quality range from approximately study photo 1 (the condition visitors would prefer) to approximately photo 6 (the condition at which visitors would no longer stargaze [Table 18.3]). Of course this study applies specifically to Acadia, but it could be replicated in other parks or regions.

As described earlier, management of night skies will also require monitoring the brightness of celestial bodies and the amount of light pollution, as well actions designed to maintain standards of quality by controlling light pollution in and around national parks. The NPS Night Sky Team is engaged in a program of monitoring the condition of night skies in the national park system (Albers and Duriscoe 2001; Chadwick Moore 2001). However, controlling light pollution is likely to be more challenging. Of course, the NPS can and should adopt best lighting practices designed to minimize light pollution within national parks (L. Chan and Clark 2001). But controlling light pollution outside park boundaries will require a proactive approach of working with surrounding communities. Acadia offers a good example of this approach, working with the gateway town of Bar Harbor, which recently adopted a new lighting ordinance for the town designed to encourage efficiency and reduce light pollution (Maine Association of Conservation Commissions 2010). Chaco Culture National Historical Park offers another good example, successfully working with stakeholder groups to encourage the state legislature to pass the New Mexico Night Sky Protection Act, regulating outdoor lighting throughout the state (Rogers and Sovick 2001b; Manning et al. 2017).

Controlling light pollution in and around national parks might further be promoted by "astronomical tourism" (Bemus 2001; Collison and Poe 2013). Paradoxically, as the opportunity for high-quality stargazing has diminished; its value may be increasing. In this way, the economic benefits of tourism based on stargazing (and other elements of natural darkness) may encourage communities in and around national parks to help reduce light pollution.

Fortunately, natural darkness, particularly the night sky, is a renewable resource; light pollution is largely transitory in both space and time. Though light pollution may have already had irreversible biological and ecological impacts, it can be controlled and even reduced, thus restoring the brightness of the night sky. The national parks, with their emphasis on protection of natural and cultural resources and the quality of visitor experiences, are a good place to advance this cause.

19

VISITOR EVALUATION OF NIGHT SKY INTERPRETATION

Britton Mace and Jocelyn McDaniel

How much do national park visitors know about the night sky? Are interpretive programs about the night sky well attended, and do they increase knowledge about the night sky? This survey of visitors to two units of the national park system found that visitors who participate in night sky interpretive programs gained a significant amount of knowledge about night sky issues and that visitors have strongly held attitudes about light pollution (perhaps at least partially related to their experience with night sky interpretive programs). These programs can enhance the quality of visitors' experience and can contribute to the protection of natural darkness.

INTRODUCTION

In a growing number of locations around the planet, observing the Milky Way galaxy amidst a tapestry of stars has become a rarity. Increases in population, urbanization, and suburban sprawl have created a blanket of outdoor lights enshrouding many continents (Cinzano et al. 2001). Yet within us there are deep historical and cultural connections with the stars and night sky. We try to gain understanding of the universe and contemplate our place within it as we gaze at the vast expanse of the stars above. With a reduction in night sky visibility where the majority of us reside, many are left to seek out the few remaining areas of the planet that are void of anthropogenic lighting to stargaze. In the United States, many of these remaining dark areas are found within parks and other protected areas.

The National Park Service (NPS) classifies dark night skies as a natural resource and has taken the initiative in documenting levels of light pollution and educating the public about night sky issues. As part of meeting this initiative, parks have educational programs, including exhibits, ranger-led talks, and other interpretive materials and opportunities. Two parks highly involved in night sky interpretation are Bryce Canyon National Park and Cedar Breaks National Monument, both in southern Utah. Because of their high elevations and dry, relatively clean air, Bryce Canyon and Cedar Breaks are ideal locations for stargazing, especially with their distance from major cities. This study seeks to gain an understanding about the effectiveness of night sky interpretation by surveying visitors at Bryce Canyon and Cedar Breaks who attended a ranger-led program. Visitor studies of night sky related issues in national parks are limited, so park managers and interpreters are lacking needed information to help guide their programs. Attitudes about the night sky and light pollution are also compared with a sample of day visitors at both locations. The results of this study will provide insight into the knowledge, attitudes, benefits, and behaviors related to the night sky of day visitors and night sky interpretive program participants.

A great deal of research has documented the environmental impacts of visitors in parks and protected areas (Hammitt and Cole 1998; Leung and Marion 2001; Manning 2011), including harm to the land and wildlife. Anthropogenic impacts to park resources can also be caused by non-visitors, hundreds of miles away in such forms as air and light pollution. While the effects of air pollution on parks and protected areas has received substantial research attention (see Mace et al. 2004 for a review), the same is not true for light pollution. Research on the effects of light pollution on park natural lightscapes, especially from a social scientific orientation, is just beginning.

Bryce Canyon National Park began the first night sky interpretive programs in the national park system over four decades ago, and has since become a leader in night sky protection and interpretation. While the expanse of land viewable during the day is quite large (referred to as a viewshed), on a clear dark night visibility can stretch as far as the Andromeda galaxy, 2.2 million light years away, or 527,000,000,000,000,000,000 miles (Bryce Canyon National Park 2012). Both Bryce and Cedar Breaks have a limiting magnitude of seven, which translates into a night panorama of 7,500 stars being visible. By comparison, in most urban areas about 500 stars can be seen and the limiting magnitude is four or less. There are full-time and seasonal rangers who have contributed to the growth of the night sky interpretive programs in Bryce and Cedar Breaks over the past 40 years. Educational lectures, stargazing, eclipse viewing, astronomy festivals, ranger-led telescope viewing, night hikes, and constellation tours are some of the interpretive programs offered from April through October. During these months anywhere from 100 to 300 visitors a day will partake of these interpretive opportunities offered by the "Dark Rangers." The programs have become so popular that Bryce reports having over 27,000 visitors a year engaging in these interpretive programs, essentially equaling the popularity of all other interpretive programs combined (Bryce Canyon National Park 2012). Furthermore, many visitors anecdotally report that seeing starlit scenery is one of the reasons for choosing their destination. Clearly, night skies are an important resource and part of a quality visitor experience at Bryce and Cedar Breaks.

Parks and protected areas need to understand what drives visitor interest and what the expectations are regarding night skies. Furthermore, park managers and interpreters need to assess if their programs are effective in communicating the importance of the resource to the public. Standardized questions related to night skies and light pollution are seldom included on visitor surveys in parks and protected areas. A set of standards and indicators of quality for night skies does not exist from a social scientific perspective. This lack of social scientific information can be frustrating, particularly since groups such as the National Parks Conservation Association have criticized the NPS for not adequately managing this visitor and natural resource. On the other hand, an opportunity exists to begin building a database and set of standards to provide useable and informative research for decision makers and stakeholders.

Given the impact of light pollution and its projected increase in the future, it is important to better understand the attitudes and recreational activities of nighttime park visitors. Unfortunately, research from a social scientific perspective on the topic is limited at best. While the majority of visitors to parks and protected areas go to these areas during the day, it is not clearly understood what types of night sky related activities are desirable to visitors. This research examines the relationship between lightscapes and visitor enjoyment by comparing the responses of day-use visitors with those attending night sky related interpretive programs at Bryce Canyon and Cedar Breaks. The primary objectives of the research are to evaluate the effectiveness of night sky interpretive programs, and identify important night sky related factors including attitudes, benefits, and behaviors. The findings of this study will benefit park managers and interpretive personnel by providing needed data to help fine-tune interpretive messaging and identify topics of interest to park visitors related to the night sky.

STUDY METHODS

A total of 1,650 visitors (day-use and night sky program attendees) to Bryce Canyon and Cedar Breaks were approached and asked to complete a short survey regarding their park experience. Of these 1,650, a total of 1,179 visitors agreed to participate, reflecting a response rate of 71.5%. The majority (67%) of visitors who declined to participate were with family (based on the surveyor's visual assessment or visitors' verbal cues), and did not want to make the rest of their group wait. Others who refused participation also stated they were short on time (21%). Non-response bias was checked by statistically comparing results from the different sampling sessions on several demographic variables. Further descriptive statistics of the sample are presented in the results.

A 36-item survey was developed with input from a team of social scientists, park personnel from Bryce Canyon, and the NPS Night Sky Team. Attitude items assessed light pollution, night sky protection, and the role of gateway communities in addressing these issues. Behavioral based questions asked visitors whether they recently have engaged in night sky viewing to notice the phases of the moon, planets and meteor showers, and stargazing. A set of questions also assessed the perceived benefits of night skies, including connecting with nature, the universe, and the past; inspiring an interest in science; creativity; and solitude. Visitors were also asked about their activities and time spent in the park, and if night sky programs played into their travel plans. Basic demographic questions such as age, sex, place of residence, and night sky visibility at their place of residence were also included in the survey. The primary motive of this study was to assess the effectiveness of night sky related interpretive programs in the park by surveying program attendees and comparing their data with those of day-use visitors. To this end a set of scale items probed visitors' knowledge of five particular night sky issues to see if they gained new information about the night sky as a result of visiting the park. For those who took part in a night sky interpretive program, this set of questions evaluated the level of awareness conveyed in the ranger-led sessions. Finally, visitors were asked if they would visit another park based on night sky quality and interpretive opportunities.

NIGHT SKY INTERPRETIVE PROGRAM FINDINGS

Knowledge of night sky issues was assessed with a set of Likert-type scale items asking visitors to rate their knowledge of these issues prior to visiting the park and after engaging in a night-sky interpretive program. The items ranged on a one to five scale, with five representing "very aware" and one being "completely unaware." Specific items assessed knowledge of the impacts of ground lights on wildlife, human health, and visibility; the type of lighting that contributes to pollution; the cultural and historical components of the night sky; and general astronomy. The mean responses of program participants showed increased knowledge in all five night sky related issues. Mean scores of knowledge before visiting the park ranged from 2.55 to 3.69, corresponding to somewhat unaware to somewhat aware levels of night sky knowledge. The issues which program participants knew the least about prior to the interpretive program were the impact of ground lighting in creating sky glow and glare that affects human health and wildlife ($M = 2.55$). Interpretive program participants reported being most aware of the impact of ground lights on night sky visibility ($M = 3.69$). A multivariate analysis of variance (MANOVA) was run on the dependent variables using before and after as the fixed factor, with sex, age, and place of residence as covariates. Results show the interpretive program participants gained a significant amount of night sky knowledge as a result of participating in a ranger-led interpretive program. Univariate analyses found the interpretive programs significantly improved visitor knowledge about the types of lighting that help reduce light pollution, the impact light has on human health and wildlife, and the effect ground light has on night sky visibility. The interpretive program visitors also gained more knowledge about astronomy, and the cultural and historical connections with the night sky, as a result of their participation in a ranger-led night sky program.

Day Visitors

Those visitors who went to the park during the day and did not participate in a ranger-led night sky interpretive program also gained knowledge about night sky related issues from other sources in the park. Univariate analyses found day visitors improved their knowledge about the effects of light on wildlife and human health, lighting that reduces pollution, and the impact of ground lights on night visibility. Day visitors also became more knowledgeable about cultural and historical connections with the night sky and astronomy, as a result of visiting the park. Daytime visitors gained this knowledge primarily through a park brochure, newspaper, or handout (28%), and visitor center exhibits (24.3%). Some daytime visitors (21.2%) indicated they did not encounter any night sky information while visiting the park.

Differences between Visitors

Statistical comparisons were conducted between night interpretive program attendees and day visitors. While both types of visitors became more knowledgeable about the night sky while visiting the park, interpretive program respondents gained more knowledge than daytime visitors. Specifically, interpretive program participants became more knowledgeable than daytime visitors about how to reduce light pollution, how lights impact the night sky, cultural and historical aspects of the night sky, astronomy, and the effects of light on human health and wildlife.

Visitors to the parks come with nearly the same level of knowledge about night sky issues, as daytime visitors and interpretive program participants were found to differ on only one knowledge variable. Night sky program attendees knew less about the effects of light on wildlife and human health before visiting the park than daytime visitors.

Interpretive program visitors placed more importance on the park's night sky and stargazing programs than day users. In addition, 39.7% of interpretive program participants marked night sky interpretive opportunities as "very important" or "somewhat important" to their travel plans, compared to 23.3% of day visitors. Anecdotal evidence, based on verbal and open-ended comments made by interpretive program attendees, show that some made the trip to Bryce Canyon specifically for the park's night sky. Yet, many visitors were simply unaware of the night sky and stargazing opportunities at the parks, with 62.5% of day visitors and 41.8% of interpretive program attendees reporting they had no knowledge of these available programs prior to visiting the park. Visitors learned about interpretive program opportunities from the park brochure and newspaper and from the visitor center.

Night Sky Related Behaviors

To gain an understanding of the frequency which visitors engaged in night sky related behaviors, a set of six Likert-type five-point scale items (always to never) was used. Visitors were asked to rate the frequency with which they noticed the phases of the moon, the planets, the night sky, meteor showers; stargazed with family; and took night walks. A majority of day respondents (57.5%) and program visitors (64.2%) said they always (once a week) or often (once a month) notice the phases of the moon or observe the night sky. Furthermore, 21.2% of day and 32.8% of program users take night walks at least once a month. A MANOVA found interpretive program visitors engaged in night sky related behaviors more frequently than day visitors. Univariate analyses found night program attendees notice the phases of the moon, and take night walks, more often than day visitors.

Benefits of Night Sky Viewing

A series of seven scale items measured the types of benefits associated with viewing the night sky. These items were rated on a five-point scale ranging from very unimportant (1) to very important (5). Day users of the parks rated the night sky as a chance to connect with nature (75.5%) and to experience solitude (68.5%) as very or somewhat important. Night interpretive program respondents perceived the night sky as very or somewhat important to providing a better understanding of the universe (87.8%) and creating curiosity about science (87.6%). Both user groups reported creativity, a connection to the past, and a spiritual connection as less important than the other potential benefits.

A MANOVA found interpretive program visitors to have a greater degree of perceived benefits associated with the night sky. A series of univariate analyses found interpretive program respondents rated all potential night sky benefits, with the exception of solitude, as more important than day visitors did. Specifically, night program visitors viewed the nocturnal cosmos as creating curiosity about science and offering more opportunities to understand the universe, inspire creativity, connect with nature, provide spiritual inspiration, and connect with the past, to a greater degree than day visitors.

Light Pollution Attitudes

Day and night visitors believe that light pollution is a problem, at least in some areas, and that steps should be taken to preserve night visibility, especially in parks. When asked if light pollution is a problem in urban areas, 84.4% of respondents agreed or strongly agreed that it was a problem. Interestingly, 55.6% of those surveyed agreed or strongly agreed that light pollution is a problem in rural and remote areas. A majority of visitors (79.1%) stated it was important or very important to be able to see stars in their backyards. This percentage is put in perspective when considering 54% of respondents indicated they could see stars at their place of residence. Most respondents (89.1%) agreed or strongly agreed that places such as national parks should be preserved for their dark skies, and that areas around the parks (especially gateway communities) should help protect night skies (90.3%). More than half (59.2%) agreed or strongly agreed they would visit other parks because of their dark skies.

Based on a multivariate analysis of variance, those attending a night sky interpretive program show more concern for light pollution and night sky preservation than day users. Univariate analyses found differences between the two types of user groups on all seven light pollution and dark sky attitude items. Nocturnal program attendees agreed more strongly than day visitors that they would visit other parks because of their dark skies. Day visitors felt more strongly than night visitors that light pollution was an inevitable consequence of economic growth. Interpretive program respondents believed more strongly that light pollution is a problem in urban areas, as well as rural and remote locations, than daytime visitors. Night visitors also felt it was more important to be able to view the stars in their backyard. Finally, night program attendees want parks preserved for their night skies, and feel areas around the parks should help protect night skies, more than day visitors.

DISCUSSION

Visitors to Bryce Canyon and Cedar Breaks view parks and wilderness areas as the most preferred locations for stargazing, with 99.4% of interpretive program visitors and 79.9% of day users marking this choice. In comparison, 19.4% of day and 30% of night visitors identified a planetarium or observatory as one of the preferred locations for stargazing. This shows parks and other protected areas have inherent night sky resources visitors expect to be able to experience. This may be the case even if visitors are unaware of the night sky related activities a park has to offer. In this study, 62.5% of day users and 41.8% of program visitors were unaware of existing night sky and stargazing opportunities at the parks prior to visiting. Once in the parks, however, visitors can learn about night sky interpretive resources by frequenting the visitor center or reading the information provided to them when they enter the park. Yet, based on the percentages of visitors who were unaware of night sky opportunities at Bryce Canyon and Cedar Breaks, there remains room for improvement. More information about night sky recreational opportunities needs to be imparted to day visitors through other outlets so more become aware of the resource and the available interpretive programs.

While Bryce Canyon and Cedar Breaks make it a priority to communicate the importance of the night sky as a resource and the impact humans are having with their lighting (especially through their interpretive programs), other parks may not recognize that visitors arrive expecting to view and learn about the night sky. Parks and other protected areas need to recognize this expectation and include night sky related information and opportunities for their visitors, even if they are located in areas that experience light pollution. Moreover, a concerted effort on a national and international level would help to educate potential park visitors and the general public about the importance of the night sky and the impacts of light pollution, as many seem simply unaware.

In order to gauge the effectiveness of the night sky interpretive programs, attendees were asked a series of knowledge based questions before and after participating in the program. Naturally, how much visitors knew about the night sky played a role in their experiences at the park. The topic program participants said they knew the least about prior to visiting the park was the impact of ground lights (a term also referring to "skyglow" and "glare") on wildlife and human health. In fact, day visitors reported knowing more about this topic than night visitors before coming to the park, although the mean for both groups shows a relatively limited amount of knowledge. This may be due to a general lack of informational focus in society on the health effects of light pollution. The night sky interpretive programs highlight the health effects of light pollution, perhaps making attendees aware of how little they knew about the topic, a result that may be an artifact of the post-program design of the study. Yet, by attending a ranger-led interpretive program, visitors improved their knowledge on several night sky related issues. For example, the topic which showed the greatest increase in knowledge was the type of lighting that can reduce light pollution. All interpretive programs emphasize this issue, providing information not readily available elsewhere in the park, and visitors are clearly learning from this experience. Eighty-six percent of interpretive program participants reported the program was the most informative park resource on night sky topics. It appears the night sky interpretive programs are doing a good job of fulfilling the mandate of providing for the visitor experience while also communicating valuable information about the importance of night sky resource protection.

Results for day visitors are not as striking. Day users come to the park with nearly the same amount of knowledge about night sky topics as interpretive program participants. Daytime visitors, however, gained less knowledge about the night sky than did interpretive program attendees. Day users knew the least about the type of lights that reduce light pollution and the most about the impact of ground lights on night sky visibility. While day users did learn about visibility as a result of visiting the park, the increase in knowledge was not as great as for those who took part in an interpretive program. Indeed, the differences between night sky knowledge before and after park visitation are much smaller for day visitors than program participants, indicating that outside the ranger programs, the parks may not be effectively communicating information on night sky subjects to all visitors.

The challenge becomes how to best communicate night sky information to all types of visitors. Bryce Canyon and Cedar Breaks emphasize their night skies by offering interpretive programs and displaying times and locations of those programs in the visitor center and in the newspaper distributed when visitors enter the park. In addition, both parks have exhibits and informational placards distributed throughout the park to educate visitors about night sky visibility resources. Still, as indicated by the small amount of knowledge gained by day users, the park needs to find new ways to reach out to this segment—the largest—of its visitors. Given that day users spend less time at the parks than program visitors, and with 54% of day visitors indicating they had toured the visitor center, the parks should look first to the visitor centers to help increase night sky awareness. Day users see the night sky as a chance to enjoy nature and solitude, while program participants view the night sky as an opportunity to better understand the universe and create curiosity about science. Furthermore, since 76% of day and 86% of program respondents rated the connection to nature as a very or somewhat important benefit when viewing the night sky, targeting these themes could enable the parks to align exhibits and programs with visitor interests. It would also be advantageous for the parks to find other means of spreading information on night sky issues, including web based resources and the development of mobile applications that are available prior to and during visitation. With the increased dependence on technology, this would be a good way to communicate the importance of the night sky to younger visitors, who were found to view the protection of the night sky as less important than older visitors. In addition, based on the demographic differences found in this study, educational programming should communicate the importance of protecting the viewshed at night and reducing light pollution, specifically appealing to male urban residents, perhaps through historical and cultural connections.

Results from this study can also prove useful to other parks with night sky interpretive opportunities and those areas seeking to educate their visitors about the importance of the resource. The responses to attitude, behavioral, and night sky benefit questions in this study demonstrate that park guests want to learn more about the night sky. In order to continue educating the public on this resource, parks need to expand their astronomy and night sky interpretive programs, even in parks that are affected by light pollution. Results show that visitors agree that light pollution is a problem, especially in urban areas, and a majority do not believe that light pollution can be avoided.

Demonstrations showing how light pollution can be reduced in affected parks would be one way of demonstrating how individuals, businesses, and communities can make a difference with their lighting choices. Nearly 90% of all respondents believe that some places need to be preserved solely for their night visibility and that areas near national parks should assist in maintaining dark night skies. Parks affected by skyglow and ground based light pollution offer an excellent opportunity to illustrate human impact and the degradation of the dark night sky, even in remote areas. The NPS could also tailor its programs to the most frequent night sky–related activities reported by respondents in this study (noticing the phases of the moons, observing the night sky, and taking night walks). Increasing ranger-led night hikes, giving frequent astronomy and telescope presentations, and discussing lunar cycles and their effects are possible methods parks can use to capitalize on their visitors' interests and expectations. National parks are excellent places for informal learning, and if certain areas have a night sky that people cannot view at home, that visibility, as well as the information contained in interpretive programs, will lead to increased knowledge and also positively influence night sky attitudes, intentions, and behaviors.

20

"ASTRONOMICAL TOURISM"

Fredrick Collison and Kevin Poe

Because of its high elevation, low humidity, and remote location, Bryce Canyon National Park is one of the premier night sky viewing parks in the nation. The park began offering interpretive programs on night skies in 1969 and now has what may be the largest suite of visitor programs on night skies in the national park system. Visitor counts at night sky programs and a survey of a representative sample of park visitors suggest that a relatively large number of visitors take advance of opportunities to view the night sky at the park. This, in turn, suggests that "astronomical tourism" may have a large economic impact in the park and the surrounding region. However, for astronomical tourism to be successful, the quality of the night sky must be protected.

INTRODUCTION

Astronomical tourism is a potential attraction for visitors to destination areas where night skies free from artificial light pollution can be enjoyed. This is unlike most developed locations where visitors might live and presents an opportunity for places such as Bryce Canyon National Park (Bryce Canyon). Astronomical tourism goes back many centuries throughout the world. Examples of this phenomenon include sites as widely dispersed as Nabta Playa in the Sahara Desert, Stonehenge and Woodhenge in the United Kingdom, Newgrange in Ireland, Chichen Itza in Mexico, Machu Picchu in Peru, and the pyramids of Giza in Egypt. Ancient astronomical sites in the U.S. include Cahokia (American Woodhenge) in Illinois, the Bighorn Medicine Wheel in Wyoming, Mesa Verde in Colorado, and Chaco Canyon in New Mexico (Malville 2008; Williamson 1984).

The significance of night skies in the above locations had specific purposes related to agriculture and other traditional practices (e.g., timing of planting

and harvesting, timing of solar equinoxes and solstices) (Malville 2008; Mc-
Coy 1992; Richman 2004; Shattuck and Cornucopia 2001). Astronomical
tourism has changed in recent times to include traveling to locations to enjoy
the beauty of the night sky. For many people living in light polluted areas, the
night sky as seen from their home locations is that of a few bright stars, the
brightest planets, and a ubiquitous sky glow from artificial illumination
(Longcore and Rich 2004; National Park Service n.d.a.; Nordgren 2010).

Astronomical tourism has several market segments, including visitation
to observatories (Robson 2005; Weaver 2011); locations with aurora displays
(Weaver 2011); national/state/local parks with dark skies; and amateur astronomy organizations that offer public programs. In the U.S., visitors go to
astronomical observatories for tours such as at McDonald Observatory in
Texas and Kitt Peak and Lowell Observatories in Arizona (Kitt Peak National
Observatory n.d.; Lowell Observatory 2017; McDonald Observatory 2017).
Tours are held during the day when an observatory is idle or, if at night, to
an area near an observatory. Some observatories offer Web-based outreach
whereby the tourist does not even have to be physically present to take a tour
and be educated (Robson 2005). Some observatories in other parts of the
world also offer such programs for visitors.

Aurora tourism is particularly found in northern latitudes of North
America and Europe, where tourists may go to view auroral displays. Typically, the aurora borealis is easier seen during the northern winter when
many hours of darkness exist. For North America, the principal geographic
area providing this type of astronomical tourism is Alaska (McDowell Group
with DataPath Systems and Davis, Hibbits, and Midghall, Inc. 2007; Milner,
Collins, Tachibana, and Hiser 2000). Canada's provinces of the Northwest
Territories (Government of Northwest Territories Department of Industry, Tourism and Investment 2007) and the Yukon (Government of Yukon
Department of Tourism and Culture 2008) are other iconic destinations.
Indications are that this tourism sector is usually packaged with other components such as dog sledding and snow coach tours, with the Japanese as a
principal geographic market (Government of Northwest Territories Department of Industry, Tourism and Investment 2007; Milner et al. 2000).

Other providers of astronomical tourism sites such as bed and breakfasts
(B and Bs) and private observatories offer resources for observation of the
night sky, and sometimes solar observing (Astronomy at Starry Nights Bed
& Breakfast n.d; San Pedro Valley Observatory 2017; Sedona Star Gazing
2017). Also included are amateur astronomy organizations that operate their
own observatories which may be open to the public. For example, the Astro-

nomical Society of Kansas City (ASKC; n.d.) operates two observatories, one at a local university open on Friday nights and one at a public park open on Saturday nights, mid-spring to mid-fall.

In the case of B and Bs and private observatories, the demand for these facilities would be classified as tourism related, since visitors are expected to stay at least one night in the accommodations provided along with the astronomical observing. Private observatory (only) visitors would need overnight accommodations elsewhere. For astronomical organizations such as the ASKC, some users of its facilities would be classified as visitors, especially for situations when an extended "star party" is provided (Heart of America Star Party 2011).

National/state/local parks also offer astronomy programs. A number of national park system units offer some form of astronomy-related tourism, especially in the southwestern states where good night sky observing is possible due to the high altitude, low levels of relative humidity and air pollution, and widespread areas with little artificial illumination at night (C. A. Moore 2001). Nearly all of these NPS sites have some form of organized astronomy program in which rangers offer programs followed by night sky observing. Grand Canyon National Park has a one-week star party while Bryce Canyon offers astronomy programs three nights per week at the park and two nights per week off-site during the visitor use season (May through October), along with a multi-day star party in the summer (National Park Service n.d.a; Tucson Amateur Astronomy Association 2017).

THE NIGHT SKY AT BRYCE CANYON

An important attribute of Bryce Canyon is the quality of the night sky for viewing solar system bodies, stars, constellations, and many deep sky wonders. The dark sky at the park represents a resource few other U.S. areas have, especially outside the Southwest. On a typical night at Bryce Canyon when the moon or extensive cloud cover is not featured in the night sky, seeing is excellent with the Milky Way and Andromeda (if in the night sky) galaxies clearly visible. Up to 7,500 individual stars may be visible to the unaided eye. Threats to this astronomical sight are just beyond the horizon, including artificial light from nearby towns and even those as far away as Las Vegas (National Parks Conservation Association 2005). Some local artificial illumination in the park at various facilities and in nearby towns also creates some night sky visual pollution, although it tends to be no more than minimal at present.

The first form of an astronomy program at Bryce Canyon was initiated in 1969, and for the next few years, the program was maintained by seasonal rangers. Beginning in 1984 and continuing, the park has offered monthly programs focused on current astronomical developments in our solar system (National Park Service n.d.a). Since approximately 2000, one or more permanent rangers have led the Astronomy and Dark Sky Program at the park, thus enhancing the long-term viability of the program.

Short-term volunteers also have played a role in the Bryce Canyon program during the peak months, beginning in 2002. A formal application process helps to determine who will serve as volunteers (National Park Service 2016a; National Park Service 2016b). The first permanent telescope at the park was received from the NPS Air Resources Division in 1985 and a second telescope was received in 1987 from the Bryce Canyon Natural History Association, enhancing night-time observing sessions for visitors (Bryce Canyon National Park-A canyon alight with stars: A brief history of astronomy at Bryce Canyon, 2007; Bryce Canyon National Park-Astronomy volunteer, 2009). Astronomy volunteers are typically based at the park from April through October, with numbers based on the amount of visitation expected. Since then, additional telescopes have been donated to the park, including a solar telescope that allows visitors daytime viewing of the surface of the sun in a safe manner.

Bryce Canyon, with assistance from the Salt Lake Astronomical Society, put on the first Bryce Canyon Astronomy Festival in 2001, an event that continues to the present day. This annual event attracts hundreds of astronomy enthusiasts to the park. In the summer of 2011, during and directly following the Astronomy Festival, was the annual convention of the Astronomical League—which, combined with the Astronomy Festival, produced substantial astronomy visitation to the park and the surrounding area (National Park Service 2017a).

From 2004 to 2008, Bryce Canyon hosted the NPS Night Sky Team, which is focused on protection and enjoyment of dark skies. Additional astronomy staff were also brought on board temporarily to help increase the offerings by the Astronomy and Dark Sky Program (National Park Service 2016a). During this period the astronomy festival was expanded, and in 2005, 4,500 visitor contacts were registered as participating in at least part of the festival, while for 2010 the visitor number was about 3,600 (many of whom had multiple contacts during the festival).

The Astronomy and Dark Sky Program at Bryce Canyon consists of four primary components that are regularly offered to visitors: solar observing during the early afternoon, multimedia presentations of about 60 minutes in the early evening, star gazing at night through the program's telescopes after the multimedia presentations, and full moon hikes into the Bryce Canyon amphitheater during the weekend's nearest full moon. Visitors may take part in one or more of these activities.

The night sky viewing sessions are used as a time to further educate viewers about the need to reduce artificial lighting at night. For many visitors it is the first time to see the Milky Way in the night sky as well as planets, deep sky objects, and the host of fainter stars that are lost in the sky glow of the urbanized (and not so urbanized) areas where they live (Nordgren 2010). The reaction of most visitors is one of appreciation for being able to see a night sky free of nearly all light pollution. Although the astronomical views cannot compare to what some visitors have seen in the media (e.g., results from the Hubble space telescope), the sights they are seeing are seen directly with their own eyes at that moment in time.

ASTRONOMY AND DARK SKY PROGRAM ANNUAL VISITOR STATISTICS

For 2010, detailed data on monthly astronomy visitors were available. Data were collected on visitation by specific activity, both at Bryce Canyon and at off-site locations where park staff offer programs. Determining total recreational visitors for the Astronomy and Dark Sky Program is somewhat difficult as one visitor might have anywhere from one to three or more program contacts. For example, a visitor on a given day might take part in solar observing, attend an evening multimedia presentation, and participate in night sky viewing, representing three program contacts but only one visitor. A conservative approach was taken with the data due to the potential multiple counting of some visitors to ensure an accurate count of actual individual visitors. This yielded a total of 17,002 verifiable individual visitors at the Astronomy and Dark Sky Program in 2010.

VISITOR SURVEY

A survey of visitors to Bryce Canyon National Park in 2009 used a systematic random sampling procedure and found that 10% reported participating in "stargazing activities/astronomy" (N. Holmes, Schuett, and Hollenhorst 2010). This survey controlled for the possibility of visitors completing the survey form more than once (M. Littlejohn, personal communication, 3 May

2012). This resulting estimate of the number of participants in the park's Astronomy and Dark Sky Program is higher than that found as described above. This higher number may be due to the fact that 10% can include visitors who did "stargazing/astronomy" on their own or simply viewed the astronomy exhibits at the visitor center or astronomy information signs found in the park, rather than as part of the formal programs presented by park staff. However, both the Astronomy and Dark Sky Program participant numbers and the 2009 visitor survey numbers indicate that the Astronomy and Dark Sky Program is an important component of Bryce Canyon's offerings to visitors.

In terms of visitor learning while at the park, 67% of visitors indicated that they learned about one or more park topics. Of all the visitor respondents indicating learning, the largest percentage was from the topic of "geology" (94%) with 56% for "night skies/astronomy" (multiple reasons were allowed). For the topic category of "night skies/astronomy" 21% of respondents reported that their learning level improved a lot, while 38% indicated their level improved somewhat. For learning topics of interest during future visits, "geology" (80%) was the most popular, with "night skies/astronomy" at 56% (N. Holmes et al. 2010).

Visitors were also asked to rate the importance of protecting park attributes and resources; 47% indicated "dark, starry night sky." Related attributes were scenic vistas (96%) and clean air (85%) (N. Holmes et al. 2010). The above visitation data suggest that the Astronomy and Dark Sky Program has an important impact on Bryce Canyon visitors, although there may be more visitors who would participate, but did not know about the program, did not stay overnight at or near the park, or visited the park on a day that the program was not offered.

CONCLUSION

The Astronomy and Dark Sky Program represents a significant resource for Bryce Canyon and the NPS. It interacts with a substantial number of visitors to the park. The program enables visitors to see the night sky absent nearly all light pollution and other impediments that might diminish the beauty of that sky. Various interpretive multimedia programs are presented to many visitors and nearly all of them contain a component that addresses the impacts of artificial light pollution on the visible night sky, humans, and other animals and plants. In this latter sense, the Astronomy and Dark Sky Program serves particularly as an educational program, some of whose components can be implemented when visitors return home, such as improving home and business lighting to reduce the impact of artificial lighting on the night sky where they live and work.

One of the issues facing the Astronomy and Dark Sky Program is the need to have a better understanding of the visitors who participate in such programs. Knowing such information might enable the Astronomy and Dark Sky Program and similar programs at other parks to increase the demand for astronomy/night sky programs and to better tailor them to visitor demand. Visitor data could be collected via a survey administered to program participants to include items such as visitor characteristics, where visitors learned about the program including where they found information, and what information they considered relevant. More importantly, data are needed on the decisions visitors make to participate in the Astronomy and Dark Sky Program and what their evaluations were of the program after participation. Knowing whether such as visitors would return to participate again and whether they would recommend the program to others would also be of value.

Another important topic that needs examining with regard to the Astronomy and Dark Sky Program at Bryce Canyon is the economic impact of the program. For visitors to participate in the evening components of the program, they must of necessity stay overnight within the park or in the near vicinity. As a result, those visitors need to purchase lodging, food, and other products during their overnight stay. At present, only preliminary estimates of this economic impact are available. More precise estimates are needed through surveys of program participants and/or through more general studies of visitor economic impacts of Bryce Canyon. The results of this research could be of assistance in the management of the park and the Astronomy and Dark Sky Program, as well as allowing the program to more actively seek additional resources, especially those of a financial nature.

The role of the Astronomy and Dark Sky Program at Bryce Canyon will become increasingly important, especially if the quality of the night sky (and day sky too) can be protected at the park and if the sky quality further declines in other areas. The NPS has recognized the importance of dark sky resources at parks such as Bryce Canyon by calling for establishment of a Dark Sky Preserve on the Colorado Plateau of which Bryce Canyon is a part (National Park Service 2011b). This effort will become increasingly important as a way to protect one of the few remaining dark sky areas in the continental U.S. This effort can have a positive impact on the program at Bryce Canyon since potential visitors may seek out dark sky parks even more so than they have done in the past (C. Moore, Hoffman, Fields, and Mastroguiseppe n.d.). Astronomical tourism represents an opportunity to expand dark sky–related programs to visitors at a number of national parks that have dark skies as a natural resource. Additionally, visitors may return home with personal commitments to more actively engage in astronomical activities and to work to improve the darkness of the night sky where they live.

CONCLUSION

PRINCIPLES OF STUDYING AND MANAGING NATURAL QUIET AND NATURAL DARKNESS IN THE NATIONAL PARKS

Robert Manning, Peter Newman, Jesse Barber,
Christopher Monz, Jeffrey Hallo, and Steven Lawson

INTRODUCTION

The conclusion of this book synthesizes much of the information presented in the preceding chapters and the larger scientific and professional literature. It begins by reaffirming the premise of the book that natural quiet and natural darkness have evolved as important "new" resources of the national parks and related protected areas. The focus of the conclusion is the development and presentation of twenty-two principles or best practices for studying and managing natural quiet and natural darkness. The book concludes with some observations on the increasing urgency of advancing our knowledge about natural quiet and natural darkness and applying this knowledge to manage and protect these new resources in the national parks, protected areas more broadly, and the communities in which we live.

THE IMPORTANCE OF NATURAL QUIET AND NATURAL DARKNESS

This book began with the assertion that natural quiet and natural darkness are the newest resources of the national parks (and protected areas more broadly). Natural quiet is generally defined as the sounds of nature uninterrupted by human-caused noise, and natural darkness is darkness unaffected by human-caused light. It is important to note that natural quiet and natural darkness do not necessarily mean absolute quiet or darkness, as the natural world often generates sounds of its own (for example, birds calling, wind blowing, and rivers rushing) and the natural world has sources of illumination (such as the glow of celestial bodies and the fluorescence of some plants and animals). Moreover, there are certain park contexts in which "cultural" sounds and light are appropriate and can even enhance the quality of the visitor experience (for example, the sounds of cannon firing at historical battlefields and church bells ringing at historical religious sites, and the lighting of dark caves and historical monuments). We believe that the chapters of this book and the larger scientific and professional literature from which they are drawn support our assertions that natural quiet and natural darkness are vital to many species of plants and animals and to natural processes, and that they are increasingly appreciated by park visitors and the public in general.

The chapters in this book are representative of the growing body of scientific and professional literature on natural quiet and natural darkness. The studies described in these chapters were intended to answer a number of questions. For example, what is the condition of natural quiet and natural darkness in national parks and related areas? What roles do natural quiet and natural darkness play in the natural environment and the experience of visiting national parks? What are the ecological and experiential impacts of human-caused noise and light? How can human-caused noise and light inside and outside national parks best be managed to protect natural quiet and natural darkness?

We have looked closely at these studies and many of the other scientific and professional papers that are listed in the bibliography of this book. Our intent was to inform ourselves about the state of the science and associated professional literature on natural quiet and natural darkness, with the goal of synthesizing this work into a series of principles that might help guide future research on and practice applied to natural quiet and natural darkness. These principles are presented in the following section. As we noted in the introduction, much of the book focuses on national parks in the United States, but we feel that the principles we have derived from our review of the scientific and professional literature on natural quiet and natural darkness apply equally well to a variety of parks and protected areas in the United States and elsewhere.

PRINCIPLES FOR STUDYING AND MANAGING NATURAL QUIET AND NATURAL DARKNESS

Principle 1. Natural quiet and natural darkness are the "new" resources of the national parks.

We began the book with the assertion above, and we conclude the book by noting that the study findings from the preceding chapters firmly support it. This evidence is manifested in three ways. First, the history of the national parks illustrates a clear expansion—in fact, an evolution—of the resources and values of the growing national park system, and of protected areas around the world more broadly. Figure i.1 in the introduction presents an outline of this evolution, starting with appreciation of the grand scenery of the national parks and moving through a host of other valued aspects of the parks: history, cultural heritage, wildlife, education, ecology and biodiversity, recreation, wilderness, native subsistence, cities and urban areas, science, and ecosystem services. Natural quiet and natural darkness can be seen as the most recent additions to this growing list of resources. It is important to note that these resources and values are additive: one does not replace another, but each new one extends the increasingly rich reservoir of resources that are appreciated and protected in national parks and related reserves. We expect this evolution to continue as a result of our growing knowledge about and appreciation of natural and cultural history.

Second, natural quiet and natural darkness are increasingly recognized as important resources. The classic definition of a "resource" is something that has value but is becoming increasingly scarce. A growing body of scientific and professional literature, of which the chapters in this book are examples, clearly illustrates the ways in which natural quiet and natural darkness are vital components of our natural environment (for example, they make natural processes such as predator-prey relationships and mating behaviors possible) and contribute to the quality of our lives (for example, they enhance our enjoyment and appreciation of national parks and contribute to our health and well-being).

Third, human-caused sound and light have made these resources increasingly scarce, even in the national parks. For example, nearly all national parks include roads and other transportation corridors that produce a great deal of noise, and this noise is growing louder with the increase in the numbers of visits to the national parks—now measured in the hundreds of millions annually. The study in Rocky Mountain National Park reported in chapter 10 illustrates that much of this noise spills over into the wilderness portions of parks. Moreover, the study by Weinzimmer et al. reported in chapter 6 demonstrates the excessive noise produced by snowmobiles in Yellowstone National Park, air tours in a growing number of national parks, and the increasingly ubiquitous motorcycles inside and outside of the parks. The landmark study by Cinzano et al. (2001) cited in several chapters mapped the lighted portions of the planet and found that almost a fifth of the earth's land surface was polluted by light. Moreover, light pollution is increasing at a rate of 6 percent annually. The vast majority of people in the United States (excluding Alaska and Hawaii) live in areas that are polluted by light. Consequently, most Americans can not see the Milky Way from their homes. Natural quiet and natural darkness are important resources in the classic sense of this term: they are valuable but becoming scarcer, and they are increasingly recognized as vital components of the US national parks and protected areas globally.

Principle 2. Natural quiet can be defined as the sounds of nature without interference by human-caused noise, and natural darkness can be defined as darkness without interference by human-caused light.

As noted earlier, natural quiet and natural darkness do not necessarily mean absolute quiet or darkness, as nature often generates sounds of its own (for example, birds calling, wind blowing, rivers rushing) as well as light of its own (such as the glow of celestial bodies, the bioluminescence and fluorescence of some plants and animals). The need to protect natural quiet and natural darkness is based on the growing recognition and appreciation of the roles that these resources play (protecting plants, animals, and ecological services; promoting a deeper appreciation of parks and protected areas; and improving human health and well-being) and a greater understanding of the detrimental effects of excessive, unwanted, and unneeded human-caused noise and light. As noted above, in certain contexts, "cultural" sounds and light may contribute to the quality of visitor experiences, and this topic is addressed in principle 13.

Principle 3. The emergence of natural quiet and natural darkness in the scientific and public conscience has been accompanied by a revealing set of concepts and terminology.

Many of the chapters in this book refer to human-caused sound and light as noise and light *pollution*, by-products of our economy and culture that are unwanted and damaging in the same ways as other more conventional forms of pollution, such as air and water pollution. Excessive noise and light that flow onto another's property are increasingly referred to as noise and light *trespass* and as *nuisance* noise and light. These terms reflect a growing appreciation of the ecological and experiential values of natural quiet and natural darkness and increasing concern about the detrimental effects of unwanted and damaging human-caused noise and light.

Principle 4. Noise pollution can have substantial impacts on the ecological and biodiversity values of national parks.

Many of the studies reported in the book, especially those of chapters 1–5, document the multiple and wide-ranging impacts of human-caused noise on park resources. The literature review by Francis and Barber in chapter 1 illustrates several mechanisms by which these impacts occur. For example, chronic and frequent noise interferes with animals' ability to detect important sounds, whereas intermittent and unpredictable noise is often perceived as a threat. These effects can lead to "fitness costs" to animals, either directly or indirectly—for example, by compromising predator or prey detection or mating signals, altering temporal movement patterns, increasing psychological stress, altering spatial distributions of animals, decreasing foraging or provisioning efficiency, and altering mate attraction and territorial defense. The chapters in the first part of the book offer examples of these impacts. For instance, in chapter 2, Ware et al. found that nearly a third of migrating birds avoided areas with experimentally broadcast traffic noise, and that some of the migrating species that remained in the exposed area had reduced body condition and "stopover efficiency" (the ability to gain weight). Based on these and other findings, the authors characterized noise pollution as an "invisible source of habitat degradation." In chapter 3, Francis et al. found that noise pollution altered the community of animals that feed on and disperse seeds in selected species of trees. In chapter 4, Simpson et al. found that the impacts of noise extend to fish. In this study, the target prey species was found to have elevated metabolic rates with exposure to motorboat noise and to respond less often and less rapidly to predator strikes, thus suffering higher mortality rates under noise exposure. More recent research has studied the interaction of noise and light pollution. For example, in chapter 5, McMahon et al. found that both of these forms of pollution affect frog midges and their túngara frog hosts, disrupting this parasitic relationship. In addition to the impacts on individual species described in these chapters, the authors note that these impacts can extend from individuals and species to communities of species and ecological services. For example, altered patterns of species that pollinate and disperse seeds can affect foundational habitat structure.

Principle 5. Light pollution can have substantial impacts on the ecological and biodiversity values of national parks.

Many of the studies reported in this book, especially those in chapters 11 through 15, document the multiple and wide-ranging impacts of human-caused light on park resources. The literature review by Gaston et al. in chapter 11 documents the mechanisms by which these impacts occur. The authors note that the biological world is organized largely by light and darkness: the rotation of the earth partitions time into a regular cycle of day and night. But the spread of electric lighting causes major perturbations in these natural light and darkness regimes, affecting life from the cellular to the ecosystem levels. The effects can include foundational alteration of natural processes, including primary productivity; partitioning of temporal niches; repair and recovery of physiological function; measurement of time through circadian, lunar, and seasonal cycles; detection of resources and natural enemies; and navigation. The following chapters provide examples of these and other effects. In chapter 12, Altermatt and Ebert found reduced natural flight-to-light behavior in moths that had been exposed to human-caused light. In chapter 13, Kempenaers et al. found that compared to birds at control sites, male birds near streetlights started singing earlier in the day and were more successful in obtaining extra-pair mates, while female birds near streetlights started laying eggs an average of 1.5 days earlier. In chapter 14, Lewanzik et al. found that artificial light reduced the feeding behavior of fruit-eating bats and that bats were less likely to use habitats that were light polluted. In chapter 15, ffrench-Constant et al. found that light pollution causes several species of trees to move budburst up to 7.5 days earlier. As with research on noise pollution, these studies suggest that these effects on individuals and species have broader impacts on communities of species and ecological services. For example, reduced flight-to-light behavior of moths may reduce pollination and the food source for bats and spiders, altered behavior of birds may lead to both a mismatch between time of peak demand for food and peak food availability and females' maladaptive choices of mates, and reduced feeding behavior in bats results in reductions in pollination and seed dispersal.

Principle 6. Natural quiet can contribute to the quality of the visitor experience in national parks; correspondingly, noise pollution can substantially diminish the quality of the visitor experience.

Many of the studies reported in this book, especially those in chapters 6–10, document both the ways in which natural quiet can contribute to the quality of visiting national parks and the impacts of human-caused noise on the experience of visiting national parks. For example, in chapter 6, Weinzimmer et al. note the importance of natural quiet to park visitors, citing a survey conducted in the national parks in which "more than 90% of visitors reported the enjoyment of natural quiet and the sounds of nature as primary motivators for their visit." They also note that excessive environmental noise poses threats to human well-being. In chapter 7, Benfield et al. reported on the results of a laboratory experiment they conducted on the effects of noise in natural environments. They found that natural sound had restorative benefits independent of the well-documented benefits produced by the visual qualities of natural environments. Of course, as natural sounds are masked by human-caused noise, the restorative benefits of natural quiet are likely to be diminished. In chapter 8, Manning et al. found that natural quiet is considered an important indicator of quality for visitors to Muir Woods National Monument and identified minimum acceptable thresholds for noise pollution. In chapter 9, Taff et al. found that messages to visitors about the cause of noise pollution (in this case, military flights over a national park) can reduce the impacts of such noise, at least to some degree. In chapter 10, Park et al. note the importance of natural quiet to visitors to the national parks and demonstrate how the consequences of such noise (in this case, transportation-related noise in Rocky Mountain National Park) can be better understood through simulation modeling—including the extent to which this noise penetrates into the park. It should be noted that there is often a strong relationship between the impacts of noise pollution on the natural environment and on the experience of visiting the national parks. For example, alterations in wildlife distributions and behaviors caused by noise pollution may diminish visitors' opportunity to see and hear iconic wildlife. Moreover, natural quiet and lack of noise pollution can contribute not only to the quality of visiting parks, but also to human health and well-being.

Principle 7. Natural darkness can contribute to the quality of the visitor experience in national parks; correspondingly, light pollution can substantially diminish the quality of the visitor experience.

Many of the studies reported in this book, especially those in chapters 16 through 20, show how natural darkness can contribute to the quality of the national park visitors' experience and how human-caused light can diminish that experience. For example, in chapter 16, Duriscoe notes that "an unbroken carpet of stars extending nearly to the horizon in all directions with no evidence of artificial light is one of the more impressive features of a wilderness landscape." Of course, human-caused light can and does substantially diminish the opportunity to see those stars. In chapter 17, Smith and Hallo call attention to the suite—and growing popularity—of nighttime recreation opportunities in national parks, but light pollution can damage the quality of these activities. In chapter 18, Manning et al. found that dark night skies are an important indicator of quality for visitors to Acadia National Park and identified minimum acceptable thresholds for light pollution. In chapter 19, Mace and McDaniel found that there was substantial interest in night sky–related educational and interpretive programing at Bryce Canyon National Park and that these programs were enjoyable and effective in educating visitors about night skies and natural darkness more broadly. In chapter 20, Collison and Poe describe the night sky programming at Bryce Canyon and other national parks and suggest both that "astronomical tourism" is emerging in the national parks and elsewhere and that it has important economic benefits. However, this form of recreation ultimately relies on protecting the quality of dark night skies. It should be noted that, as is the case with the impacts of noise pollution (as noted above), light pollution's impacts on the natural environment can be strongly related to its impacts on the experience of visiting the national parks. For example, changes in distributions and behaviors of selected species caused by light pollution can affect visitors' ability to see and hear iconic nocturnal wildlife.

Principle 8. National parks are some of the last refuges of natural quiet and natural darkness and must be protected.

Many national parks and other protected areas are located far from large metropolitan areas and have therefore been somewhat insulated from the noise and light generated outside these places. Examples include the national parks in Alaska (for example, Denali National Park and Gates of the Arctic National Park), in the northern Rocky Mountains (such as Yellowstone National Park and Glacier National Park), and on the Colorado Plateau (for example, Grand Canyon National Park and Natural Bridges National Monument). Moreover, in the visitor survey conducted at Bryce Canyon National Park by Mace and McDaniel reported in chapter 19, many respondents reported that national parks are the preferred sites to stargaze. However, natural quiet and natural darkness are threatened in all national parks by increasing numbers of visits and human-caused noise and light—especially since both noise and light can travel substantial distances. For example, light can travel up to 200 miles from its source.

Principle 9. The National Park Service (NPS) recognizes the value of natural quiet and natural darkness and has implemented policies and programs to help protect these resources.

The NPS has been a leader in recognizing and protecting natural quiet and natural darkness, especially in the many national parks and protected areas the agency manages. Much of this activity has been conducted under the guidance and direction of the Congress. Perhaps the most foundational guidance is the Organic Act of 1916 that created the NPS, which directed the agency to manage the national parks to "conserve the scenery and the natural and historic objects and the wild life [sic] therein." Of course, there was little thought about natural quiet and darkness at that time, but the act has provided a framework to protect all new park resources as scientists, managers, and the public become conscious of them. In chapter 16, Duriscoe makes a strong argument that natural darkness as manifested in night skies should be protected under the auspices of the Organic Act and the Wilderness Act of 1964. Analogous arguments can be made for natural quiet. More contemporary legislation includes the National Parks Overflight Act of 1987 and the National Parks Air Tour Management Act of 2000, both intended to restore and protect natural quiet in the national parks from the impacts of the air tours over many national parks—a growing business. The NPS has taken the initiative with its creation of the Night Sky Team in 1999, designed to monitor light pollution in national parks and advise park staff members on how to help reduce this pollution. And in 2000 the NPS established its Natural Sounds and Night Skies Division, whose mission is to "restore, maintain, and protect acoustical environments and natural dark skies throughout the national park system" (National Park Service 2017b). Furthermore, in 2006 the NPS added the protection of natural quiet and natural darkness to its agencywide Management Policies, the directives that guide the work of the agency. Many parks participate in this work, developing and presenting interpretive programming to park visitors, hosting star parties to engage the public in discussions of natural darkness, and implementing management actions to protect natural quiet and natural darkness. For example, Mace and McDaniel in chapter 19 and Collison and Poe in chapter 20 describe interpretive programs conducted at Bryce Canyon National Park, one of the premier night sky parks in the country, and Manning et al. in chapter 8 describe how managers at Muir Woods National Monument have successfully worked to reduce visitor-caused noise at this park.

Principle 10. More research is needed on natural quiet and natural darkness.

The scientific and professional literature on natural quiet and natural darkness is expanding. The chapters in this book and the references in its bibliography are encouraging indications of the growing interest in these resources by both scientists and managers. However, much more remains unknown than known. Coverage of species is spotty at best: we are only beginning to uncover the interrelationships among species and the mechanisms by which these interrelationships work; and the important and often cascading ecological effects of noise and light pollution on plant and animal communities—as well as vital ecological services such as pollination—are only now becoming apparent. More research is needed to help guide the formulation of ecological and experiential indicators and thresholds for both noise and light pollution. Research is needed to develop effective and efficient monitoring protocols and assess the effectiveness of management actions. Generally, the literature on noise pollution is more fully developed than that on light pollution, suggesting the need for a greater emphasis on the latter than has been the case. Research on noise and light pollution's impacts on terrestrial animals and plants is more developed than research on the impacts on marine environments. Knowledge of the economic impacts of noise and light pollution (including the benefits of protecting the resources of natural quiet and natural darkness) is limited. These and many more topics warrant an expanded program of research to help ensure that the management of these resources is as informed as possible.

Principle 11. Research on natural quiet and natural darkness can and should draw on a range of scholarly disciplines and incorporate a variety of research methods.

The studies in this book come from both the natural and social sciences, recognizing the multi- and interdisciplinary character of natural quiet and natural darkness. The chapters in parts 1 and 3 of the book use the natural sciences to focus on the biological and ecological impacts of human-caused noise and light, respectively, while the chapters in parts 2 and 4 use the social sciences to focus on the experiential impacts of human-caused noise and light and the manifold benefits of natural quiet and natural darkness. Where appropriate, work in the natural and social sciences should be integrated. For example, the natural sciences can help identify the ecological impacts of human-caused noise and light, while the social sciences can help determine societal norms for the maximum acceptable levels of this pollution, as research in chapters 8 and 18 suggests. The studies in this book also employ a range of research methods. For example, studies in the natural sciences use literature reviews and synthesis, transects, field and laboratory experimental designs, ecological monitoring, and spatial analysis, while studies in the social sciences use literature reviews and synthesis, visitor surveys, social norms, laboratory experiments, spatial analysis, and simulation modeling. Scientists should think broadly about the range of disciplines and methods that can be brought to bear on understanding and managing natural quiet and natural darkness.

Principle 12. Many of the impacts of human-caused noise and light are reversible, though it is becoming evident that some may not be.

Excessive noise and light can be reduced (sometimes with little cost or effort) or even eliminated, and thus their impacts can be substantially diminished. For example, many traditional forms of transportation—including automobiles, aircraft, and motorized recreational equipment—employ new, quieter technology. Quieter snowmobiles are now required in Yellowstone National Park. Moreover, an increasing number of national parks are using forms of public transit such as shuttle buses to reduce the number of cars and associated noise (Manning et al. 2014). Similarly, new lighting technologies can reduce the amount of light needed in national parks, including lights that are oriented more directly down at their target—which allows less light to escape into the atmosphere. The use of lights that are activated by motion sensors, so they come on only when needed, is another way to reduce light pollution. These technologies and best management practices can substantially reduce human-caused noise and light in the national parks and elsewhere, suggesting that natural quiet and natural darkness can be considered as renewable resources. In chapter 16 Duriscoe writes that "unlike losing a species to extinction, topsoil to erosion, or yet-to-be-explored virgin lands to development, the night sky is 100% recoverable." While recent ecological research suggests that such recovery may be overly optimistic, it is a worthy goal. However, some impacts of human-caused noise and light may not be reversible. For example, in the study of the effects of light pollution whose results are reported in chapter 12, Altermatt and Ebert found decreased moth flight-to-light behavior that may be the product of microevolutionary response to contemporary light environments and thus likely difficult to reverse. Nevertheless, there are substantial opportunities to reduce noise and light pollution, many of which could reduce or even eliminate the impacts of that pollution and that also make sense economically. For example, reducing the use of excessive and inefficient lighting saves money.

Principle 13. Human-caused sound and light can be important cultural resources in certain contexts.

In the context of some national parks and related areas, selected forms of human-caused sound and light can be considered valuable cultural resources. For example, at San Antonio Missions National Historical Park, the sounds of church bells add an authentic layer of sound to the park experience. And in national park sites related to the Civil War, such as Gettysburg National Military Park, booming cannons can add to the authenticity of the visitor experience. Similarly, light is not inherently deleterious in the national parks, at least from an experiential perspective: well-designed and appropriate lighting allows some park resources to be viewed so as to be better appreciated. An obvious example is the caves of Carlsbad Caverns National Park (though these lights and visitors' use of the area has reduced the presence of bats). And well-designed lighting can add drama and aesthetic appeal to cultural sites, such as the American flag flying over Fort McHenry National Monument and Historic Shrine and the monuments and memorials of the National Mall and Memorial Parks in Washington, D.C.

Principle 14. Protecting natural quiet and natural darkness is an important manifestation of the twofold mission of the NPS.

As outlined in the introduction to this book, national parks are to be managed in ways that provide appropriate public access and enjoyment but also protect park resources and the quality of visitors' experiences. But these objectives can often conflict when demand for access to the national parks is high and increasing. For example, very large numbers of visitors or even a small number of noisy visitors can degrade natural quiet. Visitors can also diminish natural darkness, as can the often extensive facilities and services in the parks (such as roads, hotels, and visitor centers) that are the infrastructure for visitor use. National parks should be managed to limit the impacts of visitor use on all important park resources, including natural quiet and natural darkness. An effective program of reducing visitor-caused noise at Muir Woods National Monument, described in chapter 8, is a good example. Moreover, as reported in chapter 18, research at Acadia National Park has identified indicators and thresholds for the minimum acceptable condition of dark night skies, and the park should be managed to help ensure that these standards are maintained.

Principle 15. Natural quiet and natural darkness are common property resources that should be managed to protect the quality of park resources and visitors' experiences.

The conceptual framework of common property resources was described in the introduction to this book. Natural quiet and natural darkness are examples of commons—resources that are owned by all but that are inherently vulnerable to use-related impacts caused by individuals. Common property resources are subject to overuse and degradation if they are not managed appropriately. In his classic paper "The Tragedy of the Commons," Garrett Hardin suggested that "mutual coercion, mutually agreed upon" (1968, 1247) was needed to manage common property resources. In other words, society must agree on rules that govern the amount and types of uses of common property resources to protect them, and it must then abide by these rules. Government agencies responsible for managing common property resources—the NPS, in the case of the national parks—must establish and enforce such rules. Of course, these rules must be established in consultation with the public.

Principle 16. The conceptual frameworks of carrying or visitor capacity, limits of acceptable change, and indicators and thresholds should be used to help protect important park resources, including natural quiet and natural darkness.

The conceptual frameworks of the dual mission of national parks and related protected areas (outlined in principle 14) and of common property resources (outlined in principle 15) suggest that rules and other management actions are needed to protect important park resources, including natural quiet and natural darkness. Several other conceptual frameworks described in the introduction to this book (the closely related frameworks of carrying capacity, limits of acceptable change, and indicators and thresholds) can be used to help protect park resources and visitors' experiences. "Carrying capacity" is defined as the amount and type of visitor use that can be accommodated in a park or related area without having unacceptable impacts on park resources and the quality of visitors' experiences. The concept of limits of acceptable change recognizes that visitors' use of national parks will inevitably have some impacts, but these impacts must be limited to maintain the quality of park resources and experiences. Indicators and thresholds empirically define the limits of acceptable change. These three conceptual frameworks—carrying capacity, limits of acceptable change, and indicators and thresholds—can be integrated and made operational in the form of an adaptive management-by-objectives framework as illustrated in Figure i.3 in the introduction of this book. First, management objectives are formulated that describe the desired conditions of park resources and experiences. These objectives are then expressed in empirical terms in the form of indicators and thresholds. Indicators are measurable, manageable variables that serve as quantitative proxies for management objectives, and thresholds define the minimum acceptable condition of indicator variables. Second, indicators are periodically monitored to assess their condition. Third, management actions (as outlined in principle 17) are applied to help ensure that the condition of indicator variables does not fall below defined thresholds. Application of this management-by-objectives framework is an ongoing and adaptive process: it provides periodic information on the condition of indicator variables as well as the effectiveness of management actions, and managers can adjust their management regime as warranted and guided by monitoring data. This management-by-objectives approach addresses the concept of carrying capacity by helping ensure that management objectives of parks and protected areas are attained. Chapter 8, on protecting natural quiet at Muir Woods National Monument, is a good example of the application of a management-by-objectives framework.

Principle 17. A range of management strategies and practices can be used to protect natural quiet and natural darkness in the national parks, but new approaches will also be required.

Several chapters in this book suggest ways in which the NPS can help protect natural quiet and natural darkness. In chapter 18, Manning et al. describe designation of a "quiet zone" and "quiet days" at Muir Woods National Monument. An educational program was implemented at the park to communicate to visitors the importance of protecting natural quiet, and this program was found to be effective, substantially reducing the level of visitor-caused noise, and was well received by visitors. In chapter 9, Taff et al. used educational messages about military flights over Sequoia National Park, which were found to be effective in reducing the impacts of these flights on the visitor experience. However, these are only very limited examples. The professional literature suggests that there are four basic strategies for reducing recreation-related impacts in parks and related reserves: limit use, increase the supply of opportunities, reduce the impacts of use, and harden park resources and experiences (that is, make them more resistant to potential impacts) (Manning et al. 2017). In addition, six basic practices can be used to carry out these strategies: information or education, zoning, rules and regulations, law enforcement, rationing and allocation, and site design and maintenance. Relatively little is known about the effectiveness of these management strategies and practices in preserving natural quiet and natural darkness and related visitor experiences, and this topic deserves more study. Furthermore, the protection of natural quiet and natural darkness will require the application of specific management strategies and practices that are aimed at this more specialized context.

Principle 18. A landscape-scale approach is needed to protecting natural quiet and natural darkness in the national parks.

In the introduction, we noted that it is becoming increasingly apparent that many conservation issues require a landscape-scale approach, because some of the impacts on parks are caused by sources outside their boundaries. This is especially the case with natural quiet and natural darkness. Human-generated noise that is produced outside the parks may drift into parks, disturbing natural quiet there. Transportation-related noise is a good example. Human-caused light generated outside parks can also intrude into the parks, disturbing natural darkness. In fact, in chapter 17 Smith and Hallo note that light from large cities can be seen as far as 200 miles from its origin, diminishing the darkness of the night skies of otherwise relatively remote national parks. Most national parks and other protected areas are not large enough to be fully insulated from these types of external threats. Progressive national park management must include reaching out to neighboring land owners—including other public land management agencies, surrounding communities, nonprofit organizations, and private land owners—and working with them to limit impacts on natural quiet and natural darkness in national parks. This outreach has succeeded at Acadia National Park and Chaco Culture National Historical Park, for example. NPS staff members at Acadia collaborated with officials in Bar Harbor, the gateway town to the park, and this resulted in passage of a progressive ordinance that reduces the amount of light that drifts into the park, thereby helping protect the park's natural darkness and night skies. And park managers at Chaco Canyon worked with stakeholders in New Mexico to convince the state legislature to pass an act that limits light pollution throughout the state. In chapter 20, Collison and Poe note that viewing the night sky is becoming increasingly popular and that this can have substantial economic implications in the form of "astronomical tourism," which offers an economic incentive to local communities to help protect the darkness of night skies. An innovative and promising landscape-scale initiative called the Dark Sky Preserve is now under way on the Colorado Plateau, a large geographic region in the American Southwest. This is an NPS-led effort to coordinate management of the many national parks in this region to help preserve the remarkable quality of their natural darkness. This landscape-scale approach to managing natural darkness is strongly supported by park visitors: for example, in their survey of visitors to Bryce Canyon National Park whose results are reported in chapter 19, Mace and McDaniel found that nearly 90 percent of respondents believe "that areas near national parks should assist in maintaining dark night skies."

Principle 19. A large network of partners and stakeholders will be required to protect natural quiet and natural darkness in the national parks.

Many partners are needed to advance the protection of natural quiet and natural darkness for two basic reasons. First, as noted in principle 18, many threats to these resources emanate from outside the parks. Second, the NPS has limited funding and staffing to do all the work that is needed. In the case of Chaco Culture National Historical Park noted in principle 18, many stakeholder groups participated in successfully lobbying the New Mexico state legislature to pass a statewide outdoor lighting ordinance. Local astronomers and related organizations, along with a host of volunteers, are instrumental in hosting star parties at many national parks, offering night sky–related programming and setting up their telescopes for park visitors. In Sequoia National Park, as Taff et al. reported in chapter 9, managers are collaborating with the local military base to help minimize the impacts of aircraft overflights on the park's natural quiet. The International Dark Sky Association is instrumental in advancing the cause of night sky protection, certifying parks and local communities that meet best management practices. Local communities are interested in advancing tourism-based economic development through protecting high-quality natural quiet and natural darkness in the parks near them. As shown by the chapters in this book and the references in the bibliography, natural and social scientists are interested in increasing our knowledge about natural quiet and natural darkness, which can inform effective park management. The NPS has been a leader among US land management agencies, including the protection of natural quiet and natural darkness in its policies and creating and funding organizational units to carry out this mandate. Perhaps the strongest advocates of natural quiet and natural darkness are park visitors, many of whom visit the parks, at least in part, to experience those resources. For example, in chapter 17 Smith and Hallo found substantial visitor interest in night-related recreation activities. And as noted in principle 18, Mace and McDaniel reported in chapter 19 that a large majority of their survey respondents believed that areas near national parks should help maintain dark night skies. Furthermore, in their study of visitors to Bryce Canyon National Park, whose results are reported in chapter 20, Collison and Poe found substantial visitor interest in interpretive programs on the night sky. These stakeholders appreciate the values of natural quiet and natural darkness and can offer many types of support to help reduce noise and light pollution in the national parks and related areas.

Principle 20. Natural quiet and natural darkness should be incorporated into comprehensive park management plans.

As is clear from several of the preceding principles, natural quiet and natural darkness are clearly important natural, cultural, and recreational resources in the national parks. The NPS has a planning process in which foundation statements are periodically prepared for all units of the national park system, followed by a portfolio of specific plans based on the operational and management needs of each park. Natural quiet and natural darkness should receive explicit consideration in this planning process. What resources related to natural quiet and natural darkness are found in the park? What are the challenges and threats to protecting these resources? Do these resources need to be restored? Many parks include specific plans that address resources or operations that are especially important to them—natural and cultural resources, wilderness, transportation, visitor use management, and so on. At parks with iconic natural quiet and natural darkness—such as Bryce Canyon National Park, Muir Woods National Monument, Acadia National Park, many of the national parks in Alaska, and many more parks—plans focused on natural quiet and natural darkness may be justified. Alternatively, natural quiet and natural darkness should be explicitly considered within the foundation statements and more specific types of plans noted above.

Principle 21. More information and educational programs are needed on natural quiet and natural darkness.

The NPS has a long history of providing high-quality education to visitors about the natural, cultural, and recreational resources of the national parks. In NPS lingo, this is called interpretive programming because it "interprets" technical information about the parks in ways that are engaging and aimed at a nontechnical audience. Classic examples are nature walks and campfire talks, though the form of these programs has been substantially updated to include collaborative relationships with local schools and a strong presence on the Internet and through social media (Manning et al. 2016). More programing is needed on the newly recognized resources of natural quiet and natural darkness. The studies by Mace and McDaniel in chapter 19 and Collison and Poe in chapter 20 suggest the growing interest in these resources and associated issues on the part of many park visitors. For example, Mace and McDaniel's findings demonstrate that participation in natural darkness–related interpretive programs increases knowledge and improves attitudes about this resource, enhances the quality of visitor experiences, can lead to support for the protection of natural darkness, and stimulates visitors' interest in supporting protection of natural darkness in their home communities. Moreover, educational programming about natural quiet and natural darkness can be a management tool, teaching park visitors how to limit their impact on these resources, both inside and outside the national parks. And as Taff et al. noted in chapter 9, educational messages can help visitors cope with human-caused noise when it is beyond the control of park managers.

Principle 22. Management of natural quiet and natural darkness should be conducted in a proactive manner.

Like all good park management, potential problems should be addressed before they reach crisis proportions. Unfortunately, human-caused noise and light have already affected park resources and the quality of the visitor experience. The chapters of this book describe many of the current ecological and experiential impacts on natural quiet and natural darkness, both of which have already undergone substantial degradation. While the growing body of research advances our understanding of natural quiet and natural darkness, much remains to be learned (principle 10). But this should not be an impediment to an aggressive program of management. We will never have complete knowledge of the ecological and experiential impacts on natural quiet and natural darkness, the causes and consequences of these impacts, and the effectiveness of alternative management practices. After making a good faith effort to inform themselves, park managers must ultimately exercise their professional judgment. Managers should find courage in their knowledge of the burgeoning scientific and professional literature (such as that presented in this book), the conceptual and management frameworks that have emerged from this literature (as outlined in the introduction, the adaptive character of the management-by-objectives approach for park management, formulating indicators and associated thresholds, making a commitment to monitor indicator variables, and adjusting management accordingly), and the trust the public has placed in them to protect park resources and experiences. A general recognition that natural quiet and natural darkness are vital park resources will require strong and deliberate management action and a promise to manage by design, not by default.

THE SUSTAINABILITY OF NATURAL QUIET AND NATURAL DARKNESS IN THE NATIONAL PARKS

The concept of sustainability—much like carrying capacity—has emerged as a vital concept for the contemporary world, and for good reason: we must learn to live within the constraints posed by our environment or face the possibility of grave consequences in the form of a degraded planet and diminished quality of life. What better place to address sustainability than in the national parks, iconic symbols of our commitment to protecting the environment? What better issues to focus on than natural quiet and natural darkness, resources that have important implications for the ecological and experiential integrity of the national parks? And what better time to do so than now, as these resources are rapidly disappearing? The sustainability of these resources in the national parks makes good, and common, sense.

In important ways, the national parks have been at the forefront of sustainability for decades. As outlined in the introduction to this book and highlighted in principle 15, national park management has historically been based on the NPS's foundational twofold mission—to foster public use and appreciation of the parks while protecting their environmental and experiential integrity. This is at the heart of sustainability: how much and in what ways can we use the world without spoiling what is so valuable about it? In the context of the national parks, this issue is often called carrying capacity and addressed in terms of how much and what kinds of use can be accommodated in the national parks (and also, in the case of natural quiet and natural darkness in the national parks, how much and what kinds of use can be accommodated outside the national parks) without having unacceptable impacts on park resources and the quality of the visitor experience? With more than 300 million annual visits to the national parks, this is an increasingly urgent question.

As described throughout this book, carrying capacity—and sustainability more broadly—can be addressed through a management-by-objectives framework that has three steps: formulating management objectives and associated indicators and thresholds, monitoring indicator variables, and implementing management actions designed to prevent violation of thresholds. Management objectives describe the desired condition of park resources and the quality of the visitor experience, while indicators and thresholds express these objectives in quantitative terms that can (and should) be measured to assess how well objectives are being met. Monitoring indicator variables is a long-term commitment to this approach to carrying capacity or sustainability and assesses the extent to which management actions are effective in reaching and maintaining management objectives. This is an iterative, adaptive process in which sustainability is defined and measured over time; management effectiveness is continuously assessed, tested, revised, and refined; and the public is regularly consulted.

The introduction to this book also described the three main dimensions of carrying capacity: concern for the quality of the environment, concern for the quality of the visitor experience, and attention to the opportunities and constraints of management to address carrying capacity or sustainability. This multidimensional framework is in keeping with the emerging body of scientific and professional literature on sustainability in general. For example, the earliest discussion of sustainability in the contemporary environmental literature suggested that it has two important dimensions: ecological and social (Brundtland Commission 1987). More recent treatments of sustainability are based on what are often called the "three pillars" of sustainability, the "three Es," or the "triple bottom line" (Elkington 1998; Ekins 2000; de Mooij and van Den Bergh 2002). All of these frameworks suggest that comprehensive consideration of sustainability must address matters of environment, society, and economy. In the case of natural quiet and natural darkness in the national parks, the scientific and professional literature is beginning to address all of these dimensions. Most of the chapters in this book address the relationship between natural quiet or natural darkness, the environment, and the quality of the visitor experience. However, relatively little is known about the economic dimension of sustainability, though tourism based on natural quiet and natural darkness suggests the potential for substantial economic benefits, as does the energy saving from more efficient lighting and transportation practices.

This book illustrates ways in which the management of natural quiet and natural darkness is becoming more sustainable. There is a growing understanding of the potential impacts of noise and light pollution on park resources and the quality of the visitor experience, as well as of the ways in which natural quiet and natural darkness can contribute to human health and well-being. Programs of natural and social science research in the national parks are providing a stronger theoretical and empirical foundation for formulating indicators and thresholds for defining and measuring the sustainability of natural quiet and natural darkness in the national parks. And this research is also testing the effectiveness of a range of management actions designed to maintain thresholds. This growing body of work draws on the literature in the natural and social sciences and the fields of conservation and park and outdoor recreation management, and integrates the work in multiple fields where possible. While a great deal more research on natural quiet and natural darkness in the national parks and protected areas more broadly is warranted, it is beginning to reach a critical mass—as reflected in the principles described above.

REFERENCES

Absher, J. D., and A. D. Bright. 2004. "Communication Research in Outdoor Recreation and Natural Resources Management." In *Society and Natural Resources—A Summary of Knowledge,* edited by M. Manfredo, J. Vaske, B. Bruyere, D. Field, and P. Brown, 117–26. Jefferson, MO: Modern Litho.

Acharya, L., and M. B. Fenton. 1999. "Bat Attacks and Moth Defensive Behaviour around Street Lights." *Canadian Journal of Zoology* 77 (1): 27–33.

Aihara, I., P. d. Silva, and X. E. Bernal. 2015. "Acoustic Preference of Frog-Biting Midges (*Corethrella* spp) Attacking Tungara Frogs in Their Natural Habitat." *Ethology* 122 (2): 105–13.

Albers, S., and D. Duriscoe. 2001. "Modeling Light Pollution from Population Data and Implications for National Park Service Lands." In *George Wright Forum* 18 (4): 56–68. Accessed June 18, 2017. http://weather.cavemanastronomy.com/dup/practice/184albers_DARK_SKY_MODELLING.pdf.

Allen, J. J., L. M. Maethger, K. C. Buresch, T. Fetchko, M. Gardner, and R. T. Hanlon. 2010. "Night Vision by Cuttlefish Enables Changeable Camouflage." *Journal of Experimental Biology* 213 (23): 3953–60.

Almany, G. R., and M. S. Webster. 2006. "The Predation Gauntlet: Early Post-Settlement Mortality in Reef Fishes." *Coral Reefs* 25 (1): 19–22.

Altermatt, F., A. Baumeyer, and D. Ebert. 2009. "Experimental Evidence for Male Biased Flight-to-Light Behavior in Two Moth Species." *Entomologia experimentalis et applicata* 130 (3): 259–65.

Altermatt, F. and D. Ebert. 2016. "Reduced Flight-to-Light Behaviour of Moth Populations Exposed to Long-Term Urban Light Pollution." *Biology Letters* 12 (4) Accessed August 20, 2017. http://rsbl.royalsocietypublishing.org/content/12/4/20160111.

Alvarez del Castillo, E. M., D. L. Crawford, and D. R. Davis. 2003. "Preserving Our Nighttime Environment: A Global Approach." In *Light Pollution: The Global View,* edited by H. E. Schwarz, 49–67. Dordrecht, the Netherlands: Springer.

Ambrose, S., and S. Burson. 2004. "Soundscape Studies in National Parks." *George Wright Forum* 21 (1): 29–38. Accessed June 18, 2017. http://www.georgewright.org/211ambrose.pdf.

Anderson, C. A., A. J. Lindsay, and B. J. Bushman. 1999. "Research in the Psychological Laboratory: Truth or Triviality?" *Current Directions in Psychological Science* 8 (1): 3–9.

Anderson, L., B. Mulligan, L. Goodman, and H. Regen. 1983. "Effects of Sounds on Preferences for Outdoor Settings." *Environment and Behavior* 15 (5): 539–66.

Arnfield, A. J. 2003. "Two Decades of Urban Climate Research: A Review of Turbulence, Exchanges of Energy and Water, and the Urban Heat Island." *International Journal of Climatology* 23 (1): 1–26.

Ashmore, M. R. 2005. "Assessing the Future Global Impacts of Ozone on Vegetation." *Plant Cell and Environment* 28 (8): 949–64.

Astronomical Society of Kansas City. n.d. "Astronomical Society of Kansas City." Accessed August 15, 2017. http://www.askc.org/index.php.

Astronomy at Starry Nights Bed & Breakfast. n.d. "Astronomy at Starry Nights Bed & Breakfast." Accessed August 15, 2017. http://starrynight.homestead .com/Astronomy.html.

Aydin, Y., and M. Kaltenbach. 2007. "Noise Perception, Heart Rate and Blood Pressure in Relation to Aircraft Noise in the Vicinity of the Frankfurt Airport." *Clinical Research in Cardiology* 96 (6): 347–58.

Babisch, W., H. Fromme, A. Beyer, and H. Ising. 2001. "Increased Catecholamine Levels in Urine in Subjects Exposed to Road Traffic Noise—The Role of Stress Hormones in Noise Research." *Environment International* 26 (7–8): 475–81.

Bachleitner, W., L. Kempinger, C. Wuelbeck, D. Rieger, and C. Helfrich-Foerster. 2007. "Moonlight Shifts the Endogenous Clock of Drosophila Melanogaster." *Proceedings of the National Academy of Sciences of the United States of America* 104 (9): 3538–43.

Baird, E., E. Kreiss, W. Wcislo, Eric Warrant, and Marie Dacke. 2011. "Nocturnal Insects Use Optic Flow for Flight Control." *Biology Letters* 7 (4): 499–501.

Baker, B. J., and J. M. L. Richardson. 2006. "The Effect of Artificial Light on Male Breeding-Season Behaviour in Green Frogs, Rana Clamitans Melanota." *Canadian Journal of Zoology—Revue Canadienne de zoologie* 84 (10): 1528–32.

Baker, G. C., and Rwrj Dekker. 2000. "Lunar Synchrony in the Reproduction of the Moluccan Megapode *Megapodius Wallacei*." *Ibis* 142 (3): 382–88.

Baker, P. J., C. V. Dowding, S. E. Molony, P. C. L. White, and S. Harris. 2007. "Activity Patterns of Urban Red Foxes (*Vulpes Vulpes*) Reduce the Risk of Traffic-Induced Mortality." *Behavioral Ecology* 18 (4): 716–24.

Baker, R., and Y. Sadovy. 1978. "Distance and Nature of Light-Trap Response of Moths." *Nature* 276 (5690): 818–21.

Balda, R. P. 1987. "Avian Impacts on Pinyon-Juniper Woodlands." General Technical Report INT-US Department of Agriculture, Forest Service, Intermountain Research Station.

Barber, J. R., C. L. Burdett, S. E. Reed, K. A. Warner, C. Formichella, K. R. Crooks, D. M. Theobald, and K. M. Fristrup. 2011. "Anthropogenic Noise Exposure in Protected Natural Areas: Estimating the Scale of Ecological Consequences." *Landscape Ecology* 26 (9): 1281–95.

Barber, J. R., K. R. Crooks, and K. M. Fristrup. 2010. "The Costs of Chronic Noise Exposure for Terrestrial Organisms." *Trends in Ecology & Evolution* 25 (3): 180–89.

Barber, J. R., K. M. Fristrup, C. L. Brown, A. R. Hardy, L. M. Angeloni, and K. R.

Crooks. 2009. "Conserving the Wild Life Therein—Protecting Park Fauna from Anthropogenic Noise." *Park Science* 26 (3): 26–31.

Barta, A., and G. Horvath. 2004. "Why Is It Advantageous for Animals to Detect Celestial Polarization in the Ultraviolet? Skylight Polarization under Clouds and Canopies Is Strongest in the UV." *Journal of Theoretical Biology* 226 (4): 429–37.

Barton, B. A. 2002. "Stress in Fishes: A Diversity of Responses with Particular Reference to Changes in Circulating Corticosteroids." *Integrative and Comparative Biology* 42 (3): 517–25.

Bates, D. M., M. Mächler, and B. Dai. "Welcome to lme4—Mixed Effects Model Project!" Accessed August 26, 2017. http://lme4.r-forge.r-project.org/.

Baugh, A., and M. Ryan. 2010. "Ambient Light Alters Temporal Updating Behaviour During Mate Choice in a Neotropical Frog." *Canadian Journal of Zoology* 88 (5): 448–53.

Baugh, K., C. D. Elvidge, T. Ghosh, and D. Ziskin. 2010. "Development of a 2009 Stable Lights Product Using DMSP-OLS Data." *Proceedings of the Asia-Pacific Advanced Network* 30:114–30.

Baum, K. A., and W. E. Grant. 2001. "Hummingbird Foraging Behavior in Different Patch Types: Simulation of Alternative Strategies." *Ecological Modelling* 137 (2–3): 201–9.

Baumgartner, T., F. Wunderlich, D. Wee, A. Jaunich, T. Sato, U. Erxleben, G. Bundy, and R. Bundsgaard, 2011. *Lighting the Way: Perspectives on the Global Lighting Market.* New York: McKinsey and Company.

Bayne, E. M., L. Habib, and S. Boutin. 2008. "Impacts of Chronic Anthropogenic Noise from Energy-Sector Activity on Abundance of Songbirds in the Boreal Forest." *Conservation Biology* 22 (5): 1186–93.

Bedrosian, T. A., L. K. Fonken, J. C. Walton, A. Haim, and R. J. Nelson. 2011a. "Dim Light at Night Provokes Depression-Like Behaviors and Reduces CA1 Dendritic Spine Density in Female Hamsters." *Psychoneuroendocrinology* 36 (7): 1062–69.

Bedrosian, T. A., L. K. Fonken, J. C. Walton, and R. J. Nelson. 2011b. "Chronic Exposure to Dim Light at Night Suppresses Immune Responses in Siberian Hamsters." *Biology Letters* 7 (3): 468–71.

Beier, P. 1995. "Dispersal of Juvenile Cougars in Fragmented Habitat." *Journal of Wildlife Management* 59 (2): 228–37.

———. 2006. "Effects of Artificial Night Lighting on Terrestrial Mammals." In *Ecological Consequences of Artificial Night Lighting* edited by Catherine Rich and Travis Longcore, 19–42. Washington: Island Press.

Bejder, L., A. Samuels, H. Whitehead, and N. Gales. 2006. "Interpreting Short-Term Behavioural Responses to Disturbance within a Longitudinal Perspective." *Animal Behaviour* 72 (November): 1149–58.

Bell, P. A., B. L. Mace, and J. A. Benfield. 2009. "Aircraft Overflights at National Parks: Conflict and Its Potential Resolution." *Park Science* 26 (3): 65–67.

Bemus, T. 2001. "Stargazing: A Driving Force in Ecotourism at Cherry Springs State Park." *George Wright Forum*, 18 (4): 69–71. Accessed June 18, 2017. http://www.georgewright.org/184bemus.pdf.

Benfield, J. A., P. A. Bell, L. J. Troup, and N. C. Soderstrom. 2010a. "Aesthetic and Affective Effects of Vocal and Traffic Noise on Natural Landscape Assessment." *Journal of Environmental Psychology* 30 (1): 103–11.

———. 2010b. "Does Anthropogenic Noise in National Parks Impair Memory?" *Environment and Behavior* 42 (5): 693–706.

Benfield, J. A., G. Rainbolt, P. Bell, and G. Donovan. 2015. "Classrooms with Nature Views: Evidence of Differing Student Perceptions and Behaviors." *Environment and Behavior* 47 (2): 140–57.

Benítez-López, A., R. Alkemade, and P. A. Verweij. 2010. "The Impacts of Roads and Other Infrastructure on Mammal and Bird Populations: A Meta-Analysis." *Biological Conservation* 143 (6): 1307–16.

Bennie, J., T. W. Davies, D. Cruse, and K. J. Gaston. 2016. "Ecological Effects of Artificial Light at Night on Wild Plants." *Journal of Ecology* 104 (3): 611–20.

Bennie, J., T. W. Davies, J. P. Duffy, R. Inger, and K. J. Gaston. 2014. "Contrasting Trends in Light Pollution across Europe Based on Satellite Observed Night Time Lights." *Scientific Reports* 4 (January): 3789.

Bennie, J., J. P. Duffy, T. W. Davies, M. E. Correa-Cano, and K. J. Gaston. 2015. "Global Trends in Exposure to Light Pollution in Natural Terrestrial Ecosystems." *Remote Sensing* 7 (3): 2715–30.

Bennie, J., E. Kubin, A. Wiltshire, B. Huntley, and R. Baxter. 2010. "Predicting Spatial and Temporal Patterns of Bud-Burst and Spring Frost Risk in North-West Europe: The Implications of Local Adaptation to Climate." *Global Change Biology* 16 (5): 1503–14.

Berge, J., F. Cottier, K. S. Last, O. Varpe, E. Leu, J. Soreide, K. Eiane, S. Falk-Peterson, K. Willis, H. Nygard, et al. 2009. "Diel Vertical Migration of Arctic Zooplankton during the Polar Night." *Biology Letters* 5 (1): 69–72.

Berger, D., and K. Gotthard. 2008. "Time Stress, Predation Risk and Diurnal-Nocturnal Foraging Trade-Offs in Larval Prey." *Behavioral Ecology and Sociobiology* 62 (10): 1655–634.

Berman, M. G., J. Jonides, and S. Kaplan. 2008. "The Cognitive Benefits of Interacting with Nature." *Psychological Science* 19 (12): 1207–12.

Bernal, X. E., and P. de Silva. 2015. "Cues Used in Host-Seeking Behavior by Frog-Biting Midges (*Corethrella spp Coquillet*)." *Journal of Vector Ecology* 40 (1): 122–28.

Bernal, X. E., and C. M. Pinto. 2016. "Sexual Differences in Prevalence of a New Species of Trypanosome Infecting Túngara Frogs." *International Journal of Parasitology: Parasites and Wildlife* 5 (1): 40–47.

Bernal, X. E., A. S. Rand, and M. J. Ryan. 2006. "AcousticPpreferences and Localization Performance of Blood-Sucking Flies (*Corethrella coquillett*) to Tungara Frog Calls." *Behavioral Ecology* 17 (5): 709–15.

Bernard, E., and M. B. Fenton. 2003. "Bat Mobility and Roosts in a Fragmented Landscape in Central Amazonia, Brazil." *Biotropica* 35 (2): 262–77.

Berson, D. M., F. A. Dunn, and M. Takao. 2002. "Phototransduction by Retinal Ganglion Cells That Set the Circadian Clock." *Science* 295 (5557): 1070–73.

Berthold, P. 1996. *Control of Bird Migration*. London: Chapman and Hall.

Biebouw, K., and D. T. Blumstein. 2003. "Tammar Wallabies (*Macropus Eugenii*) Associate Safety with Higher Levels of Nocturnal Illumination." *Ethology Ecology & Evolution* 15 (2): 159–72.

Bird, B. L., L. C. Branch, and D. L. Miller. 2004. "Effects of Coastal Lighting on Foraging Behavior of Beach Mice." *Conservation Biology* 18 (5): 1435–39.

Bishop, Ian D., and H. R. Gimblett. 2000. "Management of Recreational Areas: GIS, Autonomous Agents, and Virtual Reality." *Environment and Planning B: Planning and Design* 27 (3): 423–35.

Black, A. 2005. "Light Induced Seabird Mortality on Vessels Operating in the Southern Ocean: Incidents and Mitigation Measures." *Antarctic Science* 17 (1): 67–68.

Blake, D., A. Hutson, P. Racey, J. Rydell, and J. Speakman. 1994. "Use of Lamplit Roads by Foraging Bats in Southern England." *Journal of Zoology* 234 (November): 453–62.

Blickley, J. L., D. Blackwood, and G. L. Patricelli. 2012a. "Experimental Evidence for the Effects of Chronic Anthropogenic Noise on Abundance of Greater Sage-Grouse at Leks." *Conservation Biology* 26 (3): 461–71.

Blickley, J. L., and G. L. Patricelli. 2012. "Potential Acoustic Masking of Greater Sage-Grouse (*Centrocercus Urophasianus*) Display Components by Chronic Industrial Noise." *Ornithological Monographs* 2012 (74): 23–35.

Blickley, J. L., K. R. Word, A. H. Krakauer, J. L. Phillips, S. N. Sells, C. C. Taff, J. C. Wingfield, and G. L. Patricelli. 2012b. "Experimental Chronic Noise Is Related to Elevated Fecal Corticosteroid Metabolites in Lekking Male Greater Sage-Grouse (*Centrocercus Urophasianus*)." *PLoS One* 7 (11): e50462.

Bogard, P. 2013. *The End of Night: Searching for Natural Darkness in an Age of Artificial Light*. New York: Little, Brown.

Booth-Butterfield, S., and J. Welbourne. 2002. "The Elaboration Likelihood Model: Its Impact on Persuasion Theory and Research." In *The Persuasion Handbook: Developments in Theory and Practice*, edited by J. Dillard and M. Pfau, 135–73. Thousand Oaks, CA: Sage Publications.

Borkent, A. 2008. *The Frog-Biting Midges of the World (Corethrellidae:Diptera)*. Auckland, New Zealand: Magnolia Press.

Boscarino, B. T., L. G. Rudstam, J. L. Eillenberger, and R. O'Gorman. 2009. "Importance of Light, Temperature, Zooplankton and Fish in Predicting the Nighttime Vertical Distribution of *Mysis Diluviana*." *Aquatic Biology* 5 (3): 263–79.

Bosch, J. A., Eco J. C. Geus, A. Kelder, E. C. I. Veerman, J. Hoogstraten, and A. V. Amerongen. 2001. "Differential Effects of Active versus Passive Coping on Secretory Immunity." *Psychophysiology* 38 (5): 836–46.

Bouskila, A. 1995. "Interactions Between Predation Risk and Competition—A Field Study of Kangaroo Rats and Snakes." *Ecology* 76 (1): 165–78.

Boyce, Peter Robert. 2003. *Human Factors in Lighting*. New York: Taylor and Francis.

Bradley, C., and S. Altizer. 2007. "Urbanization and the Ecology of Wildlife Diseases." *Trends in Ecology and Evolution* 22 (2) :95–102.

Bradshaw, W. E. 1976. "Geography of Photoperiodic Response in Diapausing Mosquito." *Nature* 262 (5567): 384–86.

———— and C. M. Holzapfel. 2010. "Light, Time, and the Physiology of Biotic Response to Rapid Climate Change in Animals." *Annual Review of Physiology* 72:147–66.

Bradshaw, W. E., P. A. Zani, and C. M. Holzapfel. 2004. "Adaptation to Temperate Climates." *Evolution* 58 (8): 1748–62.

Brainard, G., B. Richardson, T. King, S. Matthews, and R. Reiter. 1983. "The Suppression of Pineal Melatonin Content and N-Acetyltransferase Activity by Different Light Irradiances in the Syrian Hamster—A Dose-Response Relationship." *Endocrinology* 113 (1): 293–96.

Brainard, G., B. Richardson, L. Petterborg, and R. Reiter. 1982. "The Effect of Different Light Intensities on Pineal Melatonin Content." *Brain Research* 233 (1): 75–81.

Brasler, D., and C. Körner. 2012. "Photoperiod Sensitivity of Bud Burst in 14 Temperate Forest Tree Species." *Agricultural and Forest Meteorology* 165:73–81.

————. 2014. "Photoperiod and Temperature Responses of Bud Swelling and Bud Burst in Four Temperate Forest Tree Species." *Tree Physiology* 34 (4): 377–88.

British Standards Institute. 2003. *Road Lighting Part 2: Performance Requirements*. London: British Standards Institute.

Britt, A. B. 1996. "DNA Damage and Repair in Plants." *Annual Review of Plant Physiology and Plant Molecular Biology* 47:75–100.

Brons, J. A., J. D. Bullough, and M. S. Rea. 2008. "Outdoor Site-Lighting Performance: A Comprehensive and Quantitative Framework for Assessing Light Pollution." *Lighting Research & Technology* 40 (3): 201–24.

Brown, J. S. 1999. "Vigilance, Patch Use and Habitat Selection: Foraging under Predation Risk." *Evolutionary Ecology Research* 1 (1): 49–71.

Bruce-White, C., and M. Shardlow. 2011. *A Review of the Impact of Artificial Light on Invertebrates*. Peterborough, UK: Buglife.

Brumm, H. 2013. *Animal Communication and Noise*. New York: Springer.

———— and H. Slabbekoorn. 2005. "Acoustic Communication in Noise." In *Advances in the Study of Behavior*, vol. 35, edited by P. J. B. Slater, C. T. Snowdon, H. J. Brockmann, T. J. Roper, and M. Naguib, 151–209. San Diego, CA: Academic Press.

Brundtland Commission. 1987. *Report of the World Commission on Environment and Development: Our Common Future*. Accessed August 27, 2017. http://www.un-documents.net/our-common-future.pdf.

Bryce Canyon National Park. 2012. "Lightscape/Night Sky." http://www.nps.gov /brca/naturescience/lightscape.htm.

Buchanan, B. W. 1993. "Effects of Enhanced Lighting on the Behavior of Nocturnal Frogs." *Animal Behaviour* 45 (5): 893–99.

⸺. 2006. "Observed and Potential Effects of Artificial Night Lighting on Anuran Amphibians." In *Ecological Consequences of Artificial Night Lighting* edited by C. Rich and T. Longcore, 192–220. Washington: Island Press.

Bunkley, J. P., C. J. W. McClure, N. J. Kleist, C. D. Francis, and J. R. Barber. 2015. "Anthropogenic Noise Alters Bat Activity Levels and Echolocation Calls." *Global Ecology and Conservation* 3: 62–71.

Burke, L., K. Reytar, M. Spalding, and A. Perry. 2011. *Reefs at Risk Revisited.* Washington: World Resources Institute. Accessed June 17, 2017. http://www.vliz.be /en/imis?refid=223234.

Buse, A., S. J. Dury, R. J. W. Woodburn, C. M. Perrins, and J. E. G. Good. 1999. "Effects of Elevated Temperature on Multi-Species Interactions: The Case of Pedunculate Oak, Winter Moth and Tits." *Functional Ecology* 13 (June): 74–82.

Buse, A., and J. E. G. Good. 1996. "Synchronization of Larval Emergence in Winter Moth (*Operophtera Brumata L*) and Budburst in Pedunculate Oak (*Quercus Robur L*) under Simulated Climate Change." *Ecological Entomology* 21 (4): 335–43.

Buxton, R. T., M. F. McKenna, D. Mennitt, K. Fristrup, K. Crooks, L. Angeloni, and G. Wittemyer. 2017. "Noise Pollution Is Pervasive in US Protected Areas." *Science* 356 (6337): 531–33.

Campbell, D., and K. Halama. 1993. "Resource and Pollen Limitations to Lifetime Seed Production in a Natural Plant Population." *Ecology* 74 (4): 1043–51.

Carlisle, J. D., K. L. Olmstead, C. H. Richart, and D. L. Swanson. 2012. "Food Availability, Foraging Behavior, and Diet of Autumn Migrant Landbirds in the Boise Foothills of Southwestern Idaho." *Condor* 114 (3): 449–61.

Carrascal, L. M., T. Santos, and J. L. Tellería. 2012. "Does Day Length Affect Winter Bird Distribution? Testing the Role of an Elusive Variable." *PLoS One* 7(2): e32733.

Cashmore, A. R., J. A. Jarillo, Y. J. Wu, and D. M. Liu. 1999. "Cryptochromes: Blue Light Receptors for Plants and Animals." *Science* 284 (5415): 760–65.

Cathey, H. M., and L. E. Campbell. 1975. "Effectiveness of Five Vision-Lighting Sources on Photo-Regulation of 22 Species of Ornamental Plants." *Journal of the American Society for Horticultural Science* 100:65–72.

Central Intelligence Agency. 2011. *The World Factbook.* Accessed August 26, 2017. https://www.cia.gov/library/publications/the-world-factbook/rankorder /2003rank.html.

Chambers, J. C., S. B. Vander Wall, and E. W. Schupp. 1999. "Seed and Seedling Ecology of Piñon and Juniper Species in the Pygmy Woodlands of Western North America." *Botanical Review* 65 (1): 1–38.

Chan, A. A. Y.-H., and D. T. Blumstein. 2011. "Attention, Noise, and Implications for Wildlife Conservation and Management." *Applied Animal Behaviour Science* 131 (1–2): 1–7.

Chan, A. A. Y.-H., P. Giraldo-Perez, S. Smith, and D. T. Blumstein. 2010a. "Anthropogenic Noise Affects Risk Assessment and Attention: The Distracted Prey Hypothesis." *Biology Letters* 6 (4): 458–61.

Chan, A. A. Y.-H., W. D. Stahlman, D. Garlick, C. D. Fast, D. T. Blumstein, and A. P. Blaisdell. 2010b. "Increased Amplitude and Duration of Acoustic Stimuli Enhance Distraction." *Animal Behaviour* 80 (6): 1075–79.

Chan, L., and E. Clark. 2001. "Yellowstone by Night." In *The George Wright Forum*, 18 (4): 83–86. Accessed June 18, 2017. http://www.georgewright.org/184chan.pdf.

Chepesiuk, R. 2009. "Missing the Dark: Health Effects of Light Pollution." *Environmental Health Perspectives* 117 (1): A20–27.

Chittka, L., and R. Menzel. 1992. "The Evolutionary Adaptation of Flower Colors and the Insect Pollinators Color-Vision." *Journal of Comparative Physiology a-Sensory Neural and Behavioral Physiology* 171 (2): 171–81.

Cinzano, P. 2003. "The Growth of the Artificial Night Sky Brightness over North America in the Period 1947–2000: A Preliminary Picture." In *Light Pollution: The Global View*, edited by H. E Schwarz, 39–47. Dordrecht, the Netherlands: Springer.

———, F. Falchi, and C. D. Elvidge. 2001. "The First World Atlas of the Artificial Night Sky Brightness." *Monthly Notices of the Royal Astronomical Society* 328 (3): 689–707.

Clark, B. A. J. 2008. "A Rationale for the Mandatory Limitation of Outdoor Lighting." Document Version 2:29. Melbourne, Australia: Astronomical Society of Victoria.

Clarke, J. A. 1983. "Moonlight's Influence on Predator-Prey Interactions Between Short-Eared Owls (*Asio-Flammeus*) and Deermice (*Peromyscus-Maniculatus*)." *Behavioral Ecology and Sociobiology* 13 (3): 205–9.

———, J. T. Chopko, and S. P. Mackessy. 1996. "The Effect of Moonlight on Activity Patterns of Adult and Juvenile Prairie Rattlesnakes (*Crotalus Viridis Viridis*)." *Journal of Herpetology* 30 (2): 192–97.

Cochran, W. W., H. Mouritsen, and M. Wikelski. 2004. "Migrating Songbirds Recalibrate Their Magnetic Compass Daily from Twilight Cues." *Science* 304 (5669): 405–8.

Cockburn, A., A. H. Dalziell, C. J. Blackmore, M. C. Double, H. Kokko, H. L. Osmond, N. R. Beck, M. L. Head, and K. Wells. 2009. "Superb Fairy-Wren Males Aggregate into Hidden Leks to Solicit Extragroup Fertilizations before Dawn." *Behavioral Ecology* 20 (3): 501–10.

Code of Federal Regulations. 2010. "Procedures for Abatement of Highway Traffic Noise and Construction Noise." 23 C.F.R. 772.17.

Cohen, J. 1988. *Statistical Power Analysis for the Behavioral Sciences*. 2nd ed. Hillsdale, NJ: Lawrence Earlbaum Associates.

Cohen, S., G. Evans, D. Krantz, and D. Stokols. 1980. "Physiological, Motivational, and Cognitive Effects of Aircraft Noise on Children—Moving from the Laboratory to the Field." *American Psychologist* 35 (3): 231–43.

Collison, F. M., and K. Poe. 2013. "'Astronomical Tourism': The Astronomy and Dark Sky Program at Bryce Canyon National Park." *Tourism Management Perspectives* 7: 1–15.

Combreau, O., and F. Launay. 1996. "Activity Rhythms of Houbara Bustards (*Chlamydotis Undulata Macqueenii*) in Relation to Some Abiotic Factors." *Journal of Arid Environments* 33 (4): 463–72.

Cooper, C. B., M. A. Voss, D. R. Ardia, S. H. Austin, and W. D. Robinson. 2011. "Light Increases the Rate of Embryonic Development: Implications for Latitudinal Trends in Incubation Period." *Functional Ecology* 25 (4): 769–76.

Cos, S., D. Mediavilla, C. Martinez-Campa, A. Gonzalez, C. Alonso-Gonzalez, and E. J. Sanchez-Barcelo. 2006. "Exposure to Light-at-Night Increases the Growth of DMBA-Induced Mammary Adenocarcinomas in Rats." *Cancer Letters* 235 (2): 266–71.

Crino, O. L., B. Klaassen Van Oorschot, E. E. Johnson, J. L. Malisch, and C. W. Breuner. 2011. "Proximity to a High Traffic Road: Glucocorticoid and Life History Consequences for Nestling White-Crowned Sparrows." *General and Comparative Endocrinology* 173 (2): 323–32.

Curry, R., A. Peterson, and T. Langen. 2002. "Western Scrub-Jay: *Aphelocoma Californica*." In *The Birds of North America*, edited by A. Poole. Vol. 712. Ithaca, NY: Cornell Lab of Ornithology.

Dacke, M., D. E. Nilsson, C. H. Scholtz, M. Byrne, and E. J. Warrant. 2003. "Insect Orientation to Polarized Moonlight." *Nature* 424 (6944): 33.

Daly, M., P. Behrends, M. Wilson, and L. Jacobs. 1992. "Behavioral Modulation of Predation Risk—Moonlight Avoidance and Crepuscular Compensation in a Nocturnal Desert Rodent, *Dipodomys Merriami*." *Animal Behaviour* 44 (1): 1–9.

Daniel, T. 1990. "Measuring the Quality of the Natural Environment—A Psychophysical Approach." *American Psychologist* 45 (5): 633–37.

Dartnall, H., J. Bowmaker, and J. Mollon. 1983. "Human Visual Pigments—Microspectrophotometric Results from the Eyes of 7 Persons." *Proceedings of the Royal Society Series B-Biological Sciences* 220 (1218): 115–30.

Dauchy, R. T., L. A. Sauer, D. E. Blask, and G. M. Vaughan. 1997. "Light Contamination during the Dark Phase in 'Photoperiodically Controlled' Animal Rooms: Effect on Tumor Growth and Metabolism in Rats." *Laboratory Animal Science* 47 (5): 511–18.

Davenport, J., and J. L. Davenport. 2006. "The Impact of Tourism and Personal Leisure Transport on Coastal Environments: A Review." *Estuarine Coastal and Shelf Science* 67 (1–2): 280–92.

Davenport, M. A., and W. T. Borrie. 2005. "The Appropriateness of Snowmobiling in National Parks: An Investigation of the Meanings of Snowmobiling Experiences in Yellowstone National Park." *Environmental Management* 35 (2): 151–60.

Davies, G., J. Hale, and J. Sadler. 2012. "Modelling Functional Connectivity Pathways for Bats in Urban Landscapes." In *Proceedings of the GIS Research UK 20th Annual Conference*, 95–100. Lancaster, UK: Lancaster University.

Davies, T. W., J. Bennie, and K. J. Gaston. 2012. "Street Lighting Changes the Composition of Invertebrate Communities." *Biology Letters* 8 (5): 764–67.

Dawson, A., V. M. King, G. E. Bentley, and G. F. Ball. 2001. "Photoperiodic Control of Seasonality in Birds." *Journal of Biological Rhythms* 16 (4): 365–80.

De Mooij, R. A., and J. C. van Den Bergh. 2002. "Growth and the Environment in Europe: A Guide to the Debate." *Empirica* 29 (2): 79–91.

De Silva, P., and X. Bernal. 2013. "First Report of the Mating Behavior of a Species of Frog-Biting Midge (*Diptera: Corethrellidae*)." *Florida Entomologist* 96 (4): 1522–29.

De Silva, P., C. Jaramillo, and X. Bernal. 2014. "Selection of Biting Sites on Anuran Hosts by *Corethrella coquillett* Species. *Journal of Insect Behavior* 27:302–16.

De Silva, P., B. Nutter, and X. E. Bernal. 2015. "Use of Acoustic Mating Signals in an Eavesdropping Frog-Biting Midge." *Animal Behaviour*, 103:45–51.

Delhey, K., A. Peters, A. Johnsen, and B. Kempenaers. 2006. "Fertilization Success and UV Ornamentation in Blue Tits *Cyanistes Caeruleus*: Correlational and Experimental Evidence." *Behavioral Ecology* 18 (2): 399–409.

Densmore, R. V. 1997. "Effect of Day Length on Germination of Seeds Collected in Alaska." *American Journal of Botany* 84 (2): 274–78.

Devictor, V., R. Julliard, D. Couvet, A. Lee, and F. Jiguet. 2007. "Functional Homogenization Effect of Urbanization on Bird Communities." *Conservation Biology* 21 (3): 741–51.

Devlin, A. S., and A. B. Arneill. 2003. "Health Care Environments and Patient Outcomes: A Review of the Literature." *Environment and Behavior* 35 (5): 665–94.

Dice, L. R. 1945. "Minimum Intensities of Illumination under Which Owls Can Find Dead Prey by Sight." *American Naturalist* 79 (784): 385–416.

Dick, Robert. 2011. "Royal Astronomical Society of Canada Guidelines for Outdoor Lighting in Urban Star Parks (RASC-USP-GOL)." Accessed June 28, 2017. http://rasc.ca/lpa/guidelines.

Dillman, D. 2007. *Mail and Internet Surveys: The Tailored Design Method*. 2nd ed. Hoboken, NJ: John Wiley and Sons.

Dingemanse, N. J., C. Both, P. J. Drent, K. Van Oers, and A. J. Van Noordwijk. 2002. "Repeatability and Heritability of Exploratory Behaviour in Great Tits from the Wild." *Animal Behaviour* 64 (December): 929–38.

Dolan, A. C., M. T. Murphy, L. J. Redmond, K. Sexton, and D. Duffield. 2007. "Extrapair Paternity and the Opportunity for Sexual Selection in a Socially Monogamous Passerine." *Behavioral Ecology* 18 (6): 985–93.

Dominoni, D., E. Carmona-Wagner, M. Hofmann, B. Kranstauber, and J. Partecke. 2013. "Individual-Based Measurements of Light Intensity Provide New Insights into the Effects of Artificial Light at Night on Daily Rhythms of Urban-Dwelling Songbirds." *Journal of Animal Ecology* 83 (3): 681–92.

Dooling, R. J., and A. N. Popper. 2007. "The Effects of Highway Noise on Birds." Sacramento: California Department of Transportation. Accessed June 12, 2017. https://www.researchgate.net/profile/Arthur_Popper/publication /228381219_The_Effects_of_Highway_Noise_on_Birds/links/02bfe510 fac89d682a000000.pdf.

Douglas, I. 1983. *The Urban Environment*. London: Edward Arnold.

Driver, B. L., P. J. Brown, G. H. Stankey, and T. G. Gregoire. 1987. "The ROS Planning System: Evolution, Basic Concepts, and Research Needed." *Leisure Sciences* 9 (3): 201–12.

Driver, B. L., R. Nash, G. Haas. 1987. "Wilderness Benefits: A State-of-Knowledge Review." USDA Forest Service General Technical Report INT-220, 294–319.

Driver, B. L., H. E. A. Tinsley, and M. J. Manfredo. 1991. "The Paragraphs about Leisure and Recreation Experience Preference Scales: Results from Two Inventories Designed to Assess the Breadth of the Perceived Psychological Benefits of Leisure." In *Benefits of Leisure*, edited by B. Driver, P. Brown, and G. Peterson, 263–86. State College, PA: Venture Publishing.

Dudash, M. 1991. "Plant Size Effects on Female and Male Function in Hermaphroditic *Sabatia-Angularis* (*Gentianaceae*)." *Ecology* 72 (3): 1004–12.

Duriscoe, D. M. 2001. "Preserving Pristine Night Skies in National Parks and the Wilderness Ethic." *George Wright Forum* 18 (4): 30–36. Accessed June 11, 2017. https://www.researchgate.net/profile/Dan_Duriscoe/publication/291116581 _Preserving_pristine_night_skies_in_national_parks_and_the_wilderness _ethic/links/5726432a08aef9c00b88f6fa.pdf.

Earth Observation Group. 2017. "Light Pollution Map." Washington, DC: National Geophysical Data Center. Accessed August 28, 2017. https://www.lightpollution map.info/.

Edman, J., and T. Scott. 1987. "Host Defensive Behavior and the Feeding success of Mosquitoes." *Insect Science and Its Application* 8 (4-5-6): 617–22.

Eigenbrod, F., S. J. Hecnar, and L. Fahrig. 2008. "The Relative Effects of Road Traffic and Forest Cover on Anuran Populations." *Biological Conservation* 141 (1): 35–46.

Ekins, P. 2000. "Costs, Benefits and Sustainability in Decision-Making, with Special Reference to Global Warming." *International Journal of Sustainable Development* 3 (4): 315–33.

Elkington, J. 1998. *Cannibals with Forks: The Triple Bottom Line of Sustainability*. Gabriola Island, BC: New Society Publishers.

Ellis, E. C. 2011. "Anthropogenic Transformation of the Terrestrial Biosphere." *Philosophical Transactions of the Royal Society a-Mathematical Physical and Engineering Sciences* 369 (1938): 1010–35.

——— and N. Ramankutty. 2008. "Putting People in the Map: Anthropogenic Biomes of the World." *Frontiers in Ecology and the Environment* 6 (8): 439–47.

Ellison, W. T., B. L. Southall, C. W. Clark, and A. S. Frankel. 2012. "A New

Context-Based Approach to Assess Marine Mammal Behavioral Responses to Anthropogenic Sounds." *Conservation Biology* 26 (1): 21–28.

Elvidge, C. D., F. C. Hsu, K. Baugh, and T. Ghosh. 2014. "National Trends in Satellite Observed Lighting: 1992–2012." In *Global Urban Monitoring and Assessment Through Earth Observation*, edited by Q. Weng, 97–118. Boca Raton, FL: CRC Press.

Elvidge, C. D., M. L. Imhoff, K. E. Baugh, V. R. Hobson, I. Nelson, J. Safran, J. B. Dietz, and B. T. Tuttle. 2001. "Night-Time Lights of the World: 1994-1995." *ISPRS Journal of Photogrammetry and Remote Sensing* 56 (2): 81–99.

Elvidge, C. D., D. M. Keith, B. T. Tuttle, and K. E. Baugh. 2010. "Spectral Identification of Lighting Type and Character." *Sensors* 10 (4): 3961–88.

enLighten. n.d. "Country Lighting Assessments." Accessed on August 12, 2017. http://www.enlighten-initiative.org/ResourcesTools/CountryLighting Assessments/WhataretheCountryLightingAssessments.aspx.

Evans, G. W., M. Bullinger, and S. Hygge. 1998. "Chronic Noise Exposure and Physiological Response: A Prospective Study of Children Living under Environmental Stress." *Psychological Science* 9 (1): 75–77.

Evans, J. A., J.A. Elliott, and M. R. Gorman. 2007. "Circadian Effects of Light No Brighter than Moonlight." *Journal of Biological Rhythms* 22 (4): 356–67.

Evans, W. R., Y. Akashi, N. S. Altman, and A. M. Manville. 2007. "Response of Night-Migrating Songbirds in Cloud to Colored and Flashing Light." *North American Birds* 60 (4): 476–488.

Faaborg, J., R. T. Holmes, A. D. Anders, K. L. Bildstein, K. M. Dugger, S. A. Gauthreaux, P. Heglund, K. Hobson, A. Jahn, D. Johnson, et al. 2010. "Conserving Migratory Land Birds in the New World: Do We Know Enough?" *Ecological Applications* 20 (2): 398–418.

Fahrig, L., and T. Rytwinski. 2009. "Effects of Roads on Animal Abundance: An Empirical Review and Synthesis." *Ecology and Society* 14 (1): 21–40.

Fahy, F. J. 2000. *Foundations of Engineering Acoustics*. London: Elsevier Academic Press.

Falchi, F., P. Cinzano, C. D. Elvidge, D. M. Keith, and A. Haim. 2011. "Limiting the Impact of Light Pollution on Human Health, Environment and Stellar Visibility." *Journal of Environmental Management* 92 (10): 2714–22.

Falkenberg, J. C., and J. A. Clarke. 1998. "Microhabitat Use of Deer Mice: Effects of Interspecific Interaction Risks." *Journal of Mammalogy* 79 (2): 558–65.

Feeny, Paul. 1970. "Seasonal Changes in Oak Leaf Tannins and Nutrients as a Cause of Spring Feeding by Winter Moth Caterpillars." *Ecology* 51 (4): 565–81.

Feinsinger, P., H. Tiebout, and B. Young. 1991. "Do Tropical Bird-Pollinated Plants Exhibit Density-Dependent Interactions—Field Experiments." *Ecology* 72 (6): 1953–63.

Fernandez-Duque, E. 2003. "Influences of Moonlight, Ambient Temperature, and Food Availability on the Diurnal and Nocturnal Activity of Owl Monkeys (*Aotus Azarai*)." *Behavioral Ecology and Sociobiology* 54 (5): 431–40.

Fernández-Juricic, E., R. Poston, K. Collibus, T. Morgan, B. Bastain, C. Martin, K. Jones, and R. Treminio. 2005. "Micro-Habitat Selection and Singing Behavior Patterns of Male House Finches (*Carpodacus mexicanus*) in Urban Arks in a heavily urbanized landscape in the Western U.S." *Urban Habitats* 3 (1):49–69.

Ferrari, M. C. O., M. I. McCormick, P. L. Munday, M. G. Meekan, D. L. Dixson, O. Lonnstedt, and D. P. Chivers. 2011. "Putting Prey and Predator into the CO2 Equation—Qualitative and Quantitative Effects of Ocean Acidification on Predator-Prey Interactions." *Ecology Letters* 14 (11): 1143–48.

Ferrari, M. C. O., P. L. Munday, J. L. Rummer, M. I. McCormick, K. Corkill, S.-A. Watson, B. J. M. Allan, M. G. Meekan, and D. P. Chivers. 2015. "Interactive Effects of Ocean Acidification and Rising Sea Temperatures Alter Predation Rate and Predator Selectivity in Reef Fish Communities." *Global Change Biology* 21 (5): 1848–55.

ffrench-Constant, R., R. Somers-Yeates, J. Bennie, T. Economou, D. Hodgson, A. Spalding, and P. K. McGregor. 2016. "Light Pollution Is Associated with Earlier Tree Budburst across the United Kingdom." *Proceedings of the Royal Society B* 283 (1833): 20160813. Accessed August 28, 2017. http://rspb.royal societypublishing.org/content/royprsb/283/1833/20160813.full.pdf.

Filadelfo, R., J. Mintz, E. Michlovich, A. D'Amico, P. L. Tyack, and D. R. Ketten. 2009. "Correlating Military Sonar Use with Beaked Whale Mass Strandings: What Do the Historical Data Show?" *Aquatic Mammals* 35 (4): 435–44.

Fischer, J., and D. Lindenmayer. 2007. "Landscape Modification and Habitat Fragmentation: A Synthesis." *Global Ecology and Biogeography* 16 (3): 265–80.

Fleming, T. H. 1988. *The Short-Tailed Fruit Bat: A Study in Plant-Animal Interactions.* Chicago: University of Chicago Press.

Floyd, M. L., M. Colyer, D. D. Hanna, and W. H. Romme. 2003. "Gnarly Old Trees: Canopy Characteristics of Old-Growth Piñon-Juniper Woodlands." In *Ancient Piñon-Juniper Woodlands: A Natural History of Mesa Verde County*, edited by L. M. Floy, 11–30. Boulder: University Press of Colorado.

Foerster, K., K. Delhey, A. Johnsen, J. T. Lifjeld, and B. Kempenaers. 2003. "Females Increase Offspring Heterozygosity and Fitness through Extra-Pair Matings." *Nature* 425 (6959): 714–17.

Foerster, K., M. Valcu, A. Johnsen, and B. Kempenaers. 2006. "A Spatial Genetic Structure and Effects of Relatedness on Mate Choice in a Wild Bird Population." *Molecular Ecology* 15 (14): 4555–67.

Foley, J. A., R. DeFries, G. P. Asner, C. Barford, G. Bonan, S. R. Carpenter, F. S. Chapin, M. T. Coe, G. C. Daily, H. K. Gibbs, et al. 2005. "Global Consequences of Land Use." *Science* 309 (5734): 570–74.

Food and Agriculture Organization of the United Nations and European Commission Joint Research Centre. 2012. *Global Forest Land-Use Change 1990–2005*. FAO Forestry Paper No. 169. Accessed August 26, 2017. http://www.fao.org /docrep/017/i3110e/i3110e.pdf.

Forman, R. T. T. 2000. "Estimate of the Area Affected Ecologically by the Road System in the United States." *Conservation Biology* 14 (1): 31–35.

Foster, R. G., and T. Roenneberg. 2008. "Human Responses to the Geophysical Daily, Annual and Lunar Cycles." *Current Biology* 18 (17): R784–94.

Fox, R. 2013. "The Decline of Moths in Great Britain: A Review of Possible Causes." *Insect Conservation and Diversity* 6 (1): 5–19.

Francis, C. D., and J. R. Barber. 2013. "A Framework for Understanding Noise Impacts on Wildlife: An Urgent Conservation Priority." *Frontiers in Ecology and the Environment* 11 (6): 305–13.

Francis, C. D., N. J. Kleist, C. P. Ortega, and A. Cruz. 2012a. "Noise Pollution Alters Ecological Services: Enhanced Pollination and Disrupted Seed Dispersal." *Proceedings of the Royal Society B-Biological Sciences* 279 (1739): 2727–35.

Francis, C. D., C. P. Ortega, and A. Cruz. 2009a. "Noise Pollution Changes Avian Communities and Species Interactions." *Current Biology* 19 (16): 1415–19.

———. 2009b. "Cumulative Consequences of Noise Pollution: Noise Changes Avian Communities and Species Interactions." *Current Biology* 19: 1415–19.

———. 2011a. "Different Behavioural Responses to Anthropogenic Noise by Two Closely Related Passerine Birds." *Biology Letters* 7 (6): 850–52.

———. 2011b. "Noise Pollution Filters Bird Communities Based on Vocal Frequency." *PLoS One* 6 (11): e27052.

———. 2011c. "Vocal Frequency Change Reflects Different Responses to Anthropogenic Noise in Two Suboscine Tyrant Flycatchers." *Proceedings of the Royal Society B-Biological Sciences* 278 (1714): 2025–31.

Francis, C, C. Ortega, R. Kennedy, and P. Nylander. 2012b. "Are Nest Predators Absent from Noisy Areas or Unable to Locate Nests." *Ornithological Monograph* 74 (101): e110.

Francis, C. D., J. Paritsis, C. P. Ortega, and A. Cruz. 2011d. "Landscape Patterns of Avian Habitat Use and Nest Success Are Affected by Chronic Gas Well Compressor Noise." *Landscape Ecology* 26 (9): 1269–80.

Frank, D. W., J. A. Evans, and M. R. Gorman. 2010. "Time-Dependent Effects of Dim Light at Night on Re-Entrainment and Masking of Hamster Activity Rhythms." *Journal of Biological Rhythms* 25 (2): 103–12.

Frank, K. D. 1988. "Impact of Outdoor Lighting on Moths: An Assessment." *Journal of the Lepidopterists' Society.*

———. 2009. "Exploitation of Artificial Light at Night by a Diurnal Jumping Spider." *Peckhamia* 78 (1): 1–3.

Freimund, W. A., J. J. Vaske, M. P. Donnelly, and T. A. Miller. 2002. "Using Video Surveys to Access Dispersed Backcountry Visitors' Norms." *Leisure Sciences* 24 (3–4): 349–62.

Frid, A., and L. Dill. 2002. "Human-Caused Disturbance Stimuli as a Form of Predation Risk." *Conservation Ecology* 6 (1): 11–27.

Frissell, S. S., Jr., and G. H. Stankey. 1972. "Wilderness Environmental Quality: Search for Social and Ecological Harmony." In *Society of American Foresters*

Annual Meeting, Hot Springs, Arkansas October, 1972, 170–83. Accessed June 11, 2017. http://nstrail.com/carrying_capacity/wilderness_environmental_quality _search_for_social_ecological_harmony_frissell_stankey_1972.pdf.

Fristrup, K. 2009. "Applying Community Noise Metrics in Parks." *Park Science* 26 (3): 21–22.

———, D. Joyce, and E. Lynch. 2010. "Measuring and Monitoring Soundscapes in the National Parks." *Park Science* 26 (3): 1–8.

Fuiman, L. A., and J. H. Cowan. 2003. "Behavior and Recruitment Success in Fish Larvae: Repeatability and Covariation of Survival Skills." *Ecology* 84 (1): 53–67.

Fuller, R. A., P. H. Warren, and K. J. Gaston. 2007. "Daytime Noise Predicts Nocturnal Singing in Urban Robins." *Biology Letters* 3 (4): 368–70.

Futsaether, C. M., A. V. Vollsnes, O. Mathis, O. Kruse, E. Otterholt, K. Kvaal, and A. B. Eriksen. 2009. "Effects of the Nordic Photoperiod on Ozone Sensitivity and Repair in Different Clover Species Studied Using Infrared Imaging." *Ambio* 38 (8): 437–42.

Gagnon, J. W., T. C. Theimer, N. L. Dodd, S. Boe, and R. E. Schweinsburg. 2007. "Traffic Volume Alters Elk Distribution and Highway Crossings in Arizona." *Journal of Wildlife Management* 71 (7): 2318–23.

Garber, S. D. 1978. "Opportunistic Feeding Behavior of Anolis Cristatellus (*Iguanidae: Reptilia*) in Puerto Rico." *Transactions of the Kansas Academy of Science (1903-)* 81 (1): 79–80.

Gaston, K. J., J. Bennie, T. W. Davies, and J. Hopkins. 2013. "The Ecological Impacts of Nighttime Light Pollution: A Mechanistic Appraisal." *Biological Reviews* 88 (4): 912–27.

Gaston, K. J., T. W. Davies, J. Bennie, and J. Hopkins. 2012. "Review: Reducing the Ecological Consequences of Night-Time Light Pollution: Options and Developments." *Journal of Applied Ecology* 49 (6): 1256–66.

Gaston, K. J., S. Gaston, J. Bennie, and J. Hopkins. 2014. "Benefits and Costs of Artificial Nighttime Lighting of the Environment." *Environmental Reviews* 23 (1): 14–23.

Gauthreaux, J., A. Sidney, and C. Belser. 2006. "Effects of Artificial Night Lighting on Migrating Birds." In *Ecological Consequences of Artificial Night Lighting*, edited by C. Rich and T. Longcore, 67–93. Washington: Island Press.

Gavin, S. D., and P. E. Komers. 2006. "Do Pronghorn (*Antilocapra Americana*) Perceive Roads as a Predation Risk?" *Canadian Journal of Zoology–Revue Canadienne de zoologie* 84 (12): 1775–80.

Gehring, J., P. Kerlinger, and A. M. Manville. 2009. "Communication Towers, Lights, and Birds: Successful Methods of Reducing the Frequency of Avian Collisions." *Ecological Applications* 19 (2): 505–14.

Gerrish, G. A., J. G. Morin, T. J. Rivers, and Z. Patrawala. 2009. "Darkness as an Ecological Resource: The Role of Light in Partitioning the Nocturnal Niche." *Oecologia* 160 (3): 525–36.

Ghanem, S. J., and C. C. Voigt. 2012. "Increasing Awareness of Ecosystem Services Provided by Bats." *Advances in the Study of Behavior* 44: 279–302.

Gibson, A. 2011. "Photograph Presentation Order and Range Effects in Visual Based Outdoor Recreation Research." PhD diss., Colorado State University.

Gibson, A. W., P. Newman, S. Lawson, K. Fristrup, J. A. Benfield, P. A. Bell, and G. A. Nurse. 2014. "Photograph Presentation Order and Range Effects in Visual-Based Outdoor Recreation Research." *Leisure Sciences* 36 (2): 183–205

Gliwicz, Z. 1986. "A Lunar Cycle in Zooplankton." *Ecology* 67 (4): 883–97.

Gomes, D., R. Page, I. Geipel, R. Taylor, M. Ryan, and W. Halfwerk. 2016. "Bats Perceptually Weight Prey Cues across Sensory Systems When Hunting in Noise." *Science* 353:1277–80.

Goodwin, S. E., and W. G. Shriver. 2011. "Effects of Traffic Noise on Occupancy Patterns of Forest Birds." *Conservation Biology* 25 (2): 406–11.

Government of Northwest Territories Department of Industry, Tourism and Investment. 2007. "Visitor Markets Strategic Overview: Northwest Territories." Accessed June 25, 2017. http://www.iti.gov.nt.ca/sites/www.iti.gov.nt.ca/files/visitor_markets_-_strategic_overview_-_2007.pdf.

Government of Yukon Department of Tourism and Culture. 2008. "Yukon Wilderness Tourism Status Report." Accessed June 25, 2017. http://www.tc.gov.yk.ca/publications/Yukon_Wilderness_Tourism_Status_Report_2008.pdf.

Graefe, A. R., J. J. Vaske, and F. R. Kuss. 1984. "Social Carrying Capacity: An Integration and Synthesis of Twenty Years of Research." *Leisure Sciences* 6 (4): 395–431.

Gramann, J. 1999. "The Effect of Mechanical Noise and Natural Sound on Visitor Experiences in Units of the National Park System." *Social Science Research Review* 1 (1): 1–16.

Great Barrier Reef Marine Park Authority. 2014. "Great Barrier Reef Outlook Report 2014." Townsville, Australia: Great Barrier Reef Marine Park Authority.

Grémillet, D., G. Kuntz, C. Gilbert, A. J. Woakes, P. J. Butler, and Y. le Maho. 2005. "Cormorants Dive through the Polar Night." *Biology Letters* 1 (4): 469–71.

Grimm, N. B., S. H. Faeth, N. E. Golubiewski, C. L. Redman, J. Wu, X. Bai, and J. M. Briggs. 2008. "Global Change and the Ecology of Cities." *Science* 319 (5864): 756–60.

Gross, K., G. Pasinelli, and H. P. Kunc. 2010. "Behavioral Plasticity Allows Short-Term Adjustment to a Novel Environment." *American Naturalist* 176 (4): 456–64.

Guadagnolo, F. 1985. "The Importance-Performance Analysis: An Evaluation and Marketing Tool." *Journal of Park and Recreation Administration* 3 (2): 13–22.

Gutman, R., and T. Dayan. 2005. "Temporal Partitioning: An Experiment with Two Species of Spiny Mice." *Ecology* 86 (1): 164–73.

Gwinner, E. 1977. "Photoperiodic Synchronization of Circannual Rhythms in European Starling (*Sturnus-Vulgaris*)." *Naturwissenschaften* 64 (1): 44–45.

Haas, G. E., and T. J. Wakefield. 1998. *National Parks and the American Public:*

A National Public Opinion Survey on the National Park System: A Summary Report. Fort Collins: Colorado State University.

Habib, Lucas, Erin M. Bayne, and Stan Boutin. 2007. "Chronic Industrial Noise Affects Pairing Success and Age Structure of Ovenbirds *Seiurus Aurocapilla.*" *Journal of Applied Ecology* 44 (1): 176–84.

Hadwen, S., and L. J. Palmer. 1922. "Reindeer in Alaska." *US Department of Agriculture Bulletin* 1089:1–74.

Haim, A., U. Shanas, A. E. S. Zubidad, and M. Scantlebury. 2005. "Seasonality and Seasons out of Time—The Thermoregulatory Effects of Light Interference." *Chronobiology International* 22 (1): 59–66.

Hainsworth, F., L. Wolf, and T. Mercier. 1985. "Pollen Limitation in a Monocarpic Species, *Ipomopsis-Aggregata.*" *Journal of Ecology* 73 (1): 263–70.

Halekoh, U., and S. Højsgaard. 2013. "Pbkrtest: Parametric Bootstrap and Kenward Roger Based Methods for Mixed Model Comparison." Accessed August 27, 2017. https://rdrr.io/cran/pbkrtest/.

Halfwerk, W., S. Bot, J. Buikx, M. van der Velde, J. Komdeur, C. ten Cate, and H. Slabbekoorn. 2011a. "Low-Frequency Songs Lose Their Potency in Noisy Urban Conditions." *Proceedings of the National Academy of Sciences of the United States of America* 108 (35): 14549–54.

Halfwerk, W., L. Holleman, C. Lessells, and H. Slabbekoorn. 2011b. "Negative Impact of Traffic Noise on Avian Reproductive Success." *Journal of Applied Ecology* 48 (1): 210–19.

Halfwerk, W., A. Lea, M. Guerra, R. Page, and M. Ryan. 2016. "Vocal Responses to Noise Reveal the Presence of the Lombard Effect in a Frog." *Behavioral Ecology* 27 (2): 669–76.

Halfwerk, W., and H. Slabbekoorn. 2009. "A Behavioural Mechanism Explaining Noise-Dependent Frequency Use in Urban Birdsong." *Animal Behaviour* 78 (6): 1301–7.

———. 2015. "Pollution Going Multimodal: The Complex Impact of the Human-Altered Sensory Environment on Animal Perception and Performance." *Biology Letters* 11 (4): 1–11.

Ham, S. H. 2009. "From Interpretation to Protection: Is There a Theoretical Basis?" *Journal of Interpretation Research* 14 (2): 49–57.

———, T. J. Brown, J. Curtis, B. Weiler, M. Hughes, and M. Poll. 2009. "Promoting Persuasion in Protected Areas: A Guide for Managers Who Want to Use Strategic Communication to Influence Visitor Behaviour." Gold Coast, Australia: Cooperative Research Centre for Sustainable Tourism. Accessed June 19, 2017. http://researchrepository.murdoch.edu.au/id/eprint/26313/1/Promoting_Persuasion_in_Protected_Areas_WEB.pdf.

Hammitt, W., and D. Cole. 1998. *Wildland Recreation: Ecology and Management.* New York: John Wiley.

Hammitt, W. E., D. N. Cole, and C. A. Monz. 2015. *Wildland Recreation: Ecology and Management.* 3rd ed. Oxford: John Wiley and Sons.

Hardin, G. 1968. "The Tragedy of the Commons." *Science* 162 (3859): 1243–48.

Harrington, M. G. 1987. *Characteristics of 1-Year-Old Natural Pinyon Seedlings.* USDA Forest Service Research Note RM-477. Fort Collins, CO: Rocky Mountain Forest and Range Experiment Station.

Hart, N. S., and D. M. Hunt. 2007. "Avian Visual Pigments: Characteristics, Spectral Tuning, and Evolution." *American Naturalist* 169 (1): S7–26.

Hartig, T., G. W. Evans, L. D. Jamner, D. S. Davis, and T. Garling. 2003. "Tracking Restoration in Natural and Urban Field Settings." *Journal of Environmental Psychology* 23 (2): 109–23.

Haven, W. 2015. *Noise Matters: The Evolution Of Communication.* Cambridge, MA: Harvard University Press.

Hayward, L. S., A. E. Bowles, J. C. Ha, and S. K. Wasser. 2011. "Impacts of Acute and Long-Term Vehicle Exposure on Physiology and Reproductive Success of the Northern Spotted Owl." *Ecosphere* 2 (6): 1–20.

Health Council of the Netherlands. 2000. *Impact of Outdoor Lighting on Man and Nature.* The Hague: Health Council of the Netherlands.

Heart of America Star Party. 2011. "Heart of America Star Party." http://www .askconline.org/HOASP.htm.

Hecht, S. 1993. "The Logic of Livestock and Deforestation in Amazonia." *Bioscience* 43 (10): 687–95.

Hedenstrom, A. 2008. "Adaptations to Migration in Birds: Behavioural Strategies, Morphology and Scaling Effects." *Philosophical Transactions of the Royal Society B-Biological Sciences* 363 (1490): 287–99.

Heerwagon, J. H. 1990. "The Psychological Aspect of Windows and Window Design." In *Proceedings of the 21st Annual Conference of the Environmental Design Research Association,* edited by K. Anthony, J. Choi, and B. Orland, 269–80. Oklahoma City: EDRA.

Heide, O. 2006. "Daylength and Thermal Time Responses of Budburst During Dormancy Release." *Physiologia Plantarum* 88 (4): 531–40.

Heiling, A. M. 1999. "Why Do Nocturnal Orb-Web Spiders (*Araneidae*) Search for Light?" *Behavioral Ecology and Sociobiology* 46 (1): 43–49.

Hoelker, F., T. Moss, B. Griefahn, W. Kloas, C. Voigt, D. Henckel, A. Haenel, P. Kappeler, S. Voelker, A. Schwope, et al. 2010a. "The Dark Side of Light: A Transdisciplinary Research Agenda for Light Pollution Policy." *Ecology and Society* 15(4) :13–24.

Hoelker, F., C. Wolter, E. Perkin, and K. Tockner. 2010b. "Light Pollution as a Biodiversity Threat." *Trends in Ecology & Evolution* 25 (12): 681–82.

Hoelker, F., C. Wurzbacher, C. Weissenborn, M. T. Monaghan, S. I. J. Holzhauer, and K. Premke. 2015. "Microbial Diversity and Community Respiration in Freshwater Sediments Influenced by Artificial Light at Night." *Philosophical Transactions of the Royal Society B-Biological Sciences* 370 (1667): 1–10.

Hoey, A. S., and M. I. McCormick. 2004. "Selective Predation for Low Body Condition at the Larval-Juvenile Transition of a Coral Reef Fish." *Oecologia* 139 (1): 23–29.

———. 2006. "Effects of Subcutaneous Fluorescent Tags on the Growth and Survival of a Newly Settled Coral Reef Fish, *Pomacentrus Amboinensis* (Pomacentridae)." In *Proceedings of the 10th International Coral Reefs Symposium*, edited by Y. Suzuki, 1:420–25. Okinawa, Japan: Japanese Coral Reef Society. Accessed June 14, 2017. http://citeseerx.ist.psu.edu/viewdoc/download?doi=10.1.1.717.4426&rep=rep1&type=pdf.

Hogg, C., M. Neveu, K.-A. Stokkan, L. Folkow, P. Cottrill, R. Douglas, D. M. Hunt, and G. Jeffery. 2011. "Arctic Reindeer Extend Their Visual Range into the Ultraviolet." *Journal of Experimental Biology* 214 (12): 2014–19.

Holden, A. 1992. "Lighting the Night—Technology, Urban Life and the Evolution of Street-Lighting." *Places* 8 (2): 56–63.

Hollenhorst, S. J., and L. Gardner. 1994. "The Indicator Performance Estimate Approach to Determining Acceptable Wilderness Conditions." *Environmental Management* 18 (6): 901–6.

Hollenhorst, S, J., D. Olson, and R. Fortney. 1992. "Use of Importance-Performance Analysis to Evaluate State Park Cabins: The Case of the West Virginia State Park System." *Journal of Park and Recreation Administration* 10 (1): 1–11.

Hollenhorst, S. J., and L. Stull-Gardner. 1992. "The Indicator Performance Estimate (IPE) Approach to Defining Acceptable Conditions in Wilderness." In *Proceedings of the Symposium on Social Aspects and Recreation Research, February 19–22, 1992, Ontario, California*, edited by D. J. Chavez, 48–49. Albany (CA): Pacific Southwest Research Station.

Holmes, N. C., M. Schuett, and S. J. Hollenhorst. 2010. *Bryce Canyon National Park Visitor Study: Summer 2009*. Washington: Department of the Interior.

Holmes, T. H., and M. I. McCormick. 2006. "Location Influences Size-Selective Predation on Newly Settled Reef Fish." *Marine Ecology Progress Series* 317:203–9.

Honour, S. L., J. N. B. Bell, T. W. Ashenden, J. N. Cape, and S. A. Power. 2009. "Responses of Herbaceous Plants to Urban Air Pollution: Effects on Growth, Phenology and Leaf Surface Characteristics." *Environmental Pollution* 157 (4): 1279–86.

Horvath, G., G. Kriska, P. Malik, and B. Robertson. 2009. "Polarized Light Pollution: A New Kind of Ecological Photopollution." *Frontiers in Ecology and the Environment* 7 (6): 317–25.

Hothorn, T., F. Bretz, and P. Westfall. 2008. "Simultaneous Inference in General Parametric Models." *Biometrical Journal* 50 (3): 346–63.

Howell, J. C., A. R. Laskey, and J. T. Tanner. 1954. "Bird Mortality at Airport Ceilometers." *Wilson Bulletin* 66 (3): 207–15.

Hsu, Y., C. Li, and O. Chuang. 2007. "Encounters, Norms and Perceived Crowding of Hikers in a Non-North American Backcountry Recreational Context." *Forest, Snow and Landscape Research* 81 (1–2): 99–108.

Hu, Y., and G. C. Cardoso. 2010. "Which Birds Adjust the Frequency of Vocalizations in Urban Noise?" *Animal Behaviour* 79 (4): 863–67.

Hudson, P. J., and A. P. Dobson. 1995. *Macroparasites: Observed Patterns in Naturally Fluctuating Animal Populations.* Cambridge: Cambridge University Press.

Hunt, K. S., D. Scott, S. Richardson. 2003. "Positioning Public Recreation and Park Offerings Using Importance-Performance Analysis." *Journal of Park and Recreation Administration* 21 (3): 1–21.

Hunt, M. 1999. "Management of the Environmental Noise Effects Associated with Sightseeing Aircraft in the Milford Sound Area, New Zealand." *Noise Control Engineering Journal* 47 (4): 133–41.

Hurly, T. A. 1996. "Spatial Memory in Rufous Hummingbirds: Memory for Rewarded and Non-Rewarded Sites." *Animal Behaviour* 51 (January): 177–83.

———, S. Franz, and S. D. Healy. 2010. "Do Rufous Hummingbirds (*Selasphorus Rufus*) Use Visual Beacons?" *Animal Cognition* 13 (2): 377–83.

Imber, M. J. 1975. "Behaviour of Petrels in Relation to the Moon and Artificial Lights." *Journal of the Ornithological Society of New Zealand*, 22:302–6.

Imhoff, M. L., W. T. Lawrence, D. C. Stutzer, and C. D. Elvidge. 1997. "A Technique for Using Composite DMSP/OLS City Lights Satellite Data to Map Urban Area." *Remote Sensing of Environment* 61 (3): 361–70.

Interagency Visitor Use Management Council. 2016. "Visitor Use Management Framework: A Guide to Providing Sustainable Outdoor Recreation." 1st ed. Accessed June 11, 2017. https://visitorusemanagement.nps.gov/Content /documents/lowres_VUM%20Framework_Edition%201_IVUMC.pdf.

Ising, H., and B. Krupp. 2004. "Health Effects Caused by Noise: Evidence in the Literature from the Past 25 Years." *Noise and Health* 6 (22): 5–13.

Jahncke, H., S. Hygge, N. Halin, A. Green, and K. Dimberg. 2011. "Open-Plan Office Noise: Cognitive Performance and Restoration." *Journal of Environmental Psychology* 31 (4): 373–82.

Jakle, J. A. 2001. *City Lights: Illuminating the American Night.* Baltimore, MD: Johns Hopkins University Press.

Jensen, M., and H. Thompson. 2004. "Natural Sounds: An Endangered Species." In *George Wright Forum*, 21 (1): 10–13. Accessed June 18, 2017. http://www .georgewright.org/211jensen1.pdf.

Jochner, S. C., M. Alves-Eigenheer, A. Menzel, and L. Patricia C. Morellato. 2013. "Using Phenology to Assess Urban Heat Islands in Tropical and Temperate Regions." *International Journal of Climatology* 33 (15): 3141–51.

Jochner, S. C., T. H. Sparks, N. Estrella, and A. Menzel. 2012. "The Influence of Altitude and Urbanisation on Trends and Mean Dates in Phenology (1980–2009)." *International Journal of Biometeorology* 56 (2): 387–94.

Johnson, K. 1979. "Control of Lampenflora at Waitomo Caves, New Zealand." In *Cave Management in Australia III: Proceedings of the 3rd Australasian Cave Tourism and Management Conference*, 105–22. Adelaide, Australia: South Australian National Parks and Australian Speleological Federation.

Johnson, R. G., and S. A. Temple. 1986. "Assessing Habitat Quality for Birds Nesting in Fragmented Tallgrass Prairies." In *Wildlife 2000: Modeling Habitat*

Relationships of Terrestrial Vertebrates, edited by J. Verner, M. L. Morrison, and C. J. Ralph. Madison: University of Wisconsin Press.

Jones, J., and C. M. Francis. 2003. "The Effects of Light Characteristics on Avian Mortality at Lighthouses." *Journal of Avian Biology* 34 (4): 328–33.

Jordano, P., P.-M. Forget, J. E. Lambert, K. Boehning-Gaese, A. Traveset, and S. J. Wright. 2011. "Frugivores and Seed Dispersal: Mechanisms and Consequences for Biodiversity of a Key Ecological Interaction." *Biology Letters* 7 (3): 321–23.

Juenger, T., and J. Bergelson. 1997. "Pollen and Resource Limitation of Compensation to Herbivory in Scarlet Gilia, *Ipomopsis Aggregata*." *Ecology* 78 (6): 1684–95.

Kaiser, K., D. Scofield, M. Alloush, R. Jones, S. Marczak, K. Martineau, M. Oliva, and P. Narins. 2011. "When Sounds Collide: The Effect of Anthropogenic Noise on a Breeding Assemblage of Frogs in Belize, Central America." *Behaviour* 148 (2): 215–32.

Kaliski, K., E. Duncan, and J. Cowan. 2007. "Community and Regional Noise Mapping in the United States." *Sound and Vibration* 41 (9): 14–17.

Kaplan, R. 1993. "The Role of Nature in the Context of the Workplace." *Landscape and Urban Planning* 26 (1–4): 193–201.

Kappeler, P. M., and H. G. Erkert. 2003. "On the Move around the Clock: Correlates and Determinants of Cathemeral Activity in Wild Redfronted Lemurs (*Eulemur Fulvus Rufus*)." *Behavioral Ecology and Sociobiology* 54 (4): 359–69.

Kariel, H. G. 1980. "Mountaineers and the General Public: A Comparison of Their Evaluation of Sounds in a Recreational Environment." *Leisure Sciences* 3 (2): 155–67.

———. 1990. "Factors Affecting Response to Noise in Outdoor Recreational Environments." *Canadian Geographer–Geographe Canadien* 34 (2): 142–49.

Kearns, C. A., and D. W. Inouye. 1993. *Techniques for Pollination Biologists*. Boulder: University Press of Colorado.

Keizer, G. 2008. "Preserving Silence in National Parks: A Battle against Noise Aims to Save Our Natural Soundscapes." *Smithsonian Magazine*.

Kelber, A., A. Balkenius, and E. J. Warrant. 2002. Scotopic Colour Vision in Nocturnal Hawkmoths." *Nature* 419 (6910): 922–25.

Kelber, A., and L. S. V. Roth. 2006. "Nocturnal Colour Vision—Not as Rare as We Might Think." *Journal of Experimental Biology* 209 (5): 781–88.

Kelber, A., M. Vorobyev, and D. Osorio. 2003. "Animal Colour Vision—Behavioural Tests and Physiological Concepts." *Biological Reviews* 78 (1): 81–118.

Kempenaers, B., P. Borgstroem, P. Loes, E. Schlicht, and M. Valcu. 2010. "Artificial Night Lighting Affects Dawn Song, Extra-Pair Siring Success, and Lay Date in Songbirds." *Current Biology* 20 (19): 1735–39.

Kempenaers, B., G. Verheyen, M. Vandenbroeck, T. Burke, C. Van Broeckhoven, and A. Dhondt. 1992. "Extra-Pair Paternity Results from Female Preference for High-Quality Males in the Blue Tit." *Nature* 357 (6378): 494–96.

Kight, C. R., M. S. Saha, and J. P. Swaddle. 2012. "Anthropogenic Noise Is Associated with Reductions in the Productivity of Breeding Eastern Bluebirds (*Sialia Sialis*)." *Ecological Applications* 22 (7): 1989–96.

Kight, C. R., and J. P. Swaddle. 2011. "How and Why Environmental Noise Impacts Animals: An Integrative, Mechanistic Review." *Ecology Letters* 14 (10): 1052–61.

Killen, S. S., M. D. Mitchell, J. L. Rummer, D. P. Chivers, M. C. O. Ferrari, M. G. Meekan, and M. I. McCormick. 2014. "Aerobic Scope Predicts Dominance during Early Life in a Tropical Damselfish." *Functional Ecology* 28 (6): 1367–76.

Kiser, B., S. Lawson, S., and R. Itami. 2007. "Assessing the Reliability of Computer Simulation for Modeling Low Use Visitor Landscapes." In *Monitoring, Simulation, and Management of Visitor Landscapes*, edited by R. Gimblett, H. Skov-Petersen, and A. Muhar, 371–87. Tucson: University of Arizona Press.

Kitt Peak National Observatory. n.d. "Kitt Peak Visitor Center." Accessed August 14, 2017. https://www.noao.edu/kpvc/.

Kleist, N., R. Guralnick, A. Cruz, and C. Francis. 2016. "Anthropogenic Noise Weakens Territorial Response to Intruder's Songs." *Ecosphere* 7 (3): 1–11.

Kolb, H. 1992. "The Effect of Moonlight on Activity in the Wild Rabbit (*Oryctolagus-Cuniculus*)." *Journal of Zoology* 228 (December): 661–65.

Kotler, B. 1984. "Effects of Illumination on the Rate of Resource Harvesting in a Community of Desert Rodents." *American Midland Naturalist* 111 (2): 383–89.

Kozlov, M. V., J. K. Eranen, and V. E. Zverev. 2007. "Budburst Phenology of White Birch in Industrially Polluted Areas." *Environmental Pollution* 148 (1): 125–31.

Kramer, K. M., and E. C. Birney. 2001. "Effect of Light Intensity on Activity Patterns of Patagonian Leaf-Eared Mice, *Phyllotis Xanthopygus*." *Journal of Mammalogy* 82 (2): 535–44.

Krause, B. 1999. "Loss of Natural Soundscapes within the Americas." Aberdeen, MD: U.S. Army Center for Health Promotion and Preventive Medicine.

Kristiansen, K. 1988. "Light Quality Regulates Flower Initiation, Differentiation and Development of *Campanula-Carpatica* Jacq Forster,karl." *Scientia horticulturae* 35 (3–4): 275–83.

Krog, N. H., and B. Engdahl. 2005. "Annoyance with Aircraft Noise in Local Recreational Areas and the Recreationists' Noise Situation at Home." *Journal of the Acoustical Society of America* 117 (1): 221–31.

Kronfeld-Schor, N., and T. Dayan. 2003. "Partitioning of Time as an Ecological Resource." *Annual Review of Ecology Evolution and Systematics* 34: 153–81.

Kryter, K. D. 1985. *The Effects of Noise on Man*. 2nd ed. New York: Academic Press.

Kuijper, D. P. J., J. Schut, D. van Dullemen, H. Toorman, N. Goossens, J. Ouwehand, and H. Limpens. 2008. "Experimental Evidence of Light Disturbance along the Commuting Routes of Pond Bats (*Myotis Dasycneme*)." *Lutra* 51 (1): 37–49.

Kyba, C. C. M., T. Ruhtz, J. Fischer, and F. Hoelker. 2011a. "Cloud Coverage Acts as an Amplifier for Ecological Light Pollution in Urban Ecosystems." *PLoS One* 6 (3): e17307.

———. 2011b. "Lunar Skylight Polarization Signal Polluted by Urban Lighting." *Journal of Geophysical Research* 116 (December): D24106.

———. 2012. "Red Is the New Black: How the Colour of Urban Skyglow Varies with Cloud Cover." *Monthly Notices of the Royal Astronomical Society* 425 (1): 701–8.

Lafferty, K., S. Allesina, M. Arim, C. Briggs, G. Leo, A. Dobson, J. Dunne, P. Johnson, A. Kuris, D. Marcogliese, et al. 2008. "Parasites in Food Webs: The Ultimate Missing Links." *Ecology Letters* 11 (6): 533–46.

Lagardere, J. 1982. "Effects of Noise on Growth and Reproduction of Crangon-Crangon in Rearing Tanks." *Marine Biology* 71 (2): 177–85.

Laiolo, P. 2010. "The Emerging Significance of Bioacoustics on Animal Species Conservation." *Biological Conservation* 143 (7): 1635–45.

Lambrechts, M. M., J. Blondel, M. Maistre, and P. Perret. 1997. "A Single Response Mechanism Is Responsible for Evolutionary Adaptive Variation in a Bird's Laying Date." *Proceedings of the National Academy of Sciences of the United States of America* 94 (10): 5153–55.

Larsen, L. O., and J. N. Pedersen. 1982. "The Snapping Response of the Toad, *Bufo Bufo*, towards Prey Dummies at Very Low Light Intensities." *Amphibia-Reptilia* 2 (4): 321–27.

Lashbrook, M. K., and R. L. Livezey. 1970. "Effects of Photoperiod on Heat Tolerance in *Sceloporus Occidentalis Occidentalis.*" *Physiological Zoology* 43 (1): 38–46.

Laumann, K., T. Gärling, and K. M. Stormark. 2003. "Selective Attention and Heart Rate Responses to Natural and Urban Environments." *Journal of Environmental Psychology* 23 (2): 125–34.

Lavine, H., and M. Snyder. 1996. "Cognitive Processing and the Functional Matching Effect in Persuasion: The Mediating Role of Subjective Perceptions of Message Quality." *Journal of Experimental Social Psychology* 32 (6): 580–604.

Lawson, S. R. 2006. "Computer Simulation as a Tool for Planning and Management of Visitor Use in Protected Natural Areas." *Journal of Sustainable Tourism* 14 (6): 600–617.

———, B. Kiser, K. Hockett, and A. Ingram. 2008. "Research to Support Backcountry Visitor Use Management and Resource Protection in Haleakala National Park." Blacksburg, VA: Virginia Polytechnic Institute and State University.

Lawson, S. R., and K. Plotkin. 2006. "Understanding and Managing Soundscapes in National Parks: Part 3–Computer Simulation." In *Exploring the Nature of Management*, edited by D. Siegrist, C. Clivaz, M. Hunziker, and S. Iten, 203–4. Rapperswil, Switzerland: Research Centre for Leisure, Tourism and Landscape.

Le Corre, M., A. Ollivier, S. Ribes, and P. Jouventin. 2002. "Light-Induced Mortality of Petrels: A 4-Year Study from Reunion Island (Indian Ocean)." *Biological Conservation* 105 (1): 93–102.

Lefèvre, Marcel. 1974. "La 'maladie verte' de Lascaux." *Studies in Conservation* 19 (3): 126–56.

Leis, J. M., and M. I. McCormick. 2002. "The Biology, Behavior, and Ecology of the Pelagic, Larval Stage of Coral Reef Fishes." In *Coral Reef Fishes: Dynamics and Diversity in a Complex Ecosystem*, edited by P. Sale, 171–99. San Diego, CA: Elsevier Science.

Leonard, M. L., and A. G. Horn. 2012. "Ambient Noise Increases Missed Detections in Nestling Birds." *Biology Letters* 8 (4): 530–32.

Leopold, A. 1933. *Game Management*. New York: Charles Scribner's Sons.

———. 1949. *A Sand County Almanac: With Other Essays on Conservation from Round River*. New York: Oxford University Press.

Leopold, A. S., S. A. Cain, C. M. Cottam, I. N. Gabrielson, and T. L. Kimball. 1963. "Wildlife Management in the National Parks." In *Transactions of the North American Wildlife and Natural Resources Conference*, vol. 28, edited by J. B. Trefethen, 28–45. Washington: Wildlife Management Institute.

Leung, Y., and J. Marion. 2001. "Recreation Impacts and Management in Wilderness: A State-of-Knowledge Review." In *Proceedings: Wilderness Science in a Time of Change*, 23–48. Fort Collins, CO: Rocky Mountain Research Station.

Lewanzik, D., and C. C. Voigt. 2014. "Artificial Light Puts Ecosystem Services of Frugivorous Bats at Risk." *Journal of Applied Ecology* 51 (2): 388–94.

Lewis, T., and L. R. Taylor. 1964. "Diurnal Periodicity of Flight by Insects." *Transactions of the Royal Entomological Society of London* 116 (15): 393–435.

Lima, S. L. 1998. "Stress and Decision Making under the Risk of Predation: Recent Developments from Behavioral, Reproductive, and Ecological Perspectives." *Advances in the Study of Behavior* 27:215–90.

Lin, C. T. 2000. "Photoreceptors and Regulation of Flowering Time." *Plant Physiology* 123 (1): 39–50.

Linkosalo, Tapio, and Martin J. Lechowicz. 2006. "Twilight Far-Red Treatment Advances Leaf Bud Burst of Silver Birch (*Betula Penduld*)." *Tree Physiology* 26 (10): 1249–56.

Lloyd, J. E. 2006. "Stray Light, Fireflies, and Fireflyers." In *Ecological Consequences of Artificial Night Lighting*, edited by C. Rich and T. Longcore, 345–64. Washington: Island Press.

Lockwood, R. 2011. *A Review of Local Authority Road Lighting Initiatives Aimed at Reducing Costs, Carbon Emissions and Light Pollution*. London: Department for Environment, Food and Rural Affairs.

Longcore, T., and C. Rich. 2004. "Ecological Light Pollution." *Frontiers in Ecology and the Environment* 2 (4): 191–98.

———. 2006. "Synthesis." In *Ecological Consequences of Artificial Night Lighting*, edited by Catherine Rich and Travis Longcore, 413–30. Washington: Island Press.

López, J. M. Garrido, and J. Díaz-Reixa Suárez. 2007. "The Right to Starlight, Another Step towards Controlling Pollution and Efficient Use of Natural Resources." In *Starlight: A Common Heritage: Proceedings of the International Astronomical Union*, edited by C. Marin, 241–47. Cambridge, MA: International Astronomical Union.

Lorne, J. K., and M. Salmon. 2007. "Effects of Exposure to Artificial Lighting on Orientation of Hatchling Sea Turtles on the Beach and in the Ocean." *Endangered Species Research* 3 (1): 23–30.

Love, E. K., and M. A. Bee. 2010. "An Experimental Test of Noise-Dependent Voice Amplitude Regulation in Cope's Grey Treefrog, *Hyla Chrysoscelis*." *Animal Behaviour* 80 (3): 509–15.

Lovett, P. N. 2005. "Shea Butter Industry Expanding in West Africa." *Inform* 16 (5): 273–5.

Lowell Observatory. 2017. Lowell Observatory." Accessed August 14, 2017. https://lowell.edu/.

Lowery, S. F., S. T. Blackman, and D. Abbate. 2009. "Urban Movement Patterns of Lesser Long-Nosed Bats (*Leptonycteris Curasoae*): Management Implications for the Habitat Conservation Plan within the City of Tucson and the Town of Marana." Phoenix, AZ: Research Branch, Arizona Game and Fish Department. Accessed June 23, 2017. http://www.maranaaz.gov/s/2007-2008-LLNB -Urban-Movement-report.PDF.

Lucas, R. C. 1964. *Recreational Capacity of the Quetico-Superior Area*. St. Paul, MN: Lake State Forest Experiment Station, US Department of Agriculture Forest Service.

Luginbuhl, C. B., C. E. Walker, and R. J. Wainscoat. 2009. "Lighting and Astronomy." *Physics Today* 62 (12): 32–37.

Luther, D. A. 2008. "Signaller: Receiver Coordination and the Timing of Communication in Amazonian Birds." *Biology Letters* 4 (6): 651–54.

Lynch, E., D. Joyce, and K. Fristrup. 2011. "An Assessment of Noise Audibility and Sound Levels in US National Parks." *Landscape Ecology* 26 (9): 1297–309.

Lyytimaki, J., P. Tapio, and T. Assmuth. 2012. "Unawareness in Environmental Protection: The Case of Light Pollution from Traffic." *Land Use Policy* 29 (3): 598–604.

Mace, B. L., P. A. Bell, and R. J. Loomis. 1999. "Aesthetic, Affective, and Cognitive Effects of Noise on Natural Landscape Assessment." *Society & Natural Resources* 12 (3): 225–42.

———. 2004. "Visibility and Natural Quiet in National Parks and Wilderness Areas—Psychological Considerations." *Environment and Behavior* 36 (1): 5–31.

Mace, B. L., P. A. Bell, R. J. Loomis, G. E. Haas. 2003. "Source Attribution of Helicopter Noise in Pristine National Park Landscapes." *Journal of Park and Recreation Administration* 21 (3): 97–119.

Mace, B. L., G. C. Corser, L. Zitting, and J. Denison. 2013. "Effects of Overflights on the National Park Experience." *Journal of Environmental Psychology* 35 (September): 30–39.

Mace, B. L., and J. McDaniel. 2013. "Visitor Evaluation of Night Sky Interpretation in Bryce Canyon National Park and Cedar Breaks National Monument." *Journal of Interpretation Research* 18 (1): 39–57.

Maine Association of Conservation Commissions. 2010. "Bar Harbor's Dark Skies Ordinance, Bar Harbor, Maine. Home Rules, Home Tools: Locally Led Conservation Achievements." *Mount Desert Islander.*

Malm, W., K. Kelley, J. Molenar, and T. Daniel. 1981. "Human Perception of Visual Air-Quality (Uniform Haze)." *Atmospheric Environment* 15 (10–1): 1875–90.

Malthus, T. 1798. *An Essay on the Principle of Population, as It Affects the Future Improvement of Society with Remarks on the Speculations of Mr. Godwin, M. Condorcet, and Other Writers.* London. Accessed June 11, 2017. http://129.237.201.53/books/malthus/population/malthus2.doc.

Malville, J. M. 2008. *Guide to Prehistoric Astronomy in the Southwest.* Boulder, CO: Johnson Books.

Manfredo, M., and A. Bright. 1991. "A Model for Assessing the Effects of Communication on Recreationists." *Journal of Leisure Research* 23 (1): 1–20.

Manning, R. E. 1998. "Report on Draft VIP Open House Focus Groups." Internal report prepared for the National Park Service.

———. 2003. "Emerging Principles for Using Information/Education in Wilderness Management." *International Journal of Wilderness* 9 (1): 20–27.

———. 2007. *Parks and Carrying Capacity: Commons without Tragedy.* Washington: Island Press.

———. 2011. *Studies in Outdoor Recreation: Search and Research for Satisfaction.* 3rd ed. Corvallis: Oregon State University Press.

———. 2013. "Social Norms and Reference Points: Integrating Sociology and Ecology." *Environmental Conservation* 40 (4): 310–17.

Manning, R. E., M. Budruk, W. Valliere, and J. Hallo. 2005. *Research to Support Visitor Management at Muir Woods National Monument and Muir Beach.* Burlington: University of Vermont, Park Studies Laboratory.

Manning, R. E., R. Diamant, N. Mitchell, and D. Harmon. 2016. *A Thinking Person's Guide to America's National Parks.* New York: George Braziller Publishers.

——— and L. E. Anderson. 2012. *Managing Outdoor Recreation: Case Studies in the National Parks.* Wallingford, UK: CABI.

——— and P. Pettengill. 2017. *Managing Outdoor Recreation: Case Studies in the National Parks.* 2nd ed. Wallingford, UK: CABI.

Manning, R. E., and W. A. Freimund. 2004. "Use of Visual Research Methods to Measure Standards of Quality for Parks and Outdoor Recreation." *Journal of Leisure Research* 36 (4): 557–79.

———, D. W. Lime, and D. G. Pitt. 1996. "Crowding Norms at Frontcountry Sites: A Visual Approach to Setting Standards of Quality." *Leisure Sciences* 18 (1): 39–59.

Manning, R. E., S. Lawson, P. Newman, J. Hallo, and C. Monz. 2014. *Sustainable Transportation in the National Parks: From Acadia to Zion.* Hanover, NH: University Press of New England.

Manning, R. E., P. Newman, K. Fristrup, D. Stack, and E. Pilcher. 2010. "A Program of Research to Support Management of Visitor-Caused Noise at Muir Woods National Monument." *Park Science* 26 (3): 54–58.

Manning, R. E., E. Rovelstad, C. Moore, J. Hallo, and B. Smith. 2015. "Indicators and Standards of Quality for Viewing the Night Sky in the National Parks." *Park Science* 32 (2): 1–9.

Manning, R. E., W. Valliere, J. Hallo, P. Newman, E. Pilcher, M. Savidge, and D. Dugan. 2007. "From Lightscapes to Soundscapes: Understanding and Managing Natural Quiet in the National Parks." In *Proceedings of the 2006 Northeastern Recreation Research Symposium,* edited by R. Burns and K. Robinson, 601–6. Newton Square, PA: US Department of Agriculture.

Marin, L. D. 2011. "An Exploration of Visitor Motivations: The Search for Silence." MS thesis, Colorado State University.

———, P. Newman, R. Manning, J. J. Vaske, and D. Stack. 2011. "Motivation and Acceptability Norms of Human-Caused Sound in Muir Woods National Monument." *Leisure Sciences* 33 (2): 147–61.

Marion, J. L., and S. E. Reid. 2007. "Minimising Visitor Impacts to Protected Areas: The Efficacy of Low Impact Education Programmes." *Journal of Sustainable Tourism* 15 (1): 5–27.

Marra, P. P., K. A. Hobson, and R. T. Holmes. 1998. "Linking Winter and Summer Events in a Migratory Bird by Using Stable-Carbon Isotopes." *Science* 282 (5395): 1884–86.

Marshall, J., and J. Oberwinkler. 1999. "The Colourful World of the Mantis Shrimp." *Nature* 401 (6756): 873–74.

Martin, G. 1990. *Birds by Night.* London: A. & C. Black Publishers.

Marzluff, J., R. Bowman, and R. Donnelly. 2001. *Avian Ecology and Conservation in an Urbanizing World.* New York: Springer Science & Business Media.

Matzke, E B. 1936. "The Effect of Street Lights in Delaying Leaf-Fall in Certain Trees." *American Journal of Botany,* 23 (6): 446–52.

Mayer, J. D., and Y. N. Gaschke. 1988. "The Experience and Meta-Experience of Mood." *Journal of Personality and Social Psychology* 55 (1): 102–11.

McClure, C. J. W., H. E. Ware, J. Carlisle, G. Kaltenecker, and J. R. Barber. 2013. "An Experimental Investigation into the Effects of Traffic Noise on Distributions of Birds: Avoiding the Phantom Road." *Proceedings of the Royal Society B-Biological Sciences* 280 (1773): 20132290.

McCool, S. F., and N. A. Christensen. 1996. "Alleviating Congestion in Parks and Recreation Areas through Direct Management of Visitor Behavior." In *Congestion and Crowding in the National Park System: Guidelines for Management and Research,* edited by D. Lime, 67–83. St. Paul, MN University of Minnesota Agriculture Experiment Station.

McCormick, M. I., and A. S. Hoey. 2004. "Larval Growth History Determines Juvenile Growth and Survival in a Tropical Marine Fish." *Oikos* 106 (2): 225–42.

McCoy, R. 1992. *Archaeoastronomy: Skywatching in the Native American Southwest.* Flagstaff, AZ: Museum of Northern Arizona.

McDonald, C., R. Baumgartner, and R. Iachan. 1995. "National Park Service Aircraft Management Studies." Denver (CO): National Park Service.

McDonald, M. A., J. A. Hildebrand, and S. M. Wiggins. 2006. "Increases in Deep Ocean Ambient Noise in the Northeast Pacific West of San Nicolas Island, California." *Journal of the Acoustical Society of America* 120 (2): 711–18.

McDonald Observatory. 2017. "McDonald Observatory." Accessed August 14, 2017. https://mcdonaldobservatory.org/.

McDowell Group with DataPath systems and Davis, Hibbits and Midghall, Inc. 2007. "Alaska Visitor Statistics Program—Alaska Visitor Volume and Profile, Fall/Winter 2006–2007." State of Alaska Department of Commerce, Community and Economic Development. http://www.dced.state.ak.us/ded/dev /toubus/pub/FW_whole_doc.pdf.

McFarlane, R. W. 1963. "Disorientation of Loggerhead Hatchlings by Artificial Road Lighting." *Copeia* 1963 (1): 153.

McKeever, S. 1977. "Observations of Corethrella Feeding on Tree Frogs (*Hyla*)." *Mosquito News* 37:522–23.

——— and W. Hartberg. 1980. "An Effective Method for Trapping Adult Female Corethrella (*Diptera: Chaoboridae*)." *Mosquito News* 40 (1): 111–12.

McKinney, M. L. 2002. "Urbanization, Biodiversity, and Conservation." *Bioscience* 52 (10): 883–90.

McMahon, T., J. Romansic, and J. Rohr. 2013. "Non-Monotonic and Monotonic Effects of Pesticides on the Pathogenic Fungus *Batrachochytrium dendrobatidis* in Culture and on Tadpoles." *Environmental Science and Technology* 47 (14): 7958–64.

Medellin, R. A., and O. Gaona. 1999. "Seed Dispersal by Bats and Birds in Forest and Disturbed Habitats of Chiapas, Mexico." *Biotropica* 31 (3): 478–85.

Meekan, M. G., S. G. Wilson, A. Halford, and A. Retzel. 2001. "A Comparison of Catches of Fishes and Invertebrates by Two Light Trap Designs, in Tropical NW Australia." *Marine Biology* 139 (2): 373–81.

Megdal, S. P., C. H. Kroenke, F. Laden, E. Pukkala, and E. S. Schernhammer. 2005. "Night Work and Breast Cancer Risk: A Systematic Review and Meta-Analysis." *European Journal of Cancer* 41 (13): 2023–32.

Menaker, M. 1968. "Extraretinal Light Perception in the Sparrow. I. Entrainment of the Biological Clock." *Proceedings of the National Academy of Sciences* 59 (2): 414–21.

Mengak, K. K. 1986. "Use of Importance-Performance Analysis to Evaluate a Visitor Center." Paper presented at the Seventeenth Annual Conference of the Travel and Tourism Research Association, Memphis, TN, June 15–18.

Menken, S., W. Herrebout, and J. Wiebes. 1992. "Small Ermine Moths (*Yponomeuta*)—Their Host Relations and Evolution." *Annual Review of Entomology* 37: 41–66.

Met Office. N.d. UKCP09: Gridded Observation Data Sets. Accessed June 24, 2017. www.metoffice.gov.uk/climatechange/science/monitoring/ukcp09/.

Metcalfe, N. B., F. Huntingford, and J. Thorpe. 1987. "The Influence of Predation Risk on the Feeding Motivation and Foraging Strategy of Juvenile Atlantic Salmon." *Animal Behaviour* 35 (June): 901–11.

Miksis-Olds, J. L., P. L. Donaghay, J. H. Miller, P. L. Tyack, and J. A. Nystuen. 2007. "Noise Level Correlates with Manatee Use of Foraging Habitats." *Journal of the Acoustical Society of America* 121 (5): 3011–20.

Miller, M. W. 2006. "Apparent Effects of Light Pollution on Singing Behavior of American Robins." *Condor* 108 (1): 130–39.

Miller N., G. Anderson, R. Horonjeff, and R. Thompson 1999. "Mitigating the Effects of Military Aircraft Overflights on Recreational Users of Parks." Lexington, MA: Harris, Miller, Miller, and Hanson.

Miller, N. P. 1995–97. "Report on Effects of Aircraft Overflights on the National Park System: Executive Summary Report to Congress." Burlington, MA: Harris, Miller, Miller, and Hanson, Inc.

———. 1999. "The Effects of Aircraft Overflights on Visitors to US National Parks." *Noise Control Engineering Journal* 47 (3): 112–17.

———. 2002. "Transportation Noise and Recreational Lands." In *Proceedings of the 2002 International Congress and Exposition on Noise Control Engineering*, 19–21 August. Dearborn, Michigan. Institute of Noise Control Engineering of the USA, Indianapolis, Indiana, USA.

———. 2004. "Understanding, Managing, and Protecting Opportunities for Visitor Experiences: Transportation Noise and the Value of Natural Quiet." In *Protecting Our Diverse Heritage: The Role of Parks, Protected Areas, and Cultural Sites. Proceedings of the 2003 George Wright Society/National Park Service Joint Conference, 14–18 April. San Diego, California*, edited by D. Harmon, 128–34. Hancock, MI: George Wright Society.

———. 2008. "US National Parks and Management of Park Soundscapes: A Review." *Applied Acoustics* 69 (2): 77–92.

Mills, A. M. 2008. "Latitudinal Gradients of Biologically Useful Semi-Darkness." *Ecography* 31 (5): 578–82.

Milner, L. M., J. M. Collins, R. Tachibana, and R. F. Hiser. 2000. "The Japanese Vacation Visitor to Alaska: A Preliminary Examination of Peak and Off-Season Traveler Demographics, Information Source Utilization, Trip Planning, and Customer Satisfaction." *Journal of Travel & Tourism Marketing* 9 (1–2): 43–56.

Milson, T. 1984. "Diurnal Behavior of Lapwings in Relation to Moon Phase During Winter." *Bird Study* 31 (July): 117–20.

Mockford, E. J., and R. C. Marshall. 2009. "Effects of Urban Noise on Song and Response Behaviour in Great Tits." *Proceedings of the Royal Society B-Biological Sciences* 276 (1669): 2979–856.

Monroe, M., P. Newman, E. Pilcher, R. Manning, and D. Stack. 2007. "Now Hear This." *Legacy* 18 (1): 18–25.

Montevecchi, W. 2006. "Influences of Artificial Light on Marine Birds." In *Ecological Consequences of Artificial Night Lighting*, edited by C. Rich and T. Longcore, 94–113. Washington: Island Press.

Montgomerie, R., and P. J. Weatherhead. 1997. "How Robins Find Worms." *Animal Behaviour* 54 (July): 143–51.

Moore, C., F. Hoffman, D. Fields, and R. Mastroguiseppe. n.d. "Promoting and Protecting Dark Night Skies in Our National Parks." National Park Trust. Accessed August 15, 2017. http://www.craterlakeinstitute.com/online-library/protecting-dark-skies/protecting-dark-skies.htm.

Moore, C. A. 2001. "Visual Estimation of Night Sky Brightness." *George Wright Forum*, 18 (4): 46–55. Accessed June 18, 2017. http://www.georgewright.org/184moore.pdf.

Moore, C. B., and T. D. Siopes. 2000. "Effects of Lighting Conditions and Melatonin Supplementation on the Cellular and Humoral Immune Responses in Japanese Quail (*Coturnix Coturnix Japonica*)." *General and Comparative Endocrinology* 119 (1): 95–104.

Moore, E. 1981. "A Prison Environment's Effect on Health Care Service Demands." *Journal of Environmental Systems* 11 (1): 17–34.

Moore, M. V., S. M. Pierce, H. M. Walsh, S. K. Kvalvik, and J. D. Lim. 2000. "Urban Light Pollution Alters the Diel Vertical Migration of Daphnia." *Internationale Vereinigung Fur Theoretische Und Angewandte Limnologie Verhandlungen* 27 (2): 779–82.

Morley, E. L., G. Jones, and A. N. Radford. 2014. "The Importance of Invertebrates When Considering the Impacts of Anthropogenic Noise." *Proceedings of the Royal Society B-Biological Sciences* 281 (1776): 20132683.

Morrison, D. 1978. "Lunar Phobia in a Neotropical Fruit Bat, *Artibeus Jamaicensis* (*Chiroptera Phyllostomidae*)." *Animal Behaviour* 26 (August): 852–55.

Mougeot, F., and V. Bretagnolle. 2000. "Predation Risk and Moonlight Avoidance in Nocturnal Seabirds." *Journal of Avian Biology* 31 (3): 376–86.

Mueller, R. C., C. M. Scudder, M. E. Porter, R. T. Trotter, C. A. Gehring, and T. G. Whitham. 2005. "Differential Tree Mortality in Response to Severe Drought: Evidence for Long-Term Vegetation Shifts." *Journal of Ecology* 93 (6): 1085–93.

Muheim, R., J. B. Phillips, and S. Akesson. 2006. "Polarized Light Cues Underlie Compass Calibration in Migratory Songbirds." *Science* 313 (5788): 837–39.

Munday, P. L., D. L. Dixson, M. I. McCormick, M. Meekan, M. C. O. Ferrari, and D. P. Chivers. 2010. "Replenishment of Fish Populations Is Threatened by Ocean Acidification." *Proceedings of the National Academy of Sciences of the United States of America* 107 (29): 12930–34.

Murphy, M. T., K. Sexton, A. C. Dolan, and L. J. Redmond. 2008. "Dawn Song of the Eastern Kingbird: An Honest Signal of Male Quality?" *Animal Behaviour* 75 (March): 1075–84.

Muscarella, R., and T. H. Fleming. 2007. "The Role of Frugivorous Bats in Tropical Forest Succession." *Biological Reviews* 82 (4): 573–90.

Nash, R. 1982. *Wilderness and the American Mind*. 3rd ed. New Haven, CT: Yale University Press.

National Marine Manufacturers Association. 2013. *Recreational Boating Statistical Abstract*. Chicago: National Marine Manufacturers Association.

National Park Service. n.d.a. Bryce Canyon: Astronomy & Night Sky Programs. Accessed August 14, 2017. https://www.nps.gov/brca/planyourvisit/astronomy programs.htm.

———. n.d.b. "Natural Sounds: A Symphony of Trees, Grasses, Birds and Streams." Accessed June 10, 2017. http://nature.nps.gov/sound/index.cfm.

———. n.d.c. "Stephen Tyng Mather." Accessed August 24, 2017. https://www .nps.gov/bestideapeople/mather.html.

———. 1997. *VERP: The Visitor Experience and Resource Protection Framework; A Handbook for Planners and Managers*. Denver, CO: National Park Service.

———. 2000. "Director's Order #47: Soundscape Preservation and Noise Management." Washington: National Park Service. Accessed August 29, 2017. https://www.nps.gov/policy/DOrders/DOrder47.html.

———. 2006. *Management Policies 2006*. Washington: National Park Service. Accessed June 11, 2017. http://www.nps.gov/policy/MP2006.pdf.

———. 2011a. *A Call to Action: Preparing for a Second Century of Stewardship and Engagement*. Accessed August 15, 2017. https://www.nps.gov/calltoaction/PDF /Directors_Call_to_Action_Report.pdf.

———. 2011b. "Healthy Parks Healthy People US Strategic Action Plan." Washington: National Park Service. Accessed June 11, 2017. http://www.nps.gov /public_health/hp/hphp/press/1012-955-WASO.pdf.

———. 2014. *A Call to Action: Preparing for a Second Century of Stewardship and Engagement*. Washington: National Park Service. Accessed June 10, 2017. https://www.nps.gov/calltoaction/PDF/C2A_2014.pdf.

———. 2016a. "A Canyon Alight With Stars: A Brief History of Astronomy at Bryce Canyon." Accessed August 15, 2017. https://www.nps.gov/brca/plan yourvisit/astrohistory.htm.

———. 2016b. "Astronomy Volunteers." Accessed August 15, 2017. https://www .nps.gov/brca/planyourvisit/astron.omyvolunteer.htm.

———. 2017a. "Bryce Canyon: Annual Astronomy Festival." Accessed August 15, 2017. https://www.nps.gov/brca/planyourvisit/astrofest.htm.

———. 2017b. "Natural Sounds and Night Skies Division: What We Do." Accessed August 15, 2017. https://www.nps.gov/orgs/1050/whatwedo.htm.

National Parks Conservation Association. 2005. *State of the Parks—Bryce Canyon National Park: A Resource Assessment*. http://www.npca.org/stateoftheparks /brycecanyon/brycecanyon.pdf.

National Research Council of the National Academies. 2005. *Marine Mammal Populations and Ocean Noise: Determining When Noise Causes Biologically Significant Effects*. Washington: National Academies Press.

Navara, K. J., and R. J. Nelson. 2007. "The Dark Side of Light at Night: Physiological, Epidemiological, and Ecological Consequences." *Journal of Pineal Research* 43 (3): 215–24.

Neff, M. M., C. Fankhauser, and J. Chory. 2000. "Light: An Indicator of Time and Place." *Genes & Development* 14 (3): 257–71.

Negro, J. J., J. Bustamante, C. Melguizo, J. L. Ruiz, and J. M. Grande. 2000. "Nocturnal Activity of Lesser Kestrels under Artificial Lighting Conditions in Seville, Spain." *Journal of Raptor Research* 34 (4): 327–29.

Newman, P., S. Lawson, D. Taff, L. Marin, and A. Gibson. 2012. *Soundscape Indicators and Standards of Quality Research in Yosemite (YOSE) and Sequoia and Kings Canyon National Parks (SEKI).* Fort Collins, CO: Colorado State University.

Newman, P., R. Manning, and K. Trevino. 2010. "From Landscapes to Soundscapes: Introduction to the Special Issue." *Park Science* 26 (3): 2–5.

Newman, P., D. Stack, and R. Manning. 2007. *Research to Inform the Development of Indicators and Standards of Quality for Soundscapes in Muir Woods National Monument.* For Collins: Warner College of Natural Resources, Colorado State University.

Nicholas, M. 2001. "Light Pollution and Marine Turtle Hatchlings: The Straw That Breaks the Camel's Back?" *George Wright Forum*, 18 (4): 77–82. Accessed June 18, 2017. http://www.georgewright.org/184nicholas.pdf.

Niva, C. C., and M. Takeda. 2003. "Effects of Photoperiod, Temperature and Melatonin on Nymphal Development, Polyphenism and Reproduction in Halyomorpha Halys (*Heteroptera: Pentatomidae*)." *Zoological Science* 20 (8): 963–70.

Niven, J. E., and S. B. Laughlin. 2008. "Energy Limitation as a Selective Pressure on the Evolution of Sensory Systems." *Journal of Experimental Biology* 211 (11): 1792–1804.

Nordgren, Tyler. 2010. *Stars Above, Earth Below: A Guide to Astronomy in the National Parks.* Berlin: Springer Science & Business Media.

Normandeau Associates, Inc. 2012. "Effects of Noise on Fish, Fisheries, and Invertebrates in the U.S. Atlantic and Arctic from Energy Industry Sound-Generating Activities: Workshop Report." Accessed August 25, 2017. https://www.cbd.int/doc/meetings/mar/mcbem-2014-01/other/mcbem-2014-01-submission-boem-04-en.pdf.

Nowinszky, L., and J. Puskás. 2010. "Light Trapping of Helicoverpa Armigera in India and Hungary in Relation with the Moon Phases." *Indian Journal of Agricultural Sciences* 81:152–55.

Odum, E. P. 1953. *Fundamentals of Ecology.* Philadelphia, PA: W. B. Saunders.

Oprea, M., P. Mendes, T. B. Vieira, and A. D. Ditchfield. 2009. "Do Wooded Streets Provide Connectivity for Bats in an Urban Landscape?" *Biodiversity and Conservation* 18 (9): 2361–717.

Orbach, D. N., and B. Fenton. 2010. "Vision Impairs the Abilities of Bats to Avoid Colliding with Stationary Obstacles. *PLoS One* 5 (11): e13912.

Ortega, C. P. 2012. "Effects of Noise Pollution on Birds: A Brief Review of Our Knowledge." *Ornithological Monographs* 74:6–22.

——— and C. D. Francis. 2012. "Effects of Gas-Well-Compressor Noise on the Ability to Detect Birds during Surveys in Northwest New Mexico." *Ornithology Monograph* 74: 78–90.

Osorio, D., and M. Vorobyev. 2008. "A Review of the Evolution of Animal Colour Vision and Visual Communication Signals." *Vision Research* 48 (20): 2042–51.

Outen, A. R. 1998. "The Possible Ecological Implications of Artificial Lighting." *Hertfordshire Environmental Records Centre.*

Packer, C., A. Swanson, D. Ikanda, and H. Kushnir. 2011. "Fear of Darkness, the Full Moon and the Nocturnal Ecology of African Lions." *PLoS One* 6 (7): e22285.

Page, R., M. Ryan, and X. Bernal. 2014. "Be Loved, Be Preyed, Be Eaten." In *Animal Behavior,* edited by K. Yasakawa. Santa Barbara, CA: BC-CLIO Editorial.

Paige, K., and T. Whitham. 1987. "Flexible Life-History Traits—Shifts by Scarlet-Gilia in Response to Pollinator Abundance." *Ecology* 68 (6): 1691–95.

Park, Logan, Steve Lawson, Ken Kaliski, Peter Newman, and Adam Gibson. 2010. "Modeling and Mapping Hikers' Exposure to Transportation Noise in Rocky Mountain National Park." *Park Science* 26 (3): 59–64.

Parris, K., M. Velik-Lord, and J. North. 2009. "Frogs Call at a Higher Pitch in Traffic Noise." *Ecology and Society* 14 (1): 25–35.

Patricelli, G. L., and J. L. Blickley. 2006. "Avian Communication in Urban Noise: Causes and Consequences of Vocal Adjustment." *Auk* 123 (3): 639–49.

Pauley, S. M. 2004. "Lighting for the Human Circadian Clock: Recent Research Indicates That Lighting Has Become a Public Health Issue." *Medical Hypotheses* 63 (4): 588–96.

Payne, S. R. 2008. "Are Perceived Soundscapes within Urban Parks Restorative?" *Proceeding of Acoustics* 8:5519–24.

———. 2013. "The Production of a Perceived Restorativeness Soundscape Scale." *Applied Acoustics* 74 (2): 255–63.

Pearson, K. M., and T. C. Theimer. 2004. "Seed-Caching Responses to Substrate and Rock Cover by Two Peromyscus Species: Implications for Pinyon Pine Establishment." *Oecologia* 141 (1): 76–83.

Penteriani, V., M. del Mar Delgado, L. Campioni, and R. Lourenco. 2010. "Moonlight Makes Owls More Chatty." *PLoS One* 5 (1): e8696.

Penteriani, V., A. Kuparinen, M. del Mar Delgado, R. Lourenco, and L. Campioni. 2011. "Individual Status, Foraging Effort and Need for Conspicuousness Shape Behavioural Responses of a Predator to Moon Phases." *Animal Behaviour* 82 (2): 413–20.

Perkin, E. K., F. Hoelker, J. S. Richardson, J. P. Sadler, C. Wolter, and K. Tockner. 2011. "The Influence of Artificial Light on Stream and Riparian Ecosystems: Questions, Challenges, and Perspectives." *Ecosphere* 2 (11): 1–16.

Perloff, R. M. 2003. *The Dynamics of Persuasion: Communication and Attitudes in the Twenty-First Century.* New York: Lawrence Erlbaum Associates.

Perry, G., and R. N. Fisher. 2006. "Night Lights and Reptiles: Observed and Potential Effects." In *Ecological Consequences of Artificial Night Lighting*, edited by C. Rich and T. Longcore, 169–91. Washington: Island Press.

Perry, M., and D. Hollis. 2005. "The Generation of Monthly Gridded Datasets for a Range of Climatic Variables over the UK." *International Journal of Climatology* 25 (8): 1041–54.

Peters, A., and K. Verhoeven. 1994. "Impact of Artificial Lighting on the Seaward Orientation of Hatchling Loggerhead Turtles." *Journal of Herpetology* 28 (1): 112–14.

Peterson, G. L., and D. W. Lime. 1979. "People and Their Behavior: A Challenge for Recreation Management." *Journal of Forestry* 77 (6): 343–46.

Petty, R. E., and J. T. Cacioppo. 1981. *Attitudes and Persuasion: Classic and Contemporary Approaches.* Dubuque, IA: William C. Brown.

———. 1986. *Central and Peripheral Routes to Persuasion: Theory and Research.* New York: Springer Verlag.

———. 1996. *Attitudes and Persuasion: Classic and Contemporary Approaches.* Boulder, CO: Westview Press.

Petty, R. E., and D. T. Wegener. 2008. "Matching versus Mismatching Attitude Functions: Implications for Scrutiny of Persuasive Messages." In *Attitudes Their Structure, Function and Consequences: Key Readings in Social Psychology*, edited by R. Fazio and R. E. Petty, 227–40. New York: Psychology Press.

Pilcher, E. J., P. Newman, and R. E. Manning. 2009. "Understanding and Managing Experiential Aspects of Soundscapes at Muir Woods National Monument." *Environmental Management* 43 (3): 425–35.

Pleasants, J. 1983. "Nectar Production Patterns in *Ipomopsis-Aggregata* (Polemoniaceae)." *American Journal of Botany* 70 (10): 1468–75.

Poesel, A., H. P. Kunc, K. Foerster, A. Johnsen, and B. Kempenaers. 2006. "Early Birds Are Sexy: Male Age, Dawn Song and Extrapair Paternity in Blue Tits, *Cyanistes* (Formerly *Parus*) *Caeruleus*." *Animal Behaviour* 72 (September): 531–38.

Polak, T., C. Korine, S. Yair, and M. W. Holderied. 2011. "Differential Effects of Artificial Lighting on Flight and Foraging Behaviour of Two Sympatric Bat Species in a Desert." *Journal of Zoology* 285 (1): 21–27.

Poole, A. 2005. *The Birds of North America Online.* Ithaca, NY: Cornell Laboratory of Ornithology. Accessed August 2, 2017. http://bna.birds.cornell.edu/BNA/.

Poot, H., B. J. Ens, H. de Vries, M. A. H. Donners, M. R. Wernand, and J. M. Marquenie. 2008. "Green Light for Nocturnally Migrating Birds." *Ecology and Society* 13 (2): 47–60.

Price, M., N. Waser, and T. Bass. 1984. "Effects of Moonlight on Microhabitat Use by Desert Rodents." *Journal of Mammalogy* 65 (2): 353–56.

Purser, J., and A. N. Radford. 2011. "Acoustic Noise Induces Attention Shifts and

Reduces Foraging Performance in Three-Spined Sticklebacks (*Gasterosteus Aculeatus*)." *PLoS One* 6 (2): e17478.

Queval, G., E. Issakidis-Bourguet, F. A. Hoeberichts, M. Vandorpe, B. Gakiere, H. Vanacker, M. Miginiac-Maslow, F. Van Breusegem, and G. Noctor. 2007. "Conditional Oxidative Stress Responses in the Arabidopsis Photorespiratory Mutant cat2 Demonstrate That Redox State Is a Key Modulator of Daylength-Dependent Gene Expression, and Define Photoperiod as a Crucial Factor in the Regulation of H^2O^2-Induced Cell Death." *Plant Journal* 52 (4): 640–57.

Quinn, J. L., M. J. Whittingham, S. J. Butler, and W. Cresswell. 2006. "Noise, Predation Risk Compensation and Vigilance in the Chaffinch Fringilla Coelebs." *Journal of Avian Biology* 37 (6): 601–8.

Rabin, L. A., R. G. Coss, and D. H. Owings. 2006. "The Effects of Wind Turbines on Antipredator Behavior in California Ground Squirrels (*Spermophilus Beecheyi*)." *Biological Conservation* 131 (3): 410–20.

Raffel, T., N. Halstead, T. McMahon, A. Davis, and J. Rohr. 2015. "Temperature Variability and Moisture Synergistically Interact to Exacerbate an Epizootic Disease." *Proceedings of the Royal Society B-Biological Sciences* 282 (1801): 2014–39.

Raffel, T., J. Romansic, N. Halstead, T. McMahon, M. Venesky, and J. Rohr. 2013. "Disease and Thermal Acclimation in a More Variable and Unpredictable Climate." *Nature Climate Change* 3 (2): 146–51.

Ragni, M., and M. R. D'Alcala. 2004. "Light as an Information Carrier Underwater." *Journal of Plankton Research* 26 (4): 433–43.

Rainbolt, G. N., J. A. Benfield, and P. A. Bell. 2012. "The Influence of Anthropogenic Sound in Historical Parks: Implications for Park Management." *Journal of Park and Recreation Administration* 30 (4): 53–65.

Rand, A., M. Bridarolli, L. Dries, and M. Ryan. 1997. "Light Levels Influence Female Choice in Tungara Frogs: Predation Risk Assessment?" *Copeia* 1997 (2):447–50.

Ratcliffe, E., B. Gatersleben, and P. Sowden. 2012. "Exploring Perceptions of Birdsong as a Restorative Stimulus." Paper presented at the 13th Congress of the International Society of Ethnobiology, Montpellier, France, May 25.

Raven, J. A., and C. S. Cockell. 2006. "Influence on Photosynthesis of Starlight, Moonlight, Planetlight, and Light Pollution (Reflections on Photosynthetically Active Radiation in the Universe)." *Astrobiology* 6 (4): 668–75.

Rea, M. S., J. D. Bullough, and Y. Akashi. 2009. "Several Views of Metal Halide and High-Pressure Sodium Lighting for Outdoor Applications." *Lighting Research & Technology* 41 (4): 297–320.

Reed, J., J. Sincock, and J. Hailman. 1985. "Light Attraction in Endangered Procellariiform Birds—Reduction by Shielding Upward Radiation." *Auk* 102 (2): 377–83.

Reijnen, R., and R. Foppen. 2006. "Impact of Road Traffic on Breeding Bird Populations." In *The Ecology of Transportation: Managing Mobility for the En-*

vironment, edited by J. Davenport and J. L. Davenport, 255–74. Dordrecht, the Netherlands: Springer.

Reiter, R. J., D.-X. Tan, A. Korkmaz, T. C. Erren, C. Piekarski, H. Tamura, and L. C. Manchester. 2007. "Light at Night, Chronodisruption, Melatonin Suppression, and Cancer Risk: A Review." *Critical Reviews in Oncogenesis* 13 (4): 303–28.

Reiter, R. J., D.-X. Tan, E. Sanchez-Barcelo, M. D. Mediavilla, E. Gitto, and A. Korkmaz. 2011. "Circadian Mechanisms in the Regulation of Melatonin Synthesis: Disruption with Light at Night and the Pathophysiological Consequences." *Journal of Experimental and Integrative Medicine* 1 (1): 13–22.

Reudink, M. W., P. P. Marra, T. K. Kyser, P. T. Boag, K. M. Langin, and L. M. Ratcliffe. 2009. "Non-Breeding Season Events Influence Sexual Selection in a Long-Distance Migratory Bird." *Proceedings of the Royal Society B-Biological Sciences* 276 (1662): 1619–26.

Rex, K., D. H. Kelm, K. Wiesner, T. H. Kunz, and C. C. Voigt. 2008. "Species Richness and Structure of Three Neotropical Bat Assemblages." *Biological Journal of the Linnean Society* 94 (3): 617–29.

Rex, K., R. Michener, T. H. Kunz, and C. C. Voigt. 2011. "Vertical Stratification of Neotropical Leaf-Nosed Bats (*Chiroptera: Phyllostomidae*) Revealed by Stable Carbon Isotopes." *Journal of Tropical Ecology* 27 (May): 211–22.

Rich, C., and T. Longcore, eds. 2006. *Ecological Consequences of Artificial Night Lighting*. Washington, DC: Island Press.

———. 2013. *Ecological Consequences of Artificial Night Lighting*. 2nd ed. Washington: Island Press.

Richman, A. 2004. "Who Will Keep the Night?" In *Protecting Our Diverse Heritage: The Role of Parks, Protected Areas, and Cultural Sites, Proceedings of the 2003 George Wright Society/National Park Service Joint Conference*, edited by D. Harmon, 152–56. Hancock, MI: George Wright Society.

Riegel, K. W. 1973. "Light Pollution." *Science* 179 (4080): 1285–91.

Rivera, G., S. Elliott, L. S. Caldas, G. Nicolossi, V. T. R. Coradin, and R. Borchert. 2002. "Increasing Day-Length Induces Spring Flushing of Tropical Dry Forest Trees in the Absence of Rain." *Trees: Structure and Function* 16 (7): 445–56.

Robbins, C., J. Sauer, R. Greenberg, and S. Droege. 1989. "Population Declines in North-American Birds That Migrate to the Neotropics." *Proceedings of the National Academy of Sciences of the United States of America* 86 (19): 7658–62.

Robertson, B. A., and R. L. Hutto. 2006. "A Framework for Understanding Ecological Traps and an Evaluation of Existing Evidence." *Ecology* 87 (5): 1075–85.

Robson, I. 2005. "The Role of the Observatories." In *IAU Commission 55: Communicating Astronomy with the Public*, edited by I. Robson and L. I. Christensen, 60–70. Munich: IAU Commission.

Rochat, J. L. 2011. "Motorcycle Noise in a Park Environment." Washington: National Park Service. Accessed June 11, 2017. https://ntl.bts.gov/lib/48000/48100/48196/NPS_Motorcycle_Noise_2013.pdf.

————. 2013. "Motorcycle Noise in a Park Environment." Natural Resource Technical Report NPS/NSNS/NRTR—2013/781. National Park Service, Fort Collins, CO. Accessed June 18, 1017. https://ntl.bts.gov/lib/48000/48100/48196/NPS_Motorcycle_Noise_2013.pdf.

Rodman, J. 1983. "Four Forms of Ecological Consciousness Reconsidered." *Ethics and the Environment* 1983:82–92.

Rodríguez, A., B. Rodríguez, and M. P. Lucas. 2012. "Trends in Numbers of Petrels Attracted to Artificial Lights Suggest Population Declines in Tenerife, Canary Islands." *Ibis* 154 (1): 167–72.

Rogers, J., and J. Sovick. 2001a. "Let There Be Dark: The National Park Service and the New Mexico Night Sky Protection Act." *George Wright Forum*, 18 (4): 37–45. Accessed June 18, 2017. http://www.georgewright.org/184rogers2.pdf.

————. 2001b. "The Ultimate Cultural Resource." *George Wright Forum*, 18 (4): 25–29. Accessed June 18, 2017. http://georgewright.org/184rogers1.pdf.

Rohr, J., T. Raffel, N. Halstead, T. McMahon, S. Johnson, R. Boughton, and L. Martin. 2013. "Early-Life Exposure to a Herbicide Has Enduring Effects on Pathogen-Induced Mortality." *Proceedings of the Royal Society B-Biological Sciences* 280 (1772): 1502–10.

Rojas, L. M., R. McNeil, T. Cabana, and P. Lachapelle. 1999. "Diurnal and Nocturnal Visual Capabilities in Shorebirds as a Function of Their Feeding Strategies." *Brain Behavior and Evolution* 53 (1): 29–43.

Romero, L. M., M. J. Dickens, and N. E. Cyr. 2009. "The Reactive Scope Model—A New Model Integrating Homeostasis, Allostasis, and Stress." *Hormones and Behavior* 55 (3): 375–89.

Roof, C. J., B. Kim, G. G. Fleming, J. Burstein, and C. S. Y. Lee. 2002. "Noise and Air Quality Implications of Alternative Transportation Systems: Zion and Acadia National Park Case Studies." Cambridge, MA: Department of Transportation. Accessed June 19, 2017. http://ntl.bts.gov/lib/47000/47200/47291/NoiseAir_QualityATS_2002.pdf.

Rotics, S., T. Dayan, and N. Kronfeld-Schor. 2011. "Effect of Artificial Night Lighting on Temporally Partitioned Spiny Mice." *Journal of Mammalogy* 92 (1): 159–68.

Rotics, S., T. Dayan, O. Levy, and N. Kronfeld-Schor. 2011. "Light Masking in the Field: An Experiment with Nocturnal and Diurnal Spiny Mice Under Semi-Natural Field Conditions." *Chronobiology International* 28 (1): 70–75.

Royal Commission on Environmental Pollution. 2009. *Artificial Light in the Environment.* London: Stationery Office.

Ryan, M. J. 1985. "Energetic Efficiency of Vocalization by the Frog *Physalaemus pustulosus.*" *Journal of Experimental Biology* 116 (1): 47–52.

Rydell, J. 1991. "Seasonal Use of Illuminated Areas by Foraging Northern Bats *Eptesicus-Nilssoni.*" *Holarctic Ecology* 14 (3): 203–7.

————. 1992. "Exploitation of Insects Around Streetlamps by Bats in Sweden." *Functional Ecology* 6 (6): 744–50.

———. 2006. "Bats and Their Insect Prey at Streetlights." *Ecological Consequences of Artificial Night Lighting* 2: 43–60.

———, A. Entwistle, and P. A. Racey. 1996. "Timing of Foraging Flights of Three Species of Bats in Relation to Insect Activity and Predation Risk." *Oikos* 76 (2): 243–52.

Salmon, M., M. G. Tolbert, D. P. Painter, M. Goff, and R. Reiners. 1995. "Behavior of Loggerhead Sea Turtles on an Urban Beach." *Journal of Herpetology* 29 (4): 568–76.

San Pedro Valley Observatory. 2017. "San Pedro Valley Observatory." Accessed August 15, 2017. http://www.go-astronomy.com/observatory.php?ID=222.

Sanderson, F. J., P. F. Donald, D. J. Pain, I. J. Burfield, and F. P. J. van Bommel. 2006. "Long-Term Population Declines in Afro-Palearctic Migrant Birds." *Biological Conservation* 131 (1): 93–105.

Santos, C. D., A. C. Miranda, J. P. Granadeiro, P. M. Lourenco, S. Saraiva, and J. M. Palmeirim. 2010. "Effects of Artificial Illumination on the Nocturnal Foraging of Waders." *Acta Oecologica* 36 (2): 166–72.

Schaaf, F. 1988. *The Starry Room: Naked Eye Astronomy in the Intimate Universe.* New York: Wiley.

Schaub, A., J. Ostwald, and B. M. Siemers. 2008. "Foraging Bats Avoid Noise." *Journal of Experimental Biology* 211 (19): 3174–80.

Schlaepfer, M. A., M. C. Runge, and P. W. Sherman. 2002. "Ecological and Evolutionary Traps." *Trends in Ecology & Evolution* 17 (10): 474–80.

Schmidt, K. A., S. R. X. Dall, and J. A. van Gils. 2010. "The Ecology of Information: An Overview on the Ecological Significance of Making Informed Decisions." *Oikos* 119 (2): 304–16.

Schreuder, D. 2010. *Outdoor Lighting: Physics, Vision and Perception.* New York: Springer-Verlag.

Schroeder, S. A., D. C. Fulton, W. Penning, and K. Doncarlos. 2012. "Using Persuasive Messages to Encourage Hunters to Support Regulation of Lead Shot." *Journal of Wildlife Management* 76 (8): 1528–39.

Sedona Star Gazing. 2017. "Sedona Star Gazing." Accessed August 15, 2017. http://www.eveningskytours.com/.

Segura, G., and L. Snook. 1992. "Stand Dynamics and Regeneration Patterns of a Pinyon Pine Forest in East Central Mexico." *Forest Ecology and Management* 47 (1–4): 175–94.

Seligman, M. E. P. 1975. *Helplessness: On Depression, Development, and Death.* New York: W. H. Freeman.

Sellars, R. W. 1997. *Preserving Nature in the National Parks: A History.* New Haven, CT: Yale University Press.

Sexton, K., M. T. Murphy, L. J. Redmond, and A. C. Dolan. 2007. "Dawn Song of Eastern Kingbirds: Intrapopulation Variability and Sociobiological Correlates." *Behaviour* 144 (October): 1273–95.

Shannon, G., L. M. Angeloni, G. Wittemyer, K. M. Fristrup, and K. R. Crooks.

2014. "Road Traffic Noise Modifies Behaviour of a Keystone Species." *Animal Behaviour* 94 (August): 135–41.

Shannon, G., M. F. McKenna, L. M. Angeloni, K. R. Crooks, K. M. Fristrup, E. Brown, K. A. Warner, M. D. Nelson, C. White, J. Briggs, et al. 2015. "A Synthesis of Two Decades of Research Documenting the Effects of Noise on Wildlife." *Biological Reviews* 91 (4): 982–1005.

Shattuck, B., and G. B. Cornucopia. 2001. "Chaco's Night Lights." *George Wright Forum* 18 (4): 72–76. http://www.georgewright.org/184shattuck.pdf.

Shelby, B., and T. A. Heberlein. 1984. "A Conceptual Framework for Carrying Capacity Determination." *Leisure Sciences* 6 (4): 433–51.

———. 1986. *Carrying Capacity in Recreation Settings*. Corvallis, OR: Oregon State University Press.

——— and M. P. Donnelly.1996. "Norms, Standards, and Natural Resources." *Leisure Sciences* 18 (2): 103–23.

Shin, W. 2007. "The Influence of Forest View through a Window on Job Satisfaction and Job Stress." *Scandinavian Journal of Forest Research* 22 (3): 248–53.

Shochat, E., P. S. Warren, S. H. Faeth, N. E. McIntyre, and D. Hope. 2006. "From Patterns to Emerging Processes in Mechanistic Urban Ecology." *Trends in Ecology and Evolution* 21 (4): 186–91.

Shuboni, D., and L. Yan. 2010. "Nighttime Dim Light Exposure Alters the Responses of the Circadian System." *Neuroscience* 170 (4): 1172–78.

Sibson, R. 1980. "The Dirichlet Tessellation as an Aid in Data-Analysis." *Scandinavian Journal of Statistics* 7 (1): 14–20.

Siemers, B. M., and A. Schaub. 2011. "Hunting at the Highway: Traffic Noise Reduces Foraging Efficiency in Acoustic Predators." *Proceedings of the Royal Society B-Biological Sciences* 278 (1712): 1646–52.

Sih, A., A. Bell, and J. C. Johnson. 2004. "Behavioral Syndromes: An Ecological and Evolutionary Overview." *Trends in Ecology & Evolution* 19 (7): 372–78.

Sillett, T. S., and R. T. Holmes. 2002. "Variation in Survivorship of a Migratory Songbird throughout Its Annual Cycle." *Journal of Animal Ecology* 71 (2): 296–308.

Simpson, S. D., J. Purser, and A. N. Radford. 2015. "Anthropogenic Noise Compromises Antipredator Behaviour in European Eels." *Global Change Biology* 21 (2): 586–93.

Simpson, S. D., A. N. Radford, S. L. Nedelec, M. C. O. Ferrari, D. P. Chivers, M. I. McCormick, and M. G. Meekan. 2016. "Anthropogenic Noise Increases Fish Mortality by Predation." *Nature Communications* 7, article no. 10544. Accessed June 13, 2017. https://www.nature.com/articles/ncomms10544.

Singmann, H. 2013. *Afex: Analysis of Factorial Experiments*. Accessed August 27, 2017. https://cran.r-project.org/web/packages/afex/afex.pdf.

Sinha, R. P., and D. P. Hader. 2002. "UV-Induced DNA Damage and Repair: A Review." *Photochemical & Photobiological Sciences* 1 (4): 225–36.

Skutelsky, O. 1996. "Predation Risk and State-Dependent Foraging in Scorpions:

Effects of Moonlight on Foraging in the Scorpion *Buthus Occitanus*." *Animal Behaviour* 52 (July): 49–57.

Slabbekoorn, H., and N. Bouton. 2008. "Soundscape Orientation: A New Field in Need of Sound Investigation." *Animal Behaviour* 76 (October): E5–8.

———, I. van Opzeeland, A. Coers, C. ten Cate, and A. N. Popper. 2010. "A Noisy Spring: The Impact of Globally Rising Underwater Sound Levels on Fish." *Trends in Ecology & Evolution* 25 (7): 419–27.

Slabbekoorn, H., and M. Peet. 2003. "Ecology: Birds Sing at a Higher Pitch in Urban Noise—Great Tits Hit the High Notes to Ensure That Their Mating Calls Are Heard above the City's Din." *Nature* 424 (6946): 267–267.

Slabbekoorn, H., and E. A. P. Ripmeester. 2008. "Birdsong and Anthropogenic Noise: Implications and Applications for Conservation." *Molecular Ecology* 17 (1): 72–83.

Small, C., and R. J. Nicholls. 2003. "A Global Analysis of Human Settlement in Coastal Zones." *Journal of Coastal Research* 19 (3): 584–99.

Smit, B., J. G. Boyles, R. M. Brigham, and A. E. McKechnie. 2011. "Torpor in Dark Times: Patterns of Heterothermy Are Associated with the Lunar Cycle in a Nocturnal Bird." *Journal of Biological Rhythms* 26 (3): 241–48.

Smith, A. D., and S. R. McWilliams. 2014. "What to Do When Stopping Over: Behavioral Decisions of a Migrating Songbird during Stopover Are Dictated by Initial Change in Their Body Condition and Mediated by Key Environmental Conditions." *Behavioral Ecology* 25 (6): 1423–35.

Smith, B., and J. Hallo. 2011. "National Park Service Unit Managers' Perceptions, Priorities, and Strategies of Night as a Unit Resource." Paper presented at the George Wright Society Conference, New Orleans, LA, March 14.

———. 2013. "A System-Wide Assessment of Night Resources and Night Recreation in the U.S. National Parks: A Case for Expanded Definitions." *Park Science* 29 (2): 54–59.

Smith, H. 2000. "Phytochromes and Light Signal Perception by Plants—An Emerging Synthesis." *Nature* 407 (6804): 585–91.

Sparks, T., and P. Carey. 1995. "The Responses of Species to Climate Over 2 Centuries—An Analysis of the Marsham Phenological Record, 1736–1947." *Journal of Ecology* 83 (2): 321–29.

Speakman, J. 1991. "Why Do Insectivorous Bats in Britain Not Fly in Daylight More Frequently?" *Functional Ecology* 5 (4): 518–24.

Stack, D. W. 2008. "Human Dimensions of Natural Soundscape Management in National Parks: Visitor Perceptions of Soundscapes and Managing for Natural Sound." PhD diss., Colorado State University.

———, P. Newman, R. E. Manning, and K. M. Fristrup. 2011. "Reducing Visitor Noise Levels at Muir Woods National Monument Using Experimental Management." *Journal of the Acoustical Society of America* 129 (3): 1375–80.

Stamps, A. 1990. "Use of Photographs to Simulate Environments—A Meta-Analysis." *Perceptual and Motor Skills* 71 (3): 907–13.

Stankey, G. H., and R. E. Manning. 1986. "Carrying Capacity of Recreational Settings. The President's Commission on Americans Outdoors." Accessed June 11, 2017. https://iucn.oscar.ncsu.edu/mediawiki/images/1/16/Stankey%281986%29.pdf.

Staples, S. L. 1996. "Human Response to Environmental Noise—Psychological Research and Public Policy." *American Psychologist* 51 (2): 143–50.

Stevens, R. G. 1987. "Electric Power Use and Breast Cancer—A Hypothesis." *American Journal of Epidemiology* 125 (4): 556–61.

———. 2009. "Light-at-Night, Circadian Disruption and Breast Cancer: Assessment of Existing Evidence." *International Journal of Epidemiology* 38 (4): 963–70.

Stone, E. L., G. Jones, and S. Harris. 2009. "Street Lighting Disturbs Commuting Bats." *Current Biology* 19 (13): 1123–27.

———. 2012. "Conserving Energy at a Cost to Biodiversity? Impacts of LED Lighting on Bats." *Global Change Biology* 18 (8): 2458–65.

Summers, P. D., G. M. Cunnington, and L. Fahrig. 2011. "Are the Negative Effects of Roads on Breeding Birds Caused by Traffic Noise?" *Journal of Applied Ecology* 48 (6): 1527–34.

Suter, S. M., D. Ermacora, N. Rieille, and D. R. Meyer. 2009. "A Distinct Reed Bunting Dawn Song and Its Relation to Extrapair Paternity." *Animal Behaviour* 77 (2): 473–80.

Sutherland, B. 1981. "Photoreactivation." *Bioscience* 31 (6): 439–44.

Sutherland, W. J., S. Armstrong-Brown, P. R. Armsworth, T. Brereton, J. Brickland, C. D. Campbell, D. E. Chamberlain. 2006. "The Identification of 100 Ecological Questions of High Policy Relevance in the UK." *Journal of Applied Ecology* 43 (4): 617–27.

Sutton, P. C. 2003. "A Scale-Adjusted Measure of 'Urban Sprawl' using Nighttime Satellite Imagery." *Remote Sensing of Environment* 86 (3): 353–69.

Svensson, A. M., and J. Rydell. 1998. "Mercury Vapour Lamps Interfere with the Bat Defence of Tympanate Moths (*Operophtera Spp.; Geometridae*)." *Animal Behaviour* 55 (January): 223–26.

Sweeney, B. M. 1963. "Biological Clocks in Plants." *Annual Review of Plant Physiology* 14 (1): 411–440.

Taff, D., P. Newman, S. R. Lawson, A. Bright, L. Marin, A. Gibson, and T. Archie. 2014. "The Role of Messaging on Acceptability of Military Aircraft Sounds in Sequoia National Park." *Applied Acoustics* 84:122–28.

Takai, N., M. Yamaguchi, T. Aragaki, K. Eto, K. Uchihashi, and Y. Nishikawa. 2004. "Effect of Psychological Stress on the Salivary Cortisol and Amylase Levels in Healthy Young Adults." *Archives of Oral Biology* 49 (12): 963–68.

Takemura, A., S. Ueda, N. Hiyakawa, and Y. Nikaido. 2006. "A Direct Influence of Moonlight Intensity on Changes in Melatonin Production by Cultured Pineal Glands of the Golden Rabbitfish, *Siganus Guttatus*." *Journal of Pineal Research* 40 (3): 236–41.

Tanner, J. E. 1996. "Seasonality and Lunar Periodicity in the Reproduction of Pocilloporid Corals." *Coral Reefs* 15 (1): 59–66.

Tarling, G. A., F. Buchholz, and J. B. L. Matthews. 1999. "The Effect of a Lunar Eclipse on the Vertical Migration Behaviour of Meganyctiphanes Norvegica (*Crustacea: Euphausiacea*) in the Ligurian Sea." *Journal of Plankton Research* 21 (8): 1475–88.

Tarrant, M., G. Haas, and M. Manfredo. 1995. "Factors Affecting Visitor Evaluations of Aircraft Overflights of Wilderness Areas." *Society & Natural Resources* 8 (4): 351–60.

Tarrant, M. A., C. Overdevest, A. D. Bright, H. K. Cordell, and D. B. K. English. 1997. "The Effect of Persuasive Communication Strategies on Rural Resident Attitudes toward Ecosystem Management." *Society & Natural Resources* 10 (6): 537–50.

Taylor, A. F., F. E. Kuo, and W. C. Sullivan. 2002. "Views of Nature and Self-Discipline: Evidence from Inner City Children." *Journal of Environmental Psychology* 22 (1–2): 49–63.

Taylor, E., T. Taylor, and N. Cuneo. 1992. "The Present Is Not the Key to the Past—A Polar Forest from the Permian." *Science* 257 (5077): 1675–77.

Taylor, M., G. Leach, L. Dimichele, W. Levitan, and W. Jacob. 1979. "Lunar Spawning Cycle in the Mummichog, Fundulus-Heteroclitus (*Pisces, Cyprinodontidae*)." *Copeia*, (2):291–97.

Tennessen, C. M., and B. Cimprich. 1995. "Views to Nature: Effects on Attention." *Journal of Environmental Psychology* 15 (1): 77–85.

Theimer, T. C. 2005. "Rodent Scatterhoarders as Conditional Mutualists." In *Seed Fate: Predation, Dispersal and Seedling Establishment*, edited by J. Lambert, P. Hulme, and S. Vander Wall, 283–96. Wallingford, UK: CABI Publishing.

Thies, W., and E. K. V. Kalko. 2004. "Phenology of Neotropical Pepper Plants (*Piperaceae*) and Their Association with Their Main Dispersers, Two Short-Tailed Fruit Bats, Carollia Perspicillata and C. Castanea (*Phyllostomidae*)." *Oikos* 104 (2): 362–76.

Thomas, D. W. 1991. "On Fruits, Seeds, and Bats." *Bats* 9 (4): 8–13.

Thomson, R. L., J. T. Forsman, Francesc Sarda-Palomera, and Mikko Monkkonen. 2006. "Fear Factor: Prey Habitat Selection and Its Consequences in a Predation Risk Landscape." *Ecography* 29 (4): 507–14.

Thorington, L. 1985. "Spectral, Irradiance, and Temporal Aspects of Natural and Artificial-Light." *Annals of the New York Academy of Sciences* 453 (September): 28–54.

Thresher, R. J., M. H. Vitaterna, Y. Miyamoto, A. Kazantsev, D. S. Hsu, C. Petit, C. P. Selby. 1998. "Role of Mouse Cryptochrome Blue-Light Photoreceptor in Circadian Photoresponses." *Science* 282 (5393): 1490–94.

Tigar, B. J., and P. E. Osborne. 1999. "The Influence of the Lunar Cycle on Ground-Dwelling Invertebrates in an Arabian Desert." *Journal of Arid Environments* 43 (2): 171–82.

Tilden, F. 1957. *Interpreting Our Heritage: Principles and Practices for Visitor Services in Parks, Museums, and Historic Places.* Durham, NC: University of North Carolina Press.

Timm, R. M., and R. K. LaVal. 1998. "A Field Key to the Bats of Costa Rica." University of Kansas Center of Latin American Studies. Occasional Publication Series No. 22.

Tin, T., Z. L. Fleming, K. A. Hughes, D. G. Ainley, P. Convey, C. A. Moreno, S. Pfeiffer, J. Scott, and I. Snape. 2009. "Impacts of Local Human Activities on the Antarctic Environment." *Antarctic Science* 21 (1): 3–33.

Topping, M. G., J. S. Millar, and J. A. Goddard. 1999. "The Effects of Moonlight on Nocturnal Activity in Bushy-Tailed Wood Rats (*Neotoma Cinerea*)." *Canadian Journal of Zoology–Revue Canadienne De Zoologie* 77 (3): 480–85.

Tranel, M. J. 2006. "Denali Air Taxis: Unique Relationships with the Park and Visitors." In *People, Places, and Parks: Proceedings of the 2005 George Wright Society Conference on Parks, Protected Areas, and Cultural Sites,* edited by D. Harmon, 242–48. Hancock (MI): George Wright Society. Accessed June 10, 2017. http://www.georgewright.org/0540tranel.pdf.

Trillo, P., X. Bernal, M. Caldwell, W. Halfwerk, M. Wessel, and R. Page. 2016. "Collateral Damage or a Shadow of Safety? The Effects of Signalling Eterospecific Neighbours on the Risks of Parasitism and Predation." *Proceedings of Biological Science B* 283 (1831):343–50.

Tucson Amateur Astronomy Association. 2017. "Grand Canyon Star Party." Accessed August 15, 2017. http://tucsonastronomy.org/upcoming-events/grand-canyon-star-party/.

Tuttle, M., and M. Ryan. 1981. "Bat Predation and the Evolution of Frog Vocalizations in the Neotropics." *Science* 214 (4521): 677–78.

Tuxbury, S. M., and M. Salmon. 2005. "Competitive Interactions between Artificial Lighting and Natural Cues during Seafinding by Hatchling Marine Turtles." *Biological Conservation* 121 (2): 311–16.

Ugolini, A., V. Boddi, L. Mercatelli, and C. Castellini. 2005. "Moon Orientation in Adult and Young Sandhoppers under Artificial Light." *Proceedings of the Royal Society B-Biological Sciences* 272 (1577): 2189–94.

Ulrich, R. 1984. "View through a Window May Influence Recovery." *Science* 224 (4647): 224–25.

United Nations Population Fund. 2011. *UNPFA State of World Population.* Accessed August 12, 2017. http://www.unfpa.org/sites/default/files/pub-pdf/EN-SWOP2011-FINAL.pdf.

Urbanski, J., M. Mogi, D. O'Donnell, M. DeCotiis, T. Toma, and P. Armbruster. 2012. "Rapid Adaptive Evolution of Photoperiodic Response during Invasion and Range Expansion across a Climatic Gradient." *American Naturalist* 179 (4): 490–500.

Van den Berg, A. E., J. Maas, R. A. Verheij, and P. P. Groenewegen. 2010. "Green Space as a Buffer between Stressful Life Events and Health." *Social Science & Medicine* 70 (8): 1203–10.

Van Langevelde, F., J. Ettema, M. Donners, M. WallisDeVries, and D. Groenendijk. 2011. "Effect of Spectral Composition of Artificial Light on the Attraction of Moths." *Biological Conservation* 144 (9): 2274–81.

Van Wagtendonk, J. W., and J. M. Benedict. 1980. "Travel Time Variation on Backcountry Trails." *Journal of Leisure Research* 12 (2): 99–117.

Vander Wall, S. B. 2000. "The Influence of Environmental Conditions on Cache Recovery and Cache Pilferage by Yellow Pine Chipmunks (*Tamias Amoenus*) and Deer Mice (*Peromyscus Maniculatus*)." *Behavioral Ecology* 11 (5): 544–49.

——— and R. Balda. 1981. "Ecology and Evolution of Food-Storage Behavior in Conifer-Seed-Catching Corvids." *Zeitschrift Fur Tierpsychologie–Journal of Comparative Ethology* 56 (3): 217–42.

Vaske, J., J. Gliner, and G. Morgan. 2002. "Communicating Judgments about Practical Significance: Effect Size, Confidence Intervals and Odds Ratios." *Human Dimensions of Wildlife* 7 (4): 287–300.

———. 2008. *Survey Research and Analysis: Applications in Parks, Recreation and Human Dimensions.* State College, PA: Venture Publishing.

Vásquez, R. 1994. "Assessment of Predation Risk Via Illumination Level— Facultative Central Place Foraging in the Cricetid Rodent *Phyllotis-Darwini.*" *Behavioral Ecology and Sociobiology* 34 (5): 375–81.

Vepsäläinen, K. 1974. "Lengthening of Illumination Period Is a Factor in Averting Diapause." *Nature* 247:385–86.

Verheijen, F. J. 1958. "The Mechanisms of the Trapping Effect of Artificial Light Sources upon Animals." *Archives Néerlandaises de Zoologie* 13 (1): 1–107.

———. 1985. "Photopollution: Artificial Light Optic Spatial Control Systems Fail to Cope with Incidents, Causation, Remedies." *Experimental Biology* 44 (1): 1–18.

Visser, M. E., and L. J. M. Holleman. 2001. "Warmer Springs Disrupt the Synchrony of Oak and Winter Moth Phenology." *Proceedings of the Royal Society B-Biological Sciences* 268 (1464): 289–94.

Vitasse, Y., and D. Basler. 2013. "What Role for Photoperiod in the Bud Burst Phenology of European Beech?" *European Journal of Forest Research* 132 (1): 1–8

Voigt, Christian C., Karin Schneeberger, Silke L. Voigt-Heucke, and Daniel Lewanzik. 2011. "Rain Increases the Energy Cost of Bat Flight." *Biology Letters* 7 (5): 793–95.

Vollsnes, A. V., A. Berglen Eriksen, E. Otterholt, K. Kvaal, U. Oxaal, and C. M. Futsaether. 2009. "Visible Foliar Injury and Infrared Imaging Show That Daylength Affects Short-Term Recovery after Ozone Stress in *Trifolium Subterraneum.*" *Journal of Experimental Botany* 60 (13): 3677–86.

Wagar, J. A. 1964. "The Carrying Capacity of Wild Lands for Recreation." Vol. 7. *Forest Science* 7 (supplement to no. 3): 1–31.

———. 1968. "The Place of Carrying Capacity in the Management of Recreation Lands." In *Third Annual Rocky Mountain-High Plains Park and Recreation Conference Proceedings,* 37–45. Fort Collins: Colorado State University.

Wale, M. A., S. D. Simpson, and A. N. Radford. 2013. "Noise Negatively Affects Foraging and Antipredator Behaviour in Shore Crabs." *Animal Behaviour* 86 (1): 111–18.

Walker, E., and J. Edman. 1985. "The Influence of Host Defensive Behavior on Mosquito (*Diptera: Culicidae*) Biting Persistence." *Journal of Medical Entomology* 22 (4): 370–72.

Ware, H. E., C. J. W. McClure, J. D. Carlisle, and J. R. Barber. 2015. "A Phantom Road Experiment Reveals Traffic Noise Is an Invisible Source of Habitat Degradation." *Proceedings of the National Academy of Sciences* 112 (39): 12105–9.

Warrant, E. 2004. "Vision in the Dimmest Habitats on Earth." *Journal of Comparative Physiology a-Neuroethology Sensory Neural and Behavioral Physiology* 190 (10): 765–89.

———— and M. Dacke. 2010. "Visual Orientation and Navigation in Nocturnal Arthropods." *Brain Behavior and Evolution* 75 (3): 156–73.

————. 2011. "Vision and Visual Navigation in Nocturnal Insects." *Annual Review of Entomology* 56:239–54.

Warren, A. D. 1990. "Predation of Five Species of Noctuidae at Ultraviolet Light by the Western Yellowjacket (*Hymenoptera: Vespidae*)." *Journal of the Lepidopterists' Society* 44 (1): 32–45.

Warren, P., M. Katti, M. Ermann, and A. Brazel. 2006. "Urban Bioacoustics: It's Not Just Noise." *Animal Behaviour* 71 (3): 491–502.

Waser, N. M., and J. A. McRobert. 1998. "Hummingbird Foraging at Experimental Patches of Flowers: Evidence for Weak Risk-Aversion." *Journal of Avian Biology* 29 (3): 305–13.

Watson, M., N. J. Aebischer, and W. Cresswell. 2007. "Vigilance and Fitness in Grey Partridges *Perdix Perdix*: The Effects of Group Size and Foraging-Vigilance Trade-Offs on Predation Mortality." *Journal of Animal Ecology* 76 (2): 211–21.

Watts, R. D., R. W. Compton, J. H. McCammon, C. L. Rich, S. M. Wright, T. Owens, and D. S. Ouren. 2007. "Roadless Space of the Conterminous United States." *Science* 316 (5825): 736–38.

Wayson, R. L. 1998. *Relationship between Pavement Surface Texture and Highway Traffic Noise.* Washington: National Academy Press.

Weaver, D. 2011. "Celestial Ecotourism: New Horizons in Nature-Based Tourism." *Journal of Ecotourism* 10 (1): 38–45.

Weinzimmer, D., P. Newman, D. Taff, J. Benfield, E. Lynch, and P. Bell. 2014. "Human Responses to Simulated Motorized Noise in National Parks." *Leisure Sciences* 36 (3): 251–67.

Welling, P., K. Koivula, and K. Lahti. 1995. "The Dawn Chorus Is Linked with Female Fertility in the Willow Tit *Parus-Montanus*." *Journal of Avian Biology* 26 (3): 241–46.

Whitfied, A. K., and A. Becker. 2014. "Impacts of Recreational Motorboats on Fishes: A Review." *Marine Pollution Bulletin* 83 (1): 24–31.

Whittaker, D., B. Shelby, R. Manning, D. Cole, and G. Haas. 2011. "Capacity Reconsidered: Finding Consensus and Clarifying Differences." *Journal of Park and Recreation Administration* 29 (1): 1–20.

Wiles, R., and T. Hall. 005. "Can Interpretive Messages Change Park Visitors' Views on Wildland Fire?" *Journal of Interpretation Research* 10 (2): 18–35.

Williamson, R. A. 1984. *Living the Sky: The Cosmos of the American Indian.* Norman: University of Oklahoma Press.

Willig, M. R., S. J. Presley, C. P. Bloch, C. L. Hice, S. P. Yanoviak, M. M. Diaz, L. Arias Chauca, V. Pacheco, and S. C. Weaver. 2007. "Phyllostomid Bats of Lowland Amazonia: Effects of Habitat Alteration on Abundance." *Biotropica* 39 (6): 737–46.

Wiltschko, R., K. Stapput, H.-J. Bischof, and W. Wiltschko. 2007. "Light-Dependent Magnetoreception in Birds: Increasing Intensity of Monochromatic Light Changes the Nature of the Response." *Frontiers in Zoology* 4:5–7.

Wiltschko, R., K. Stapput, P. Thalau, and W. Wiltschko. 2010. "Directional Orientation of Birds by the Magnetic Field under Different Light Conditions." *Journal of the Royal Society Interface* 7 (April): S163–77.

Wiltschko, W., U. Munro, H. Ford, and R. Wiltschko. 1993. "Red Light Disrupts Magnetic Orientation of Migratory Birds." *Nature* 364 (6437): 525–27.

Winker, K., D. Warner, and A. Weisbrod. 1992. "Daily Mass Gains Among Woodland Migrants at an Inland Stopover Site." *Auk* 109 (4): 853–62.

Wise, S. E., and B. W. Buchanan. 2006. "The Influence of Artificial Illumination on the Nocturnal Behavior and Physiology of Salamanders: Studies in the Laboratory and Field." In *Ecological Consequences of Artificial Night Lighting*, edited by D. Klem, 221–51. Washington: Island Press.

Witherington, B., and K. Bjorndal. 1991. "Influences of Artificial Lighting on the Seaward Orientation of Hatchling Loggerhead Turtles *Caretta-Caretta*." *Biological Conservation* 55 (2): 139–49.

Witherington, B. E., and R. E. Martin. 2000. "Understanding, Assessing, and Resolving Light-Pollution Problems on Sea Turtle Nesting Beaches." Melbourne Beach: Florida Marine Research Institute.

Wolff, R. J. 1982. "Nocturnal Activity under Artificial Lights by the Jumping Spider *Sitticus Fasciger*." *Peckhamia* 2:32.

Wood, S. N. 2006. *Generalized Additive Models: An Introduction with R.* Boca Raton, FL: CRC Press.

———. 2011. "Fast Stable Restricted Maximum Likelihood and Marginal Likelihood Estimation of Semiparametric Generalized Linear Models." *Journal of the Royal Statistical Society Series B-Statistical Methodology* 73: 3–36.

Wood, W. 2000. "Attitude Change: Persuasion and Social Influence." *Annual Review of Psychology* 51: 539–70

Woods, C. P., and R. M. Brigham. 2008. "Common Poorwill Activity and Calling Behavior in Relation to Moonlight and Predation." *Wilson Journal of Ornithology* 120 (3): 505–12.

Wright, A. J., N. A. Soto, A. L. Baldwin, M. Bateson, C. M. Beale, C. Clark, T. Deak, E. F. Edwards, A. Fernández, A. Godinho, et al. 2007. "Anthropogenic Noise as a Stressor in Animals: A Multidisciplinary Perspective." *International Journal of Comparative Psychology* 20 (2): 250–73.

Wright, G. M., J. S. Dixon, and B. H.r Thompson. 1933. *Fauna of the National Parks of the United States: A Preliminary Survey of Faunal Relations in National Parks.* Vol. 1. Washington: Government Printing Office.

Wysocki, L. E., J. P. Dittami, and F. Ladich. 2006. "Ship Noise and Cortisol Secretion in European Freshwater Fishes." *Biological Conservation* 128 (4): 501–8.

Zhang, S., J. Wei, X. Guo, T.-X. Liu, and L. Kang. 2010. "Functional Synchronization of Biological Rhythms in a Tritrophic System." *PLoS One* 5 (6): e11064.

Zhang, X. 2004. "The Footprint of Urban Climates on Vegetation Phenology." *Geophysical Research Letters* 31 (12): L12209, 1–5.

Ziegler, R., B. Dobre, and M. Diehl. 2007. "Does Matching versus Mismatching Message Content to Attitude Functions Lead to Biased Processing? The Role of Message Ambiguity." *Basic and Applied Social Psychology* 29 (3): 269–78.

Zubidat, A. E., R. Ben-Shlomo, and A. Haim. 2007. "Thermoregulatory and Endocrine Responses to Light Pulses in Short-Day Acclimated Social Voles (*Microtus Socialis*)." *Chronobiology International* 24 (2): 269–88.

ABOUT THE EDITORS

Robert Manning is a professor emeritus in the Rubenstein School of Environment and Natural Resources at the University of Vermont, where he was the Steven Rubenstein Professor of Environment and Natural Resources. He teaches the history, philosophy, and management of national parks and related areas. He is also the founding director of the University's Park Studies Laboratory, which is conducting a long-term program of research for the National Park Service and other agencies and organizations. Manning's research addresses the carrying capacity and management of parks for outdoor recreation and related matters. He has spent four year-long sabbaticals with the National Park Service at Grand Canyon National Park, Yosemite National Park, Golden Gate National Recreation Area, and the Washington, D.C., office, and he was the scholar in residence at Grand Canyon National Park. He has authored and edited several books on parks and outdoor recreation, including *Studies in Outdoor Recreation* (3rd ed.) (Oregon State University Press), *Parks and Carrying Capacity* (Island Press), *Parks and People* (University Press of New England), *Managing Outdoor Recreation* (2nd ed.) (CABI), *Sustainable Transportation in the National Parks* (University Press of New England), and *A Thinking Person's Guide to America's National Parks* (George Braziller Publishing). For the past several years, he and his students and colleagues have conducted a program of research on natural quiet and natural darkness in the national parks.

Peter Newman is department head and professor of Recreation, Park, and Tourism Management in the College of Health and Human Development and is on the faculty of the Graduate Program in Acoustics at Penn State University. He previously served as associate professor of protected areas management and associate dean of academics in the Warner College of Natural Resources at Colorado State University. He has been working in US national parks and in protected areas in Europe and Latin America for over two decades. He has experience as a field naturalist and National Park Service ranger in Yosemite National Park. Newman teaches courses in the social aspects of natural resource management, park history and policy, and conservation leadership and leads a team of researchers and graduate students who focus on systems approaches to informing visitor management (soundscapes and the acoustic environment, visitor education, integrated transportation, and visitor management) in the context of park and protected areas management. This research team also explores linkages between the acoustic environment and human health and well-being in the new Social Science Acoustic Laboratory in the College of Health and Human Development at Penn State. Newman was the recipient of the 2013 George Wright Society Social Science Achievement Award and the 2012 Cooperative Ecosystem Studios Unit National Award, and he was a Fulbright Senior Scholar in Chiapas, Mexico, in 2011.

Jesse Barber is an associate professor in the Department of Biological Sciences at Boise State University. His research program aims to understand how animals gather and use information, and how information transfer drives and constrains behavior, ecology, and evolution. Much of his lab's work focuses on how anthropogenic changes to the sensory environment alter systems. He teaches courses in sensory ecology and conservation biology. His lab's recent and current projects include landscape-scale playback experiments designed to test the role of anthropogenic and natural soundscapes in shaping the behaviors, distributions, and community structure of bats, birds, and insects. Other work in his group, centered in the US national parks, is testing the hypothesis that, as biodiversity increases with less noise pollution, human valuation of both wildlife and the acoustic environment intensifies, increasing support for soundscape conservation and ratcheting biodiversity further upward.

Christopher Monz is a professor of recreation resources management in the Department of Environment and Society at Utah State University. His research specialty is recreation ecology, and his current research interests focus on the integration of biophysical science, social science, and park planning approaches. For over twenty-five years, he has conducted research in national parks and other protected areas worldwide. Monz was principal investigator or coinvestigator on over $4.5 million in funded research from the National Science Foundation, USDA Forest Service, National Park Service, US Fish and Wildlife Service, and the Paul Sarbanes Transit in Parks Program. He has worked extensively in the United States, including Alaska, and internationally in five countries, most recently as a 2014 Fulbright Scholar in Norway. He is the coauthor of the textbook *Wildland Recreation: Ecology and Management* (John Wiley and Sons).

Jeffrey Hallo is an associate professor and the coordinator of the Park and Conservation Area Management Program at Clemson University. He is also a faculty fellow in the Clemson University Institute for Parks. His research and teaching activities are focused on understanding, planning, and managing recreational use of parks, forests, protected areas, and bodies of water. This includes a substantial emphasis on night recreation and night resources. For example, he and his doctoral students completed a National Park System–wide assessment of managers' perceptions of night recreation, a study of night hiking, and a paper on the nexus of wilderness and night recreation. His current work includes visitor management studies of waterfall users at selected Tennessee State Parks, visitors to the Delaware Water Gap National Recreation Area, and polar bear viewers at the Arctic National Wildlife Refuge. In his personal time, Hallo greatly enjoys experiencing parks and the natural world with his wife, Lisa, and his children, Bridger, Ashlyn, and Cooper.

Steven Lawson is a senior director at Resource Systems Group in White River Junction, Vermont, and leads its public lands planning and management group. He has been conducting research in national parks and related protected natural areas throughout the United States for twenty-two years, and he served as visiting social scientist in Yosemite National Park in 2007. Prior to starting a consulting practice at Resource Systems Group, Lawson was a tenured associate professor at Virginia Polytechnic Institute and State University, and he has taught courses in outdoor recreation management and public lands planning. He has expertise in using social science research to inform soundscape planning and management in national parks and protected areas, with a focus on indicator-based adaptive management. Lawson's recent work includes serving as the project team leader for studies to estimate the value of natural sounds and quiet in US national parks, and on the development and pilot field implementation of a socioeconomic monitoring program for the National Park Service. He maintains adjunct faculty status at the University of Vermont and is an associate editor of the *Journal of Park and Recreation Administration*.

INDEX

Note: Page numbers in *italics* indicate figures or tables.